General Motors Chevrolet Equinox, GMC Terrain & Pontiac Torrent Automotive Repair Manual

by Tim Imhoff, Jeff Killingsworth and John H Haynes

Member of the Guild of Motoring Writers

Models covered:

Chevrolet Equinox - 2005 through 2017
GMC Terrain - 2010 through 2017
Pontiac Torrent - 2006 through 2009

(38040 - 4Y6)

ABCDE
F

AUTOMOTIVE
PARTS &
ACCESSORIES
ASSOCIATION MEMBER

Haynes Publishing Group

Sparkford Nr Yeovil
Somerset BA22 7JJ England

Haynes North America, Inc
859 Lawrence Drive
Newbury Park
California 91320 USA
www.haynes.com

Acknowledgements

Technical writers who contributed to this project include Joe Hamilton, Jamie Sarte, Mike Stubblefield and John Wegmann. 2009 and earlier model wiring diagrams provided exclusively for Haynes North America, Inc. by Valley Forge Technical Information Services. 2010 and later model wiring diagrams provided by HaynesPro.

A book in the Haynes Automotive Repair Manual Series

Printed in Malaysia

ISBN-13: 978-1-62092-283-5
ISBN-10: 1-62092-283-5

Library of Congress Control Number: 2018934012

18-352

Contents

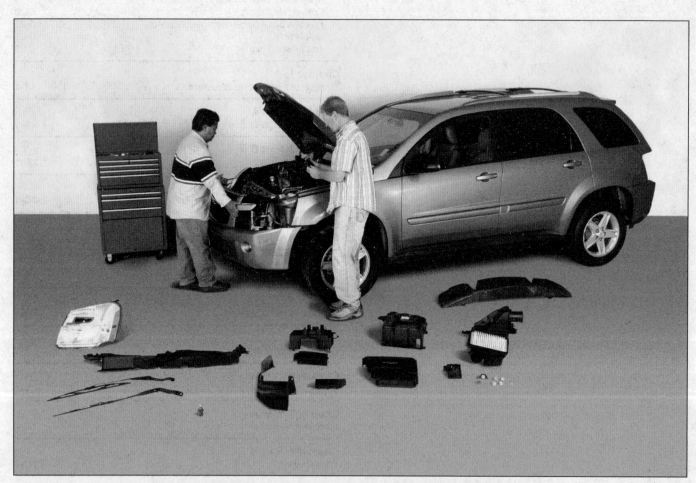

Haynes mechanic and photographer with a 2005 Chevrolet Equinox

About this manual

Its purpose

The purpose of this manual is to help you get the best value from your vehicle. It can do so in several ways. It can help you decide what work must be done, even if you choose to have it done by a dealer service department or a repair shop; it provides information and procedures for routine maintenance and servicing; and it offers diagnostic and repair procedures to follow when trouble occurs.

We hope you use the manual to tackle the work yourself. For many simpler jobs, doing it yourself may be quicker than arranging an appointment to get the vehicle into a shop and making the trips to leave it and pick it up. More importantly, a lot of money can be saved by avoiding the expense the shop must pass on to you to cover its labor and overhead

costs. An added benefit is the sense of satisfaction and accomplishment that you feel after doing the job yourself.

Using the manual

The manual is divided into Chapters. Each Chapter is divided into numbered Sections, which are headed in bold type between horizontal lines. Each Section consists of consecutively numbered paragraphs.

At the beginning of each numbered Section you will be referred to any illustrations which apply to the procedures in that Section. The reference numbers used in illustration captions pinpoint the pertinent Section and the Step within that Section. That is, illustration 3.2 means the illustration refers to Section 3 and Step (or paragraph) 2 within that Section.

Procedures, once described in the text, are not normally repeated. When it's necessary to refer to another Chapter, the reference will be given as Chapter and Section number. Cross references given without use of the word "Chapter" apply to Sections and/or paragraphs in the same Chapter. For example, "see Section 8" means in the same Chapter.

References to the left or right side of the vehicle assume you are sitting in the driver's seat, facing forward.

Even though we have prepared this manual with extreme care, neither the publisher nor the author can accept responsibility for any errors in, or omissions from, the information given.

NOTE

A **Note** provides information necessary to properly complete a procedure or information which will make the procedure easier to understand.

CAUTION

A **Caution** provides a special procedure or special steps which must be taken while completing the procedure where the Caution is found. Not heeding a Caution can result in damage to the assembly being worked on.

WARNING

A **Warning** provides a special procedure or special steps which must be taken while completing the procedure where the Warning is found. Not heeding a Warning can result in personal injury.

Introduction to the Chevrolet Equinox, GMC Terrain and Pontiac Torrent

This manual covers the Chevrolet Equinox, GMC Terrain and Pontiac Torrent. These vehicles feature a 2.4L four-cylinder engine, 3.0L, 3.4L or 3.6L V6 engine. The engine drives the front wheels via independent driveaxles. On AWD models, the rear wheels are also propelled by way of a transfer case, driveshaft, rear differential and two rear driveaxles.

Transaxles are either a five- or six-speed

automatic.

Suspension is independent at all four wheels, MacPherson struts being used at the front and coil springs with shock absorbers at the rear. The rack-and-pinion steering unit is mounted on the subframe.

Some models are equipped with electronic powering assist steering. On 2009 and earlier models, a small electric motor has

been mounted to the steering column and on 2010 and later models an electric motor has been mounted to the steering gear to help reduce steering effort.

The brakes are disc at the front and either disc or drum at the rear, with power assist standard. Some models are equipped with an Anti-Lock Brake System (ABS).

Vehicle identification numbers

Modifications are a continuing and unpublicized process in vehicle manufacturing. Since spare parts manuals and lists are compiled on a numerical basis, the individual vehicle numbers are essential to correctly identify the component required.

Vehicle Identification Number (VIN)

This very important identification number is stamped on a plate attached to the dashboard inside the windshield on the driver's side of the vehicle (see illustration). It can also be found on the certification label located on the driver's side door post. The VIN also appears on the Vehicle Certificate of Title and Registration. It contains information such as where and when the vehicle was manufactured, the model year and the body style.

On the vehicles covered by this manual, the model year codes are:

5	2005
6	2006
7	2007
8	2008
9	2009
A	2010
B	2011
C	2012
D	2013
E	2014
F	2015
G	2016
H	2017

On the vehicles covered by this manual, the engine codes are:

9	3.4L V6 engine (2005)
F	3.4L V6 engine (2006 through 2009)
7	3.6L V6 engine (2008 and 2009)
W	2.4L 4-cyl engine (2010)
C	2.4L 4-cyl engine (2011)
K	2.4L 4-cyl engine (2012 and later)
Y	3.0L V6 engine (2010)
5	3.0L V6 engine (2011 and later)
3	3.6L V6 engine (2013 and later)

Certification label

The certification label is attached to the end of the driver's door or the driver's door post. The plate contains the name of the manufacturer, the month and year of production, the Gross Vehicle Weight Rating (GVWR), the Gross Axle Weight Rating (GAWR) and the certification statement.

Engine number

The engine identification number on the 3.4L V6 engine is stamped or etched on the left rear (driver's side) end of the engine block, near the transaxle, on the firewall side of the engine. The engine identification number on the 3.6L V6 engine is stamped or etched on the left rear (driver's side) end of the engine block, near the transaxle, on the radiator side of the engine.

Transaxle number

There are two transaxles used on these models (5-speed or 6-speed). Both have important information, such as the transaxle type and build date, on a metal tag that is attached near the center top of the unit (5-speed models) (see illustration), or on the left end, near the forward upper corner (6-speed models). Some models also have a prominent sticker that gives similar information.

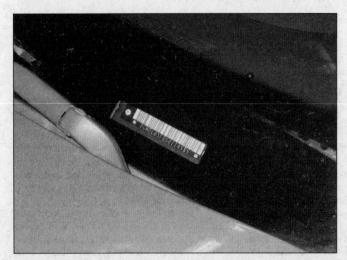

The Vehicle Identification Number (VIN) is located on a plate on top of the dash (visible through the windshield)

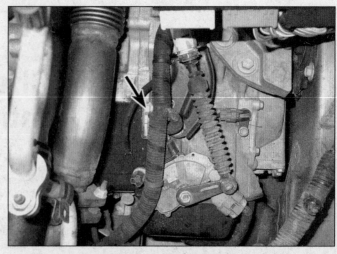

Location of the transaxle identification number (5-speed transaxle)

Recall information

Vehicle recalls are carried out by the manufacturer in the rare event of a possible safety-related defect. The vehicle's registered owner is contacted at the address on file at the Department of Motor Vehicles and given the details of the recall. Remedial work is carried out free of charge at a dealer service department.

If you are the new owner of a used vehicle which was subject to a recall and you want to be sure that the work has been carried out, it's best to contact a dealer service department and ask about your individual vehicle - you'll need to furnish them your Vehicle Identification Number (VIN).

The table below is based on information provided by the National Highway Traffic Safety Administration (NHTSA), the body which oversees vehicle recalls in the United States. The recall database is updated constantly. For the latest information on vehicle recalls, check the NHTSA website at www.nhtsa.gov, www.safercar.gov, or call the NHTSA hotline at 1-888-327-4236.

Recall date	Recall campaign number	Model(s) affected	Concern
Aug 17, 2004	04V302000	2005 Chevrolet Equinox	On certain models, the ignition key park lock cable end fitting was not fully seated on the automatic transmission shifter nail head during installation. This can prevent the cable auto-set from correctly adjusting the cable. An incorrectly adjusted cable prevents the shifter from locking in the 'Park' position when the key is removed from the key cylinder.
Jun 14, 2007	07V242000	2007 Chevrolet Equinox, Pontiac Torrent	On certain models, an incorrect primer may have been used when installing the windshield. In the event of a crash, the windshield may not be retained to the extent required by the standard, increasing the risk of personal injuries to the vehicle occupants.
Aug 06, 2007	07V344000	2007 Chevrolet Equinox, Pontiac Torrent	On certain models, the right front passenger's seat airbag sensor may have been improperly calibrated. This condition can prevent the airbag from turning off when the seat is occupied by a small child. Whenever the front passenger seat is occupied, the driver should always check the airbag indicator to see if the airbag is on or off. If it is not correct for the situation, the passenger should be moved to a different seat. This can increase the risk of injury to a seat occupant during certain crash conditions.
May 8, 2008	07V520000	2007 Chevrolet Equinox, Pontiac Torrent	On certain models not equipped with optional roof rail airbags or optional sunroof fail, the interior occupant protection test points exceeded the standard requirements. The results of this test indicated a possibility of an increased risk of head injury in a crash.
Dec 23, 2009	09V489000	2010 Equinox, GMC Terrain	The software in the center instrument panel can cause the heating, air conditioning, defrost, and radio controls as well as the panel illumination to become non-operational. Driving without a functioning defrost system can decrease your visibility under certain driving conditions and could result in a crash without warning.
Dec 15, 2010	10V623000	2011 Equinox, GMC Terrain	The driver and/or passenger side safety belt buckle anchor may fracture and separate near the seat attachment during a vehicle crash. The safety belt system may not restrain the occupant(s) as designed and could increase the risk of injury.
Oct 19, 2011	11V511000	2012 Equinox, GMC Terrain	The Tire Pressure Monitoring System (TPMS) warning light will illuminate after the recommended 25% inflation cutoff point rather than on or before 25%. Underinflated tires can result in overloading and overheating which could lead to a blowout and possible crash.
July 23, 2014	14V447000	2010, 2011, 2012 Equinox and Terrain	On some models equipped with power height adjustable driver and passenger seats, the bolt that secures the driver's and passenger's power front seat height adjuster may fall out causing the seat to drop suddenly to the lowest vertical position. If the driver's seat unexpectedly drops, the distraction and altered seat position may affect the drivers' control of the vehicle, increasing the risk of a crash.

Recall date	Recall campaign number	Model(s) affected	Concern
July 23, 2014	14V447000	2010, 2011, 2012 Equinox and Terrain	On some models equipped with power height adjustable driver and passenger seats, the bolt that secures the driver's and passenger's power front seat height adjuster may fall out causing the seat to drop suddenly to the lowest vertical position. If the driver's seat unexpectedly drops, the distraction and altered seat position may affect the drivers' control of the vehicle, increasing the risk of a crash.
Oct 16, 2015	15V666000	2015 Equinox and Terrain	Some models have front seat-mounted side impact airbags whose inflator may rupture upon its deployment. In the event of a crash necessitating deployment of one or both of the side impact airbags, the airbag's inflator may rupture and the airbag may not properly inflate. The rupture could cause metal fragments to strike the vehicle occupants, potentially resulting in serious injury or death. Additionally, if the airbag does not properly inflate, the driver or passenger is at an increased risk of injury
June 21, 2016	16V449000	2016 Equinox and Terrain	On some models the certification labels may have incorrect tire/rim size and cold tire pressure information. As such, these vehicles fail to comply with the requirements of Federal Motor Vehicle Safety Standard (FMVSS) No. 110, "Tire Selection and Rims." If the information on the certification labels is incorrect, the operator may install incorrectly sized tires or rims or may improperly inflate the tires, increasing the risk of a crash.
July 05, 2016	16V502000	2013, 2014, 2015 Equinox, 2011, 2012, 2013 Terrain	Some models may have been serviced with similar defective replacement electronic park lock levers. As such, these vehicles fail to comply with the requirements of Federal Motor Vehicle Safety Standard (FMVSS) No. 114, "Theft Protection and Rollaway Prevention" increasing the risk of a crash.
Aug 4, 2016	16V582000	2013 Equinox	On some models, the ball joints in the windshield wiper module may corrode and wear over time, possibly resulting in one or both of the windshield wipers becoming inoperative. Inoperative windshield wipers may reduce the driver's visibility, increasing the risk of a crash.

Buying parts

Replacement parts are available from many sources, which generally fall into one of two categories - authorized dealer parts departments and independent retail auto parts stores. Our advice concerning these parts is as follows:

Retail auto parts stores: Good auto parts stores will stock frequently needed components which wear out relatively fast, such as clutch components, exhaust systems, brake parts, tune-up parts, etc. These stores often supply new or reconditioned parts on an exchange basis, which can save a considerable amount of money. Discount auto parts stores are often very good places to buy materials and parts needed for general vehicle maintenance such as oil, grease, filters, spark plugs, belts, touch-up paint, bulbs, etc. They also usually sell tools and general accessories, have convenient hours, charge lower prices and can often be found not far from home.

Authorized dealer parts department: This is the best source for parts which are unique to the vehicle and not generally available elsewhere (such as major engine parts, transmission parts, trim pieces, etc.).

Warranty information: If the vehicle is still covered under warranty, be sure that any replacement parts purchased - regardless of the source - do not invalidate the warranty!

To be sure of obtaining the correct parts, have engine and chassis numbers available and, if possible, take the old parts along for positive identification.

Maintenance techniques, tools and working facilities

Maintenance techniques

There are a number of techniques involved in maintenance and repair that will be referred to throughout this manual. Application of these techniques will enable the home mechanic to be more efficient, better organized and capable of performing the various tasks properly, which will ensure that the repair job is thorough and complete.

Fasteners

Fasteners are nuts, bolts, studs and screws used to hold two or more parts together. There are a few things to keep in mind when working with fasteners. Almost all of them use a locking device of some type, either a lockwasher, locknut, locking tab or thread adhesive. All threaded fasteners should be clean and straight, with undamaged threads and undamaged corners on the hex head where the wrench fits. Develop the habit of replacing all damaged nuts and bolts with new ones. Special locknuts with nylon or fiber inserts can only be used once. If they are removed, they lose their locking ability and must be replaced with new ones.

Rusted nuts and bolts should be treated with a penetrating fluid to ease removal and prevent breakage. Some mechanics use turpentine in a spout-type oil can, which works quite well. After applying the rust penetrant, let it work for a few minutes before trying to loosen the nut or bolt. Badly rusted fasteners may have to be chiseled or sawed off or removed with a special nut breaker, available at tool stores.

If a bolt or stud breaks off in an assembly, it can be drilled and removed with a special tool commonly available for this purpose. Most automotive machine shops can perform this task, as well as other repair procedures, such as the repair of threaded holes that have been stripped out.

Flat washers and lockwashers, when removed from an assembly, should always be replaced exactly as removed. Replace any damaged washers with new ones. Never use a lockwasher on any soft metal surface (such as aluminum), thin sheet metal or plastic.

Fastener sizes

For a number of reasons, automobile manufacturers are making wider and wider use of metric fasteners. Therefore, it is important to be able to tell the difference between standard (sometimes called U.S. or SAE) and metric hardware, since they cannot be interchanged.

All bolts, whether standard or metric, are sized according to diameter, thread pitch and length. For example, a standard 1/2 - 13 x 1 bolt is 1/2 inch in diameter, has 13 threads per inch and is 1 inch long. An M12 - 1.75 x 25 metric bolt is 12 mm in diameter, has a thread pitch of 1.75 mm (the distance between threads) and is 25 mm long. The two bolts are nearly identical, and easily confused, but they are not interchangeable.

In addition to the differences in diameter, thread pitch and length, metric and standard bolts can also be distinguished by examining the bolt heads. To begin with, the distance across the flats on a standard bolt head is measured in inches, while the same dimension on a metric bolt is sized in millimeters (the same is true for nuts). As a result, a standard wrench should not be used on a metric bolt and a metric wrench should not be used on a standard bolt. Also, most standard bolts have slashes radiating out from the center of the head to denote the grade or strength of the bolt, which is an indication of the amount of torque that can be applied to it. The greater the number of slashes, the greater the strength of the bolt. Grades 0 through 5 are commonly used on automobiles. Metric bolts have a property class (grade) number, rather than a slash, molded into their heads to indicate bolt strength. In this case, the higher the number, the stronger the bolt. Property class numbers 8.8, 9.8 and 10.9 are commonly used on automobiles.

Strength markings can also be used to distinguish standard hex nuts from metric hex nuts. Many standard nuts have dots stamped into one side, while metric nuts are marked with a number. The greater the number of dots, or the higher the number, the greater the strength of the nut.

Metric studs are also marked on their ends according to property class (grade). Larger studs are numbered (the same as metric bolts), while smaller studs carry a geometric code to denote grade.

It should be noted that many fasteners, especially Grades 0 through 2, have no distinguishing marks on them. When such is the case, the only way to determine whether it is standard or metric is to measure the thread pitch or compare it to a known fastener of the same size.

Standard fasteners are often referred to as SAE, as opposed to metric. However, it should be noted that SAE technically refers to a non-metric fine thread fastener only. Coarse thread non-metric fasteners are referred to as USS sizes.

Since fasteners of the same size (both standard and metric) may have different strength ratings, be sure to reinstall any bolts, studs or nuts removed from your vehicle in their original locations. Also, when replacing a fastener with a new one, make sure that the new one has a strength rating equal to or greater than the original.

Tightening sequences and procedures

Most threaded fasteners should be tightened to a specific torque value (torque is the twisting force applied to a threaded component such as a nut or bolt). Overtightening the fastener can weaken it and cause it to break, while undertightening can cause it to eventually come loose. Bolts, screws and studs, depending on the material they are made of and their thread diameters, have specific torque values, many of which are noted in the Specifications at the beginning of each Chapter. Be sure to follow the torque recommendations closely. For fasteners not assigned a

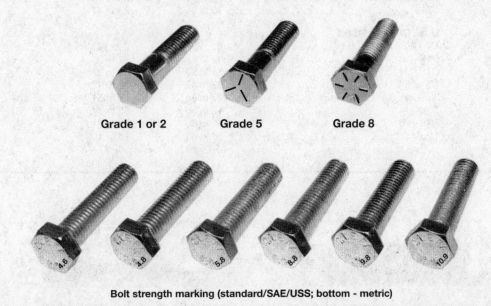

Grade 1 or 2   Grade 5   Grade 8

Bolt strength marking (standard/SAE/USS; bottom - metric)

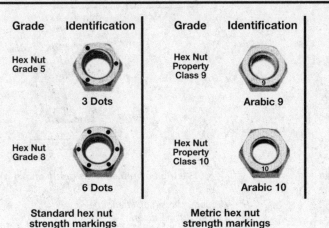

Grade	Identification
Hex Nut Grade 5	3 Dots
Hex Nut Grade 8	6 Dots

Standard hex nut
strength markings

Grade	Identification
Hex Nut Property Class 9	Arabic 9
Hex Nut Property Class 10	Arabic 10

Metric hex nut
strength markings

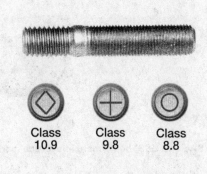

Class 10.9   Class 9.8   Class 8.8

Metric stud strength markings

specific torque, a general torque value chart is presented here as a guide. These torque values are for dry (unlubricated) fasteners threaded into steel or cast iron (not aluminum). As was previously mentioned, the size and grade of a fastener determine the amount of torque that can safely be applied to it. The figures listed here are approximate for Grade 2 and Grade 3 fasteners. Higher grades can tolerate higher torque values.

Fasteners laid out in a pattern, such as cylinder head bolts, oil pan bolts, differential cover bolts, etc., must be loosened or tightened in sequence to avoid warping the component. This sequence will normally be shown in the appropriate Chapter. If a specific pattern is not given, the following procedures can be used to prevent warping.

Initially, the bolts or nuts should be assembled finger-tight only. Next, they should be tightened one full turn each, in a crisscross or diagonal pattern. After each one has been tightened one full turn, return to the first one and tighten them all one-half turn, following the same pattern. Finally, tighten each of them one-quarter turn at a time until each fastener has been tightened to the proper torque. To loosen and remove the fasteners, the procedure would be reversed.

Component disassembly

Component disassembly should be done with care and purpose to help ensure that

Metric thread sizes	Ft-lbs	Nm
M-6	6 to 9	9 to 12
M-8	14 to 21	19 to 28
M-10	28 to 40	38 to 54
M-12	50 to 71	68 to 96
M-14	80 to 140	109 to 154

Pipe thread sizes		
1/8	5 to 8	7 to 10
1/4	12 to 18	17 to 24
3/8	22 to 33	30 to 44
1/2	25 to 35	34 to 47

U.S. thread sizes		
1/4 - 20	6 to 9	9 to 12
5/16 - 18	12 to 18	17 to 24
5/16 - 24	14 to 20	19 to 27
3/8 - 16	22 to 32	30 to 43
3/8 - 24	27 to 38	37 to 51
7/16 - 14	40 to 55	55 to 74
7/16 - 20	40 to 60	55 to 81
1/2 - 13	55 to 80	75 to 108

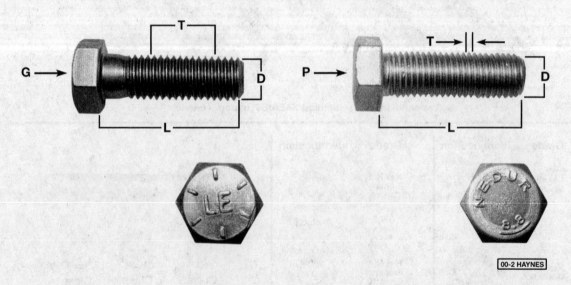

Standard (SAE and USS) bolt dimensions/grade marks

G	Grade marks (bolt strength)
L	Length (in inches)
T	Thread pitch (number of threads per inch)
D	Nominal diameter (in inches)

Metric bolt dimensions/grade marks

P	Property class (bolt strength)
L	Length (in millimeters)
T	Thread pitch (distance between threads in millimeters)
D	Diameter

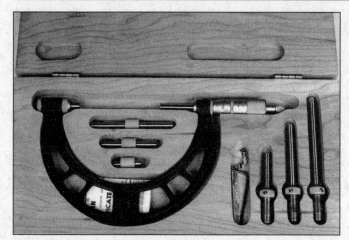

Micrometer set

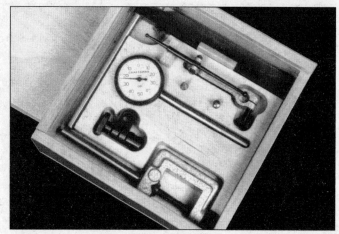

Dial indicator set

the parts go back together properly. Always keep track of the sequence in which parts are removed. Make note of special characteristics or marks on parts that can be installed more than one way, such as a grooved thrust washer on a shaft. It is a good idea to lay the disassembled parts out on a clean surface in the order that they were removed. It may also be helpful to make sketches or take instant photos of components before removal.

When removing fasteners from a component, keep track of their locations. Sometimes threading a bolt back in a part, or putting the washers and nut back on a stud, can prevent mix-ups later. If nuts and bolts cannot be returned to their original locations, they should be kept in a compartmented box or a series of small boxes. A cupcake or muffin tin is ideal for this purpose, since each cavity can hold the bolts and nuts from a particular area (i.e. oil pan bolts, valve cover bolts, engine mount bolts, etc.). A pan of this type is especially helpful when working on assemblies with very small parts, such as the carburetor, alternator, valve train or interior dash and trim pieces. The cavities can be marked with paint or tape to identify the contents.

Whenever wiring looms, harnesses or connectors are separated, it is a good idea to identify the two halves with numbered pieces of masking tape so they can be easily reconnected.

Gasket sealing surfaces

Throughout any vehicle, gaskets are used to seal the mating surfaces between two parts and keep lubricants, fluids, vacuum or pressure contained in an assembly.

Many times these gaskets are coated with a liquid or paste-type gasket sealing compound before assembly. Age, heat and pressure can sometimes cause the two parts to stick together so tightly that they are very difficult to separate. Often, the assembly can be loosened by striking it with a soft-face hammer near the mating surfaces. A regular hammer can be used if a block of wood is placed between the hammer and the part. Do

not hammer on cast parts or parts that could be easily damaged. With any particularly stubborn part, always recheck to make sure that every fastener has been removed.

Avoid using a screwdriver or bar to pry apart an assembly, as they can easily mar the gasket sealing surfaces of the parts, which must remain smooth. If prying is absolutely necessary, use an old broom handle, but keep in mind that extra clean up will be necessary if the wood splinters.

After the parts are separated, the old gasket must be carefully scraped off and the gasket surfaces cleaned. Stubborn gasket material can be soaked with rust penetrant or treated with a special chemical to soften it so it can be easily scraped off. **Caution:** *Never use gasket removal solutions or caustic chemicals on plastic or other composite components.* A scraper can be fashioned from a piece of copper tubing by flattening and sharpening one end. Copper is recommended because it is usually softer than the surfaces to be scraped, which reduces the chance of gouging the part. Some gaskets can be removed with a wire brush, but regardless of the method used, the mating surfaces must be left clean and smooth. If for some reason the gasket surface is gouged, then a gasket sealer thick enough to fill scratches will have to be used during reassembly of the components. For most applications, a non-drying (or semi-drying) gasket sealer should be used.

Hose removal tips

Warning: *If the vehicle is equipped with air conditioning, do not disconnect any of the A/C hoses without first having the system depressurized by a dealer service department or a service station.*

Hose removal precautions closely parallel gasket removal precautions. Avoid scratching or gouging the surface that the hose mates against or the connection may leak. This is especially true for radiator hoses. Because of various chemical reactions, the rubber in hoses can bond itself to the metal spigot that the hose fits over. To remove

a hose, first loosen the hose clamps that secure it to the spigot. Then, with slip-joint pliers, grab the hose at the clamp and rotate it around the spigot. Work it back and forth until it is completely free, then pull it off. Silicone or other lubricants will ease removal if they can be applied between the hose and the outside of the spigot. Apply the same lubricant to the inside of the hose and the outside of the spigot to simplify installation.

As a last resort (and if the hose is to be replaced with a new one anyway), the rubber can be slit with a knife and the hose peeled from the spigot. If this must be done, be careful that the metal connection is not damaged.

If a hose clamp is broken or damaged, do not reuse it. Wire-type clamps usually weaken with age, so it is a good idea to replace them with screw-type clamps whenever a hose is removed.

Tools

A selection of good tools is a basic requirement for anyone who plans to maintain and repair his or her own vehicle. For the owner who has few tools, the initial investment might seem high, but when compared to the spiraling costs of professional auto maintenance and repair, it is a wise one.

To help the owner decide which tools are needed to perform the tasks detailed in this manual, the following tool lists are offered: *Maintenance and minor repair, Repair/overhaul* and *Special.*

The newcomer to practical mechanics should start off with the *maintenance and minor repair* tool kit, which is adequate for the simpler jobs performed on a vehicle. Then, as confidence and experience grow, the owner can tackle more difficult tasks, buying additional tools as they are needed. Eventually the basic kit will be expanded into the *repair and overhaul* tool set. Over a period of time, the experienced do-it-yourselfer will assemble a tool set complete enough for most repair and overhaul procedures and will add tools from the special category when it is felt that the expense is justified by the frequency of use.

Dial caliper

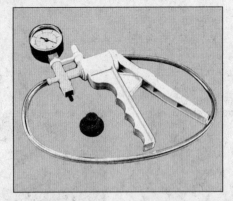

Hand-operated vacuum pump

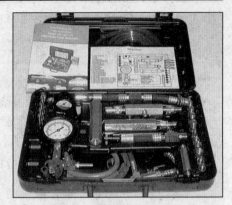

Fuel pressure gauge set

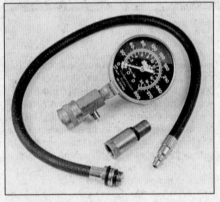

Compression gauge with spark plug hole adapter

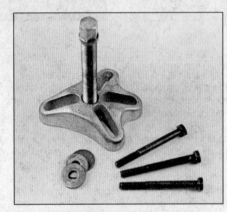

Damper/steering wheel puller

General purpose puller

Hydraulic lifter removal tool

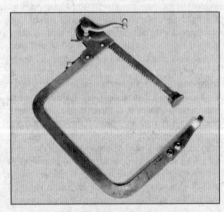

Valve spring compressor

Valve spring compressor

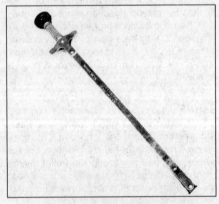

Ridge reamer

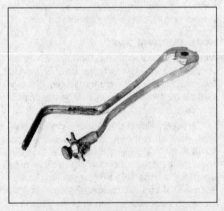

Piston ring groove cleaning tool

Ring removal/installation tool

Ring compressor

Cylinder hone

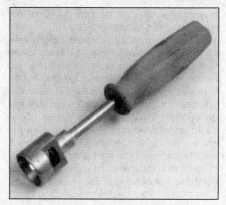

Brake hold-down spring tool

Torque angle gauge

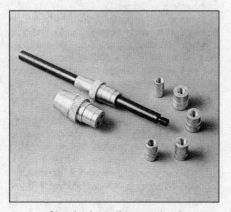

Clutch plate alignment tool

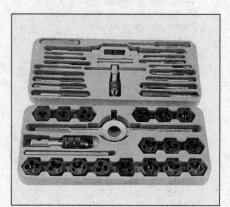

Tap and die set

Maintenance and minor repair tool kit

The tools in this list should be considered the minimum required for performance of routine maintenance, servicing and minor repair work. We recommend the purchase of combination wrenches (box-end and open-end combined in one wrench). While more expensive than open end wrenches, they offer the advantages of both types of wrench.

> Combination wrench set (1/4-inch to
>   1 inch or 6 mm to 19 mm)
> Adjustable wrench, 8 inch
> Spark plug wrench with rubber insert
> Spark plug gap adjusting tool
> Feeler gauge set
> Brake bleeder wrench
> Standard screwdriver (5/16-inch x
>   6 inch)
> Phillips screwdriver (No. 2 x 6 inch)
> Combination pliers - 6 inch
> Hacksaw and assortment of blades
> Tire pressure gauge
> Grease gun
> Oil can
> Fine emery cloth
> Wire brush
> Battery post and cable cleaning tool
> Oil filter wrench
> Funnel (medium size)
> Safety goggles
> Jackstands (2)
> Drain pan

Note: *If basic tune-ups are going to be part of routine maintenance, it will be necessary to purchase a good quality stroboscopic timing light and combination tachometer/dwell meter. Although they are included in the list of special tools, it is mentioned here because they are absolutely necessary for tuning most vehicles properly.*

Repair and overhaul tool set

These tools are essential for anyone who plans to perform major repairs and are in addition to those in the maintenance and minor repair tool kit. Included is a comprehensive set of sockets which, though expensive, are invaluable because of their versatility, especially when various extensions and drives are available. We recommend the 1/2-inch drive over the 3/8-inch drive. Although the larger drive is bulky and more expensive, it has the capacity of accepting a very wide range of large sockets. Ideally, however, the mechanic should have a 3/8-inch drive set and a 1/2-inch drive set.

> Socket set(s)
> Reversible ratchet
> Extension - 10 inch
> Universal joint
> Torque wrench (same size drive as
>   sockets)
> Ball peen hammer - 8 ounce
> Soft-face hammer (plastic/rubber)
> Standard screwdriver (1/4-inch x 6 inch)

> Standard screwdriver (stubby -
>   5/16-inch)
> Phillips screwdriver (No. 3 x 8 inch)
> Phillips screwdriver (stubby - No. 2)
> Pliers - vise grip
> Pliers - lineman's
> Pliers - needle nose
> Pliers - snap-ring (internal and external)
> Cold chisel - 1/2-inch
> Scribe
> Scraper (made from flattened copper
>   tubing)
> Centerpunch
> Pin punches (1/16, 1/8, 3/16-inch)
> Steel rule/straightedge - 12 inch
> Allen wrench set (1/8 to 3/8-inch or
>   4 mm to 10 mm)
> A selection of files
> Wire brush (large)
> Jackstands (second set)
> Jack (scissor or hydraulic type)

Note: *Another tool which is often useful is an electric drill with a chuck capacity of 3/8-inch and a set of good quality drill bits.*

Special tools

The tools in this list include those which are not used regularly, are expensive to buy, or which need to be used in accordance with their manufacturer's instructions. Unless these tools will be used frequently, it is not very economical to purchase many of them. A consideration would be to split the cost and use between yourself and a friend or friends. In addition, most of these tools can be obtained from a tool rental shop on a temporary basis.

This list primarily contains only those tools and instruments widely available to the public, and not those special tools produced by the vehicle manufacturer for distribution to dealer service departments. Occasionally, references to the manufacturer's special tools are included in the text of this manual. Generally, an alternative method of doing the job without the special tool is offered. However, sometimes there is no alternative to their use. Where this is the case, and the tool cannot be purchased or borrowed, the work should be turned over to the dealer service department or an automotive repair shop.

> Valve spring compressor
> Piston ring groove cleaning tool
> Piston ring compressor
> Piston ring installation tool
> Cylinder compression gauge
> Cylinder ridge reamer
> Cylinder surfacing hone
> Cylinder bore gauge
> Micrometers and/or dial calipers
> Hydraulic lifter removal tool
> Balljoint separator
> Universal-type puller
> Impact screwdriver
> Dial indicator set
> Stroboscopic timing light (inductive
>   pick-up)
> Hand operated vacuum/pressure pump
> Tachometer/dwell meter
> Universal electrical multimeter
> Cable hoist
> Brake spring removal and installation
>   tools
> Floor jack

Buying tools

For the do-it-yourselfer who is just starting to get involved in vehicle maintenance and repair, there are a number of options available when purchasing tools. If maintenance and minor repair is the extent of the work to be done, the purchase of individual tools is satisfactory. If, on the other hand, extensive work is planned, it would be a good idea to purchase a modest tool set from one of the large retail chain stores. A set can usually be bought at a substantial savings over the individual tool prices, and they often come with a tool box. As additional tools are needed, add-on sets, individual tools and a larger tool box can be purchased to expand the tool selection. Building a tool set gradually allows the cost of the tools to be spread over a longer period of time and gives the mechanic the freedom to choose only those tools that will actually be used.

Tool stores will often be the only source of some of the special tools that are needed, but regardless of where tools are bought, try to avoid cheap ones, especially when buying screwdrivers and sockets, because they won't last very long. The expense involved in replacing cheap tools will eventually be greater than the initial cost of quality tools.

Care and maintenance of tools

Good tools are expensive, so it makes sense to treat them with respect. Keep them clean and in usable condition and store them properly when not in use. Always wipe off any dirt, grease or metal chips before putting them away. Never leave tools lying around in the work area. Upon completion of a job, always check closely under the hood for tools that may have been left there so they won't get lost during a test drive.

Some tools, such as screwdrivers, pliers, wrenches and sockets, can be hung on a panel mounted on the garage or workshop wall, while others should be kept in a tool box or tray. Measuring instruments, gauges, meters, etc. must be carefully stored where they cannot be damaged by weather or impact from other tools.

When tools are used with care and stored properly, they will last a very long time. Even with the best of care, though, tools will wear out if used frequently. When a tool is damaged or worn out, replace it. Subsequent jobs will be safer and more enjoyable if you do.

How to repair damaged threads

Sometimes, the internal threads of a nut or bolt hole can become stripped, usually from overtightening. Stripping threads is an all-too-common occurrence, especially when working with aluminum parts, because aluminum is so soft that it easily strips out.

Usually, external or internal threads are only partially stripped. After they've been cleaned up with a tap or die, they'll still work. Sometimes, however, threads are badly damaged. When this happens, you've got three choices:

1) Drill and tap the hole to the next suitable oversize and install a larger diameter bolt, screw or stud.

2) Drill and tap the hole to accept a threaded plug, then drill and tap the plug to the original screw size. You can also buy a plug already threaded to the original size. Then you simply drill a hole to the specified size, then run the threaded plug into the hole with a bolt and jam nut. Once the plug is fully seated, remove the jam nut and bolt.

3) The third method uses a patented thread repair kit like Heli-Coil or Slimsert. These easy-to-use kits are designed to repair damaged threads in straight-through holes and blind holes. Both are available as kits which can handle a variety of sizes and thread patterns. Drill the hole, then tap it with the special included tap. Install the Heli-Coil and the hole is back to its original diameter and thread pitch.

Regardless of which method you use, be sure to proceed calmly and carefully. A little impatience or carelessness during one of these relatively simple procedures can ruin your whole day's work and cost you a bundle if you wreck an expensive part.

Working facilities

Not to be overlooked when discussing tools is the workshop. If anything more than routine maintenance is to be carried out, some sort of suitable work area is essential.

It is understood, and appreciated, that many home mechanics do not have a good workshop or garage available, and end up removing an engine or doing major repairs outside. It is recommended, however, that the overhaul or repair be completed under the cover of a roof.

A clean, flat workbench or table of comfortable working height is an absolute necessity. The workbench should be equipped with a vise that has a jaw opening of at least four inches.

As mentioned previously, some clean, dry storage space is also required for tools, as well as the lubricants, fluids, cleaning solvents, etc. which soon become necessary.

Sometimes waste oil and fluids, drained from the engine or cooling system during normal maintenance or repairs, present a disposal problem. To avoid pouring them on the ground or into a sewage system, pour the used fluids into large containers, seal them with caps and take them to an authorized disposal site or recycling center. Plastic jugs, such as old antifreeze containers, are ideal for this purpose.

Always keep a supply of old newspapers and clean rags available. Old towels are excellent for mopping up spills. Many mechanics use rolls of paper towels for most work because they are readily available and disposable. To help keep the area under the vehicle clean, a large cardboard box can be cut open and flattened to protect the garage or shop floor.

Whenever working over a painted surface, such as when leaning over a fender to service something under the hood, always cover it with an old blanket or bedspread to protect the finish. Vinyl covered pads, made especially for this purpose, are available at auto parts stores.

Jacking and towing

Jacking

The jack supplied with the vehicle should only be used for raising the vehicle for changing a tire or placing jackstands under the frame. **Warning:** *Never crawl under the vehicle or start the engine when the jack is being used as the only means of support.*

All vehicles are supplied with a scissors-type jack. When jacking the vehicle, it should be engaged with the rocker panel flange **(see illustration)**.

The vehicle should be on level ground with the wheels blocked and the transmission in Park (automatic). Pry off the hub cap (if equipped) using the tapered end of the lug wrench. Loosen the lug nuts one-half turn and leave them in place until the wheel is raised off the ground.

Place the jack under the side of the vehicle in the indicated position. Use the supplied wrench to turn the jackscrew clockwise until the wheel is raised off the ground. Remove the lug nuts, pull off the wheel and replace it with the spare.

With the beveled side in, reinstall the lug nuts and tighten them until snug. Lower the vehicle by turning the jackscrew counterclockwise. Remove the jack and tighten the nuts in a diagonal pattern to the torque listed in the Chapter 1 Specifications. If a torque wrench is not available, have the torque checked by a service station as soon as possible. Replace the hubcap by placing it in position and using the heel of your hand or a rubber mallet to seat it.

Towing

Five-speed models

Two-wheel drive models can be towed from the front with the front wheels off the ground, using a wheel lift type tow truck or a dolly. If towed from the rear, the front wheels must be placed on a dolly. All-wheel drive models must be towed with all four wheels off the ground. A sling-type tow truck cannot be used, as body damage will result. The best way to tow the vehicle is with a flat-bed car carrier.

Six-speed models

Two-wheel drive models with the six-speed transmission can be towed with all four wheels on the ground, from the front. The transmission must be in Neutral, and the ignition key in the ACC position. **Note:** *To prevent the battery from draining, remove the BATT1 (50-amp) fuse from the underhood fuse/relay box.* They may also be towed with the front wheels on a dolly. All-wheel drive models must be towed with all four wheels off the ground. A sling-type tow truck cannot be used, as body damage will result. The best way to tow the vehicle is with a flat-bed car carrier.

All models

In an emergency, the vehicle can be towed a short distance with a cable or chain attached to one of the towing eyelets located under the front or rear bumpers. The driver must remain in the vehicle to operate the steering and brakes (remember that power steering and power brakes will not work with the engine off).

The jack fits over the rocker panel flange (there are two jacking points on each side of the vehicle)

Booster battery (jump) starting

Observe these precautions when using a booster battery to start a vehicle:

a) Before connecting the booster battery, make sure the ignition switch is in the Off position.

b) Turn off the lights, heater and other electrical loads.

c) Your eyes should be shielded. Safety goggles are a good idea.

d) Make sure the booster battery is the same voltage as the dead one in the vehicle.

e) The two vehicles MUST NOT TOUCH each other!

f) Make sure the transaxle is in Park (automatic).

g) If the booster battery is not a maintenance-free type, remove the vent caps and lay a cloth over the vent holes.

Connect the red jumper lead between the positive (+) terminals of the two batteries (see illustration). Note: These vehicles are equipped with a remote positive terminal located on the side of the underhood fuse/relay box, to make jumper cable connection easier (see illustration).

Connect one end of the black jumper cable to the negative (-) terminal of the booster battery, or a good grounding point, such as a bracket on the engine. The other end of this cable should be connected to a good ground on the vehicle to be started, such as a bolt or bracket on the body.

Start the engine using the booster battery, then, with the engine running at idle speed, disconnect the jumper cables in the reverse order of connection.

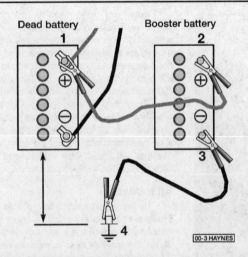

Make the booster battery connections in the numerical order shown (note that the negative cable of the booster battery is NOT attached to the negative terminal of the dead battery)

The remote positive terminal is located on the underhood fuse/relay box. To access it, remove the cover

Automotive chemicals and lubricants

A number of automotive chemicals and lubricants are available for use during vehicle maintenance and repair. They include a wide variety of products ranging from cleaning solvents and degreasers to lubricants and protective sprays for rubber, plastic and vinyl.

Cleaners

Carburetor cleaner and choke cleaner is a strong solvent for gum, varnish and carbon. Most carburetor cleaners leave a dry-type lubricant film which will not harden or gum up. Because of this film it is not recommended for use on electrical components.

Brake system cleaner is used to remove brake dust, grease and brake fluid from the brake system, where clean surfaces are absolutely necessary. It leaves no residue and often eliminates brake squeal caused by contaminants.

Electrical cleaner removes oxidation, corrosion and carbon deposits from electrical contacts, restoring full current flow. It can also be used to clean spark plugs, carburetor jets, voltage regulators and other parts where an oil-free surface is desired.

Demoisturants remove water and moisture from electrical components such as alternators, voltage regulators, electrical connectors and fuse blocks. They are non-conductive and non-corrosive.

Degreasers are heavy-duty solvents used to remove grease from the outside of the engine and from chassis components. They can be sprayed or brushed on and, depending on the type, are rinsed off either with water or solvent.

Lubricants

Motor oil is the lubricant formulated for use in engines. It normally contains a wide variety of additives to prevent corrosion and reduce foaming and wear. Motor oil comes in various weights (viscosity ratings) from 0 to 50. The recommended weight of the oil depends on the season, temperature and the demands on the engine. Light oil is used in cold climates and under light load conditions. Heavy oil is used in hot climates and where high loads are encountered. Multi-viscosity oils are designed to have characteristics of both light and heavy oils and are available in a number of weights from 0W-20 to 20W-50.

Gear oil is designed to be used in differentials, manual transmissions and other areas where high-temperature lubrication is required.

Chassis and wheel bearing grease is a heavy grease used where increased loads and friction are encountered, such as for wheel bearings, balljoints, tie-rod ends and universal joints.

High-temperature wheel bearing grease is designed to withstand the extreme temperatures encountered by wheel bearings in disc brake equipped vehicles. It usually contains molybdenum disulfide (moly), which is a dry-type lubricant.

White grease is a heavy grease for metal-to-metal applications where water is a problem. White grease stays soft under both low and high temperatures (usually from -100 to +190-degrees F), and will not wash off or dilute in the presence of water.

Assembly lube is a special extreme pressure lubricant, usually containing moly, used to lubricate high-load parts (such as main and rod bearings and cam lobes) for initial start-up of a new engine. The assembly lube lubricates the parts without being squeezed out or washed away until the engine oiling system begins to function.

Silicone lubricants are used to protect rubber, plastic, vinyl and nylon parts.

Graphite lubricants are used where oils cannot be used due to contamination problems, such as in locks. The dry graphite will lubricate metal parts while remaining uncontaminated by dirt, water, oil or acids. It is electrically conductive and will not foul electrical contacts in locks such as the ignition switch.

Moly penetrants loosen and lubricate frozen, rusted and corroded fasteners and prevent future rusting or freezing.

Heat-sink grease is a special electrically non-conductive grease that is used for mounting electronic ignition modules where it is essential that heat is transferred away from the module.

Sealants

RTV sealant is one of the most widely used gasket compounds. Made from silicone, RTV is air curing, it seals, bonds, waterproofs, fills surface irregularities, remains flexible, doesn't shrink, is relatively easy to remove, and is used as a supplementary sealer with almost all low and medium temperature gaskets.

Anaerobic sealant is much like RTV in that it can be used either to seal gaskets or to form gaskets by itself. It remains flexible, is solvent resistant and fills surface imperfections. The difference between an anaerobic sealant and an RTV-type sealant is in the curing. RTV cures when exposed to air, while an anaerobic sealant cures only in the absence of air. This means that an anaerobic sealant cures only after the assembly of parts, sealing them together.

Thread and pipe sealant is used for sealing hydraulic and pneumatic fittings and vacuum lines. It is usually made from a Teflon compound, and comes in a spray, a paint-on liquid and as a wrap-around tape.

Chemicals

Anti-seize compound prevents seizing, galling, cold welding, rust and corrosion in fasteners. High-temperature anti-seize, usually made with copper and graphite lubricants, is used for exhaust system and exhaust manifold bolts.

Anaerobic locking compounds are used to keep fasteners from vibrating or working loose and cure only after installation, in the absence of air. Medium strength locking compound is used for small nuts, bolts and screws that may be removed later. High-strength locking compound is for large nuts, bolts and studs which aren't removed on a regular basis.

Oil additives range from viscosity index improvers to chemical treatments that claim to reduce internal engine friction. It should be noted that most oil manufacturers caution against using additives with their oils.

Gas additives perform several functions, depending on their chemical makeup. They usually contain solvents that help dissolve gum and varnish that build up on carburetor, fuel injection and intake parts. They also serve to break down carbon deposits that form on the inside surfaces of the combustion chambers. Some additives contain upper cylinder lubricants for valves and piston rings, and others contain chemicals to remove condensation from the gas tank.

Miscellaneous

Brake fluid is specially formulated hydraulic fluid that can withstand the heat and pressure encountered in brake systems. Care must be taken so this fluid does not come in contact with painted surfaces or plastics. An opened container should always be resealed to prevent contamination by water or dirt.

Weatherstrip adhesive is used to bond weatherstripping around doors, windows and trunk lids. It is sometimes used to attach trim pieces.

Undercoating is a petroleum-based, tar-like substance that is designed to protect metal surfaces on the underside of the vehicle from corrosion. It also acts as a sound-deadening agent by insulating the bottom of the vehicle.

Waxes and polishes are used to help protect painted and plated surfaces from the weather. Different types of paint may require the use of different types of wax and polish. Some polishes utilize a chemical or abrasive cleaner to help remove the top layer of oxidized (dull) paint on older vehicles. In recent years many non-wax polishes that contain a wide variety of chemicals such as polymers and silicones have been introduced. These non-wax polishes are usually easier to apply and last longer than conventional waxes and polishes.

Conversion factors

Length (distance)

Inches (in)	X	25.4	= Millimeters (mm)	X 0.0394	= Inches (in)
Feet (ft)	X	0.305	= Meters (m)	X 3.281	= Feet (ft)
Miles	X	1.609	= Kilometers (km)	X 0.621	= Miles

Volume (capacity)

Cubic inches (cu in; in^3)	X	16.387	= Cubic centimeters (cc; cm^3)	X 0.061	= Cubic inches (cu in; in^3)
Imperial pints (Imp pt)	X	0.568	= Liters (l)	X 1.76	= Imperial pints (Imp pt)
Imperial quarts (Imp qt)	X	1.137	= Liters (l)	X 0.88	= Imperial quarts (Imp qt)
Imperial quarts (Imp qt)	X	1.201	= US quarts (US qt)	X 0.833	= Imperial quarts (Imp qt)
US quarts (US qt)	X	0.946	= Liters (l)	X 1.057	= US quarts (US qt)
Imperial gallons (Imp gal)	X	4.546	= Liters (l)	X 0.22	= Imperial gallons (Imp gal)
Imperial gallons (Imp gal)	X	1.201	= US gallons (US gal)	X 0.833	= Imperial gallons (Imp gal)
US gallons (US gal)	X	3.785	= Liters (l)	X 0.264	= US gallons (US gal)

Mass (weight)

Ounces (oz)	X	28.35	= Grams (g)	X 0.035	= Ounces (oz)
Pounds (lb)	X	0.454	= Kilograms (kg)	X 2.205	= Pounds (lb)

Force

Ounces-force (ozf; oz)	X	0.278	= Newtons (N)	X 3.6	= Ounces-force (ozf; oz)
Pounds-force (lbf; lb)	X	4.448	= Newtons (N)	X 0.225	= Pounds-force (lbf; lb)
Newtons (N)	X	0.1	= Kilograms-force (kgf; kg)	X 9.81	= Newtons (N)

Pressure

Pounds-force per square inch (psi; lbf/in^2; lb/in^2)	X	0.070	= Kilograms-force per square centimeter (kgf/cm^2; kg/cm^2)	X 14.223	= Pounds-force per square inch (psi; lbf/in^2; lb/in^2)
Pounds-force per square inch (psi; lbf/in^2; lb/in^2)	X	0.068	= Atmospheres (atm)	X 14.696	= Pounds-force per square inch (psi; lbf/in^2; lb/in^2)
Pounds-force per square inch (psi; lbf/in^2; lb/in^2)	X	0.069	= Bars	X 14.5	= Pounds-force per square inch (psi; lbf/in^2; lb/in^2)
Pounds-force per square inch (psi; lbf/in^2; lb/in^2)	X	6.895	= Kilopascals (kPa)	X 0.145	= Pounds-force per square inch (psi; lbf/in^2; lb/in^2)
Kilopascals (kPa)	X	0.01	= Kilograms-force per square centimeter (kgf/cm^2; kg/cm^2)	X 98.1	= Kilopascals (kPa)

Torque (moment of force)

Pounds-force inches (lbf in; lb in)	X	1.152	= Kilograms-force centimeter (kgf cm; kg cm)	X 0.868	= Pounds-force inches (lbf in; lb in)
Pounds-force inches (lbf in; lb in)	X	0.113	= Newton meters (Nm)	X 8.85	= Pounds-force inches (lbf in; lb in)
Pounds-force inches (lbf in; lb in)	X	0.083	= Pounds-force feet (lbf ft; lb ft)	X 12	= Pounds-force inches (lbf in; lb in)
Pounds-force feet (lbf ft; lb ft)	X	0.138	= Kilograms-force meters (kgf m; kg m)	X 7.233	= Pounds-force feet (lbf ft; lb ft)
Pounds-force feet (lbf ft; lb ft)	X	1.356	= Newton meters (Nm)	X 0.738	= Pounds-force feet (lbf ft; lb ft)
Newton meters (Nm)	X	0.102	= Kilograms-force meters (kgf m; kg m)	X 9.804	= Newton meters (Nm)

Vacuum

Inches mercury (in. Hg)	X	3.377	= Kilopascals (kPa)	X 0.2961	= Inches mercury
Inches mercury (in. Hg)	X	25.4	= Millimeters mercury (mm Hg)	X 0.0394	= Inches mercury

Power

Horsepower (hp)	X	745.7	= Watts (W)	X 0.0013	= Horsepower (hp)

Velocity (speed)

Miles per hour (miles/hr; mph)	X	1.609	= Kilometers per hour (km/hr; kph)	X 0.621	= Miles per hour (miles/hr; mph)

Fuel consumption*

Miles per gallon, Imperial (mpg)	X	0.354	= Kilometers per liter (km/l)	X 2.825	= Miles per gallon, Imperial (mpg)
Miles per gallon, US (mpg)	X	0.425	= Kilometers per liter (km/l)	X 2.352	= Miles per gallon, US (mpg)

Temperature

Degrees Fahrenheit = (°C x 1.8) + 32

Degrees Celsius (Degrees Centigrade; °C) = (°F - 32) x 0.56

*It is common practice to convert from miles per gallon (mpg) to liters/100 kilometers (l/100km), where mpg (Imperial) x l/100 km = 282 and mpg (US) x l/100 km = 235

DECIMALS to MILLIMETERS

Decimal	mm	Decimal	mm
0.001	0.0254	0.500	12.7000
0.002	0.0508	0.510	12.9540
0.003	0.0762	0.520	13.2080
0.004	0.1016	0.530	13.4620
0.005	0.1270	0.540	13.7160
0.006	0.1524	0.550	13.9700
0.007	0.1778	0.560	14.2240
0.008	0.2032	0.570	14.4780
0.009	0.2286	0.580	14.7320
		0.590	14.9860
0.010	0.2540		
0.020	0.5080		
0.030	0.7620		
0.040	1.0160	0.600	15.2400
0.050	1.2700	0.610	15.4940
0.060	1.5240	0.620	15.7480
0.070	1.7780	0.630	16.0020
0.080	2.0320	0.640	16.2560
0.090	2.2860	0.650	16.5100
		0.660	16.7640
0.100	2.5400	0.670	17.0180
0.110	2.7940	0.680	17.2720
0.120	3.0480	0.690	17.5260
0.130	3.3020		
0.140	3.5560		
0.150	3.8100		
0.160	4.0640	0.700	17.7800
0.170	4.3180	0.710	18.0340
0.180	4.5720	0.720	18.2880
0.190	4.8260	0.730	18.5420
		0.740	18.7960
0.200	5.0800	0.750	19.0500
0.210	5.3340	0.760	19.3040
0.220	5.5880	0.770	19.5580
0.230	5.8420	0.780	19.8120
0.240	6.0960	0.790	20.0660
0.250	6.3500		
0.260	6.6040		
0.270	6.8580	0.800	20.3200
0.280	7.1120	0.810	20.5740
0.290	7.3660	0.820	21.8280
		0.830	21.0820
0.300	7.6200	0.840	21.3360
0.310	7.8740	0.850	21.5900
0.320	8.1280	0.860	21.8440
0.330	8.3820	0.870	22.0980
0.340	8.6360	0.880	22.3520
0.350	8.8900	0.890	22.6060
0.360	9.1440		
0.370	9.3980		
0.380	9.6520		
0.390	9.9060	0.900	22.8600
0.400	10.1600	0.910	23.1140
0.410	10.4140	0.920	23.3680
0.420	10.6680	0.930	23.6220
0.430	10.9220	0.940	23.8760
0.440	11.1760	0.950	24.1300
0.450	11.4300	0.960	24.3840
0.460	11.6840	0.970	24.6380
0.470	11.9380	0.980	24.8920
0.480	12.1920	0.990	25.1460
0.490	12.4460	1.000	25.4000

FRACTIONS to DECIMALS to MILLIMETERS

Fraction	Decimal	mm	Fraction	Decimal	mm
1/64	0.0156	0.3969	33/64	0.5156	13.0969
1/32	0.0312	0.7938	17/32	0.5312	13.4938
3/64	0.0469	1.1906	35/64	0.5469	13.8906
1/16	0.0625	1.5875	9/16	0.5625	14.2875
5/64	0.0781	1.9844	37/64	0.5781	14.6844
3/32	0.0938	2.3812	19/32	0.5938	15.0812
7/64	0.1094	2.7781	39/64	0.6094	15.4781
1/8	0.1250	3.1750	5/8	0.6250	15.8750
9/64	0.1406	3.5719	41/64	0.6406	16.2719
5/32	0.1562	3.9688	21/32	0.6562	16.6688
11/64	0.1719	4.3656	43/64	0.6719	17.0656
3/16	0.1875	4.7625	11/16	0.6875	17.4625
13/64	0.2031	5.1594	45/64	0.7031	17.8594
7/32	0.2188	5.5562	23/32	0.7188	18.2562
15/64	0.2344	5.9531	47/64	0.7344	18.6531
1/4	0.2500	6.3500	3/4	0.7500	19.0500
17/64	0.2656	6.7469	49/64	0.7656	19.4469
9/32	0.2812	7.1438	25/32	0.7812	19.8438
19/64	0.2969	7.5406	51/64	0.7969	20.2406
5/16	0.3125	7.9375	13/16	0.8125	20.6375
21/64	0.3281	8.3344	53/64	0.8281	21.0344
11/32	0.3438	8.7312	27/32	0.8438	21.4312
23/64	0.3594	9.1281	55/64	0.8594	21.8281
3/8	0.3750	9.5250	7/8	0.8750	22.2250
25/64	0.3906	9.9219	57/64	0.8906	22.6219
13/32	0.4062	10.3188	29/32	0.9062	23.0188
27/64	0.4219	10.7156	59/64	0.9219	23.4156
7/16	0.4375	11.1125	15/16	0.9375	23.8125
29/64	0.4531	11.5094	61/64	0.9531	24.2094
15/32	0.4688	11.9062	31/32	0.9688	24.6062
31/64	0.4844	12.3031	63/64	0.9844	25.0031
1/2	0.5000	12.7000	1	1.0000	25.4000

Safety first!

Regardless of how enthusiastic you may be about getting on with the job at hand, take the time to ensure that your safety is not jeopardized. A moment's lack of attention can result in an accident, as can failure to observe certain simple safety precautions. The possibility of an accident will always exist, and the following points should not be considered a comprehensive list of all dangers. Rather, they are intended to make you aware of the risks and to encourage a safety conscious approach to all work you carry out on your vehicle.

Essential DOs and DON'Ts

DON'T rely on a jack when working under the vehicle. Always use approved jackstands to support the weight of the vehicle and place them under the recommended lift or support points.

DON'T attempt to loosen extremely tight fasteners (i.e. wheel lug nuts) while the vehicle is on a jack - it may fall.

DON'T start the engine without first making sure that the transmission is in Neutral (or Park where applicable) and the parking brake is set.

DON'T remove the radiator cap from a hot cooling system - let it cool or cover it with a cloth and release the pressure gradually.

DON'T attempt to drain the engine oil until you are sure it has cooled to the point that it will not burn you.

DON'T touch any part of the engine or exhaust system until it has cooled sufficiently to avoid burns.

DON'T siphon toxic liquids such as gasoline, antifreeze and brake fluid by mouth, or allow them to remain on your skin.

DON'T inhale brake lining dust - it is potentially hazardous (see *Asbestos* below).

DON'T allow spilled oil or grease to remain on the floor - wipe it up before someone slips on it.

DON'T use loose fitting wrenches or other tools which may slip and cause injury.

DON'T push on wrenches when loosening or tightening nuts or bolts. Always try to pull the wrench toward you. If the situation calls for pushing the wrench away, push with an open hand to avoid scraped knuckles if the wrench should slip.

DON'T attempt to lift a heavy component alone - get someone to help you.

DON'T *rush or take unsafe shortcuts to finish a job.*

DON'T allow children or animals in or around the vehicle while you are working on it.

DO wear eye protection when using power tools such as a drill, sander, bench grinder, etc. and when working under a vehicle.

DO keep loose clothing and long hair well out of the way of moving parts.

DO make sure that any hoist used has a safe working load rating adequate for the job.

DO get someone to check on you periodically when working alone on a vehicle.

DO carry out work in a logical sequence and make sure that everything is correctly assembled and tightened.

DO keep chemicals and fluids tightly capped and out of the reach of children and pets.

DO remember that your vehicle's safety affects that of yourself and others. If in doubt on any point, get professional advice.

Steering, suspension and brakes

These systems are essential to driving safety, so make sure you have a qualified shop or individual check your work. Also, compressed suspension springs can cause injury if released suddenly - be sure to use a spring compressor.

Airbags

Airbags are explosive devices that can **CAUSE** injury if they deploy while you're working on the vehicle. Follow the manufacturer's instructions to disable the airbag whenever you're working in the vicinity of airbag components.

Asbestos

Certain friction, insulating, sealing, and other products - such as brake linings, brake bands, clutch linings, torque converters, gaskets, etc. - may contain asbestos or other hazardous friction material. Extreme care must be taken to avoid inhalation of dust from such products, since it is hazardous to health. If in doubt, assume that they do contain asbestos.

Fire

Remember at all times that gasoline is highly flammable. Never smoke or have any kind of open flame around when working on a vehicle. But the risk does not end there. A spark caused by an electrical short circuit, by two metal surfaces contacting each other, or even by static electricity built up in your body under certain conditions, can ignite gasoline vapors, which in a confined space are highly explosive. Do not, under any circumstances, use gasoline for cleaning parts. Use an approved safety solvent.

Always disconnect the battery ground (-) cable at the battery before working on any part of the fuel system or electrical system. Never risk spilling fuel on a hot engine or exhaust component. It is strongly recommended that a fire extinguisher suitable for use on fuel and electrical fires be kept handy in the garage or workshop at all times. Never try to extinguish a fuel or electrical fire with water.

Fumes

Certain fumes are highly toxic and can quickly cause unconsciousness and even death if inhaled to any extent. Gasoline vapor falls into this category, as do the vapors from some cleaning solvents. Any draining or pouring of such volatile fluids should be done in a well ventilated area.

When using cleaning fluids and solvents, read the instructions on the container carefully. Never use materials from unmarked containers.

Never run the engine in an enclosed space, such as a garage. Exhaust fumes contain carbon monoxide, which is extremely poisonous. If you need to run the engine, always do so in the open air, or at least have the rear of the vehicle outside the work area.

The battery

Never create a spark or allow a bare light bulb near a battery. They normally give off a certain amount of hydrogen gas, which is highly explosive.

Always disconnect the battery ground (-) cable at the battery before working on the fuel or electrical systems.

If possible, loosen the filler caps or cover when charging the battery from an external source (this does not apply to sealed or maintenance-free batteries). Do not charge at an excessive rate or the battery may burst.

Take care when adding water to a non maintenance-free battery and when carrying a battery. The electrolyte, even when diluted, is very corrosive and should not be allowed to contact clothing or skin.

Always wear eye protection when cleaning the battery to prevent the caustic deposits from entering your eyes.

Household current

When using an electric power tool, inspection light, etc., which operates on household current, always make sure that the tool is correctly connected to its plug and that, where necessary, it is properly grounded. Do not use such items in damp conditions and, again, do not create a spark or apply excessive heat in the vicinity of fuel or fuel vapor.

Secondary ignition system voltage

A severe electric shock can result from touching certain parts of the ignition system (such as the spark plug wires) when the engine is running or being cranked, particularly if components are damp or the insulation is defective. In the case of an electronic ignition system, the secondary system voltage is much higher and could prove fatal.

Hydrofluoric acid

This extremely corrosive acid is formed when certain types of synthetic rubber, found in some O-rings, oil seals, fuel hoses, etc. are exposed to temperatures above 750-degrees F (400-degrees C). The rubber changes into a charred or sticky substance containing the acid. *Once formed, the acid remains dangerous for years. If it gets onto the skin, it may be necessary to amputate the limb concerned.*

When dealing with a vehicle which has suffered a fire, or with components salvaged from such a vehicle, wear protective gloves and discard them after use.

Troubleshooting

Contents

This section provides an easy reference guide to the more common problems which may occur during the operation of your vehicle. Various symptoms and their possible causes are grouped under headings denoting components or systems, such as Engine, Cooling system, etc. They also refer to the Chapter and/or Section that deals with the problem.

Remember that successful troubleshooting isn't a mysterious art practiced only by professional mechanics. It's simply the result of knowledge combined with an intelligent, systematic approach to a problem. Always use a process of elimination, starting with the simplest solution and working through to the most complex - and never overlook the obvious. Anyone can run the gas tank dry or leave the lights on overnight, so don't assume that you're exempt from such oversights.

Finally, always establish a clear idea why a problem has occurred and take steps to ensure that it doesn't happen again. If the electrical system fails because of a poor connection, check all other connections in the system to make sure they don't fail as well. If a particular fuse continues to blow, find out why - don't just go on replacing fuses. Remember, failure of a small component can often be indicative of potential failure or incorrect functioning of a more important component or system.

Engine and performance

1   Engine will not rotate when attempting to start

1   Battery terminal connections loose or corroded (Chapter 1).
2   Battery discharged or faulty (Chapter 1).
3   Automatic transaxle not completely engaged in Park (Chapter 7A).
4   Broken, loose or disconnected wiring in the starting circuit (Chapters 5 and 12).
5   Starter motor pinion jammed in flywheel ring gear (Chapter 5).
6   Starter solenoid faulty (Chapter 5).
7   Starter motor faulty (Chapter 5).
8   Ignition switch faulty (Chapter 12).
9   Transmission range switch faulty (Chapter 6).
10   Starter pinion or driveplate teeth worn or broken (Chapter 5).
11   Body Control Module (BCM) or Powertrain Control Module (PCM) faulty.

2   Engine rotates but will not start

1   Fuel tank empty.
2   Battery discharged (engine rotates slowly) (Chapter 5).
3   Battery terminal connections loose or corroded (Chapter 1).
4   Fuel not reaching fuel injection system (Chapter 4).
5   Ignition system problem (Chapter 5).

6   Worn, faulty or incorrectly gapped spark plugs (Chapter 1).

3   Engine hard to start when cold

1   Battery discharged or low (Chapter 1).
2   Fuel system malfunctioning (Chapter 4).
3   Emissions or engine control system malfunctioning (Chapter 6).

4   Engine hard to start when hot

1   Air filter clogged (Chapter 1).
2   Fuel not reaching the fuel injection system (Chapter 4).
3   Corroded battery connections, especially ground (Chapter 1).
4   Emissions or engine control system malfunctioning (Chapter 6).

5   Starter motor noisy or excessively rough in engagement

1   Pinion or driveplate gear teeth worn or broken (Chapter 5).
2   Starter motor mounting bolts loose or missing (Chapter 5).

6   Engine starts but stops immediately

1   Loose or faulty electrical connections at coil pack or alternator (Chapter 5).
2   Insufficient fuel reaching the fuel injectors (Chapter 4).
3   Vacuum leak at the gasket between the intake manifold/plenum and throttle body (Chapters 1 and 4).
4   Restricted exhaust system (most likely the catalytic converter) (Chapters 4 and 6).

7   Oil puddle under engine

1   Oil pan gasket and/or oil pan drain bolt seal leaking (Chapters 1 and 2).
2   Oil pressure sending unit leaking (Chapter 2).
3   Valve cover gaskets leaking (Chapter 2).
4   Engine oil seals leaking (Chapter 2).

8   Engine lopes while idling or idles erratically

1   Vacuum leakage (Chapter 4).
2   Leaking EGR valve or plugged PCV system (Chapter 6).
3   Air filter clogged (Chapter 1).
4   Fuel pump not delivering sufficient fuel to the fuel injection system (Chapter 4).
5   Leaking head gasket (Chapter 2).
6   Camshaft lobes worn (Chapter 2).

9   Engine misses at idle speed

1   Spark plugs worn or not gapped properly (Chapter 1).
2   Faulty ignition coil(s) (Chapter 5).
3   Vacuum leaks (Chapters 1 and 4).
4   Uneven or low compression (Chapter 2B).

10   Engine misses throughout driving speed range

1   Fuel filter clogged and/or impurities in the fuel system (Chapters 1 and 4).
2   Low fuel output at the injector (Chapter 4).
3   Faulty or incorrectly gapped spark plugs (Chapter 1).
4   Faulty emission system components (Chapter 6).
5   Low or uneven cylinder compression pressures (Chapter 2B).
6   Weak or faulty ignition coil(s) (Chapter 5).
7   Vacuum leak in fuel injection system, intake manifold or vacuum hoses (Chapter 4).

11   Engine stumbles on acceleration

1   Spark plugs worn (Chapter 1).
2   Fuel injection system problem (Chapter 4).
3   Fuel filter clogged (Chapter 1).
4   Intake manifold air leak (Chapter 4).

12   Engine surges while holding accelerator steady

1   Intake air leak (Chapter 4).
2   Fuel pump faulty (Chapter 4).
3   Defective PCM (Chapter 6).

13   Engine stalls

1   Fuel filter clogged and/or water and impurities in the fuel system (Chapters 1 and 4).
2   Ignition system problem (Chapter 5).
3   Faulty emissions system components (Chapter 6).
4   Vacuum leak in the intake manifold or vacuum hoses (Chapter 4).

14   Engine lacks power

1   Faulty or incorrectly gapped spark plugs (Chapter 1).
2   Restricted exhaust system (most likely the catalytic converter) (Chapters 4 and 6).
3   Fuel injection system malfunctioning (Chapter 4).
4   Faulty coil(s) (Chapter 5).
5   Brakes binding (Chapter 1).

6   Automatic transaxle fluid level incorrect (Chapter 1).
7   Fuel filter clogged and/or impurities in the fuel system (Chapter 1).
8   Emission control system not functioning properly (Chapter 6).
9   Low or uneven cylinder compression pressures (Chapter 2B).

15   Engine backfires

1   Emissions system not functioning properly (Chapter 6).
2   Fuel injection system malfunctioning (Chapter 4).
3   Vacuum leak at fuel injectors, intake manifold or vacuum hoses (Chapter 4).
4   Valves worn or sticking (Chapter 2).

16   Pinging or knocking engine sounds during acceleration or uphill

1   Incorrect grade of fuel.
2   Engine control system malfunctioning (Chapter 6).
3   Improper or damaged spark plugs or wires (Chapter 1).
4   Faulty emissions system (Chapter 6).
5   Vacuum leak (Chapter 4).

17   Engine runs with oil pressure light on

1   Low oil level (Chapter 1).
2   Faulty oil pressure sender (Chapter 2B).
3   Worn engine bearings and/or oil pump (Chapter 2).

18   Engine continues to run after switching off

Faulty ignition switch (Chapter 12).

Engine electrical system

19   Battery will not hold a charge

1   Drivebelt defective (Chapter 1).
2   Battery terminals loose or corroded (Chapter 1).
3   Alternator not charging properly (Chapter 5).
4   Loose, broken or faulty wiring in the charging circuit (Chapter 5).
5   Short in vehicle wiring (Chapters 5 and 12).
6   Internally defective battery (Chapters 1 and 5).

20   Voltage warning light fails to go out

1   Faulty alternator or charging circuit (Chapter 5).
2   Alternator drivebelt defective or out of adjustment (Chapter 1).
3   Alternator voltage regulator inoperative (Chapter 5).

21   Voltage warning light fails to come on when key is turned on

1   Warning light bulb defective (Chapter 12).
2   Fault in the printed circuit, dash wiring or bulb holder (Chapter 12).

Fuel system

22   Excessive fuel consumption

1   Dirty or clogged air filter element (Chapter 1).
2   Emissions system not functioning properly (Chapter 6).
3   Fuel injection system malfunctioning (Chapter 4).
4   Low tire pressure or incorrect tire size (Chapter 1).

23   Fuel leakage and/or fuel odor

1   Leak in a fuel feed or vent line (Chapter 4).
2   Tank overfilled.
3   Evaporative emissions control canister defective (Chapters 1 and 6).
4   Fuel injector seals faulty (Chapter 4).

Cooling system

24   Overheating

1   Insufficient coolant in system (Chapter 1).
2   Drivebelt defective (Chapter 1).
3   Radiator core blocked or grille restricted (Chapter 3).
4   Thermostat faulty (Chapter 3).
5   Electric cooling fan not operating (Chapter 3).
6   Expansion tank cap not maintaining proper pressure (Chapter 3).
7   Blown head gasket (Chapter 2).

25   Overcooling

Incorrect (opening temperature too low) or faulty thermostat (Chapter 3).

26   External coolant leakage

1   Deteriorated/damaged hoses or loose clamps (Chapters 1 and 3).
2   Water pump seal defective (Chapters 1 and 3).
3   Leakage from radiator core (Chapter 3).
4   Engine drain or water jacket core plugs leaking (Chapter 2).

27   Internal coolant leakage

1   Leaking cylinder head gasket (Chapter 2).
2   Cracked cylinder bore or cylinder head (Chapter 2).

28   Coolant loss

1   Too much coolant in system (Chapter 1).
2   Coolant boiling away because of overheating (Chapter 3).
3   Internal or external leakage (Chapter 2 or 3).
4   Faulty expansion tank cap (Chapter 3).

29   Poor coolant circulation

1   Inoperative water pump (Chapter 3).
2   Restriction in cooling system (Chapters 1 and 3).
3   Water pump drivebelt defective or out of adjustment (Chapter 1).
4   Thermostat sticking (Chapter 3).

Automatic transaxle

Note: *Due to the complexity of the automatic transaxle, it's difficult for the home mechanic to properly diagnose and service this component. For problems other than the following, the vehicle should be taken to a dealer service department or a transmission shop.*

30   Fluid leakage

1   Automatic transmission fluid is a deep red color. Fluid leaks should not be confused with engine oil, which can easily be blown by airflow to the transaxle.
2   To pinpoint a leak, first remove all built-up dirt and grime from the transaxle housing with degreasing agents and/or steam cleaning. Drive the vehicle at low speeds so air flow will not blow the leak far from its source. Raise the vehicle and determine where the leak is coming from. Common areas of leakage are:

a) *Fluid pan*
b) *Fill plug (Chapter 1)*
c) *Fluid cooler lines (Chapter 7)*
d) *Vehicle Speed Sensor (Chapter 6)*

31   Transaxle fluid brown or has a burned smell

Transaxle overheated. Change fluid (Chapter 1).

32   General shift mechanism problems

1   Chapter 7 deals with checking and adjusting the shift cable on automatic transaxles. Common problems which may be attributed to a poorly adjusted cable are:
a)   *Engine starting in gears other than Park or Neutral.*
b)   *Indicator on shifter pointing to a gear other than the one actually being used.*
c)   *Vehicle moves when in Park.*
2   Refer to Chapter 7 for the shift cable adjustment procedure.

33   Engine will start in gears other than Park or Neutral

Transmission range switch malfunctioning (Chapter 6).

34   Transaxle slips, shifts roughly, is noisy or has no drive in forward or reverse gears

There are many probable causes for the above problems, but the home mechanic should be concerned with only one possibility - fluid level. Before taking the vehicle to a repair shop, check the level and condition of the fluid as described in Chapter 1.

Correct the fluid level as necessary or change the fluid and filter if needed. If the problem persists, have a professional diagnose the probable cause.

Driveaxles

35   Clicking noise in turns

Worn or damaged outer CV joint. Check for cut or damaged boots (Chapter 1). Repair as necessary (Chapter 8).

36   Knock or clunk when accelerating after coasting

Worn or damaged CV joint. Check for cut or damaged boots (Chapter 1). Repair as necessary (Chapter 8).

37   Shudder or vibration during acceleration

1   Worn or damaged CV joints. Repair or replace as necessary (Chapter 8).
2   Sticking inner joint assembly. Correct or replace as necessary (Chapter 8).

Driveshaft

Note: *Refer to Chapter 8, unless otherwise specified, for service information.*

38   Leaks at front of driveshaft

Defective transfer case seal.

39   Knock or clunk when transmission is under initial load (just after transmission is put into gear)

1   Loose or disconnected rear suspension components. Check all mounting bolts and bushings (Chapter 10).
2   Loose driveshaft bolts. Inspect all bolts and nuts and tighten them securely.
3   Worn or damaged universal joint bearings.
4   Worn sleeve yoke and mainshaft spline.

40   Metallic grating sound consistent with vehicle speed

1   Pronounced wear in the universal joint bearings. Replace driveshaft.
2   Worn center support bearing. Replace driveshaft.

41   Vibration

Note: *Before blaming the driveshaft, make sure the tires are perfectly balanced and perform the following test.*
1   Install a tachometer inside the vehicle to monitor engine speed as the vehicle is driven. Drive the vehicle and note the engine speed at which the vibration (roughness) is most pronounced. Now shift the transmission to a different gear and bring the engine speed to the same point.
2   If the vibration occurs at the same engine speed (rpm) regardless of which gear the transmission is in, the driveshaft is NOT at fault since the driveshaft speed varies.
3   If the vibration decreases or is eliminated when the transmission is in a different gear at the same engine speed, refer to the following probable causes:

a)   *Bent or dented driveshaft. Inspect and replace as necessary.*
b)   *Undercoating or built-up dirt, etc. on the driveshaft. Clean the shaft thoroughly.*
c)   *Worn universal joint bearings. Replace the driveshaft.*
d)   *Driveshaft and/or companion flange out of balance. Check for missing weights on the shaft. Remove driveshaft and reinstall 180-degrees from original position, then recheck. Have the driveshaft balanced if problem persists.*
e)   *Loose driveshaft mounting bolts/nuts.*
f)   *Worn center support bearing. Replace the driveshaft.*
g)   *Worn transfer case rear bushing.*

42   Scraping noise

Make sure there is nothing, such as an exhaust heat shield or safety loop, rubbing on the driveshaft.

Rear differential

Note: *For differential servicing information, refer to Chapter 8, unless otherwise specified.*

43   Noise - same when in drive as when vehicle is coasting

1   Road noise. No corrective action available.
2   Tire noise. Inspect tires and check tire pressures (Chapter 1).
3   Hub bearings worn or damaged (Chapter 10).
4   Insufficient differential oil (Chapter 1).
5   Defective differential.

44   Knocking sound when starting or shifting gears

Worn differential.

45   Noise when turning

Worn differential.

46   Vibration

See probable causes under Driveshaft. Proceed under the guidelines listed for the driveshaft. If the problem persists, check the rear hub bearings by raising the rear of the vehicle and spinning the wheels by hand. Listen for evidence of rough (noisy) bearings. Remove and inspect.

47   Oil leaks

1   Pinion oil seal damaged.
2   Driveaxle oil seals damaged.
3   Loose filler or drain plug on differential (Chapter 1).
4   Clogged or damaged breather on differential.

Brakes

Note: *Before assuming that a brake problem exists, make sure . . .*

a)   *The tires are in good condition and properly inflated (Chapter 1).*
b)   *The front end alignment is correct (Chapter 10).*
c)   *The vehicle isn't loaded with weight in an unequal manner.*

48   Vehicle pulls to one side during braking

1   Incorrect tire pressures (Chapter 1).
2   Front end out of alignment (have the front end aligned).
3   Unmatched tires on same axle.
4   Restricted brake lines or hoses (Chapter 9).
5   Sticking caliper piston (Chapter 9).
6   Loose suspension parts (Chapter 10).
7   Contaminated brake pad material (Chapter 9).

49   Noise (grinding or high-pitched squeal) when the brakes are applied

Brake pads or shoes worn out. Replace the pads or shoes with new ones immediately (Chapter 9).

50   Brake roughness or chatter (pedal pulsates)

1   Excessive brake disc lateral runout (Chapter 9).
2   Parallelism of disc not within specifications (Chapter 9).
3   Uneven pad wear caused by caliper not sliding due to improper clearance or dirt (Chapter 9).
4   Defective brake disc (Chapter 9).

51   Excessive pedal effort required to stop vehicle

1   Malfunctioning power brake booster (Chapter 9).
2   Partial system failure (Chapter 9).
3   Excessively worn pads (Chapter 9).
4   One or more caliper pistons seized or

sticking (Chapter 9).
5   Brake pads contaminated with oil or grease (Chapter 9).
6   New pads installed and not yet seated. It will take a while for the new material to seat.

52   Excessive brake pedal travel

1   Partial brake system failure (Chapter 9).
2   Insufficient fluid in master cylinder (Chapters 1 and 9).
3   Air trapped in system (Chapter 9).
4   Faulty master cylinder (Chapter 9).

53   Dragging brakes

1   Master cylinder pistons not returning correctly (Chapter 9).
2   Restricted brake lines or hoses (Chapters 1 and 9).
3   Incorrect parking brake adjustment (Chapter 9).
4   Defective brake calipers (Chapter 9).

54   Grabbing or uneven braking action

1   Malfunction of proportioning valve (Chapter 9).
2   Malfunction of power brake booster unit (Chapter 9).
3   Binding brake pedal mechanism (Chapter 9).
4   Contaminated brake linings (Chapter 9).

55   Brake pedal feels spongy when depressed

1   Air in hydraulic lines (Chapter 9).
2   Master cylinder mounting bolts loose (Chapter 9).
3   Master cylinder defective (Chapter 9).

56   Brake pedal travels to the floor with little resistance

Little or no fluid in the master cylinder reservoir caused by leaking caliper, or loose, damaged or disconnected brake lines (Chapter 9).

57   Parking brake does not hold

Parking brake cables improperly adjusted (Chapter 9).

Suspension and steering systems

Note: *Before attempting to diagnose the suspension and steering systems, perform the following preliminary checks:*

a)   *Check the tire pressures and look for uneven wear.*
b)   *Check the steering universal joints or coupling from the column to the steering gear for loose fasteners and wear.*
c)   *Check the front and rear suspension and the steering gear assembly for loose and damaged parts.*
d)   *Look for out-of-round or out-of-balance tires, bent rims and loose and/or rough wheel bearings.*

58   Vehicle pulls to one side

1   Mismatched or uneven tires (Chapter 10).
2   Broken or sagging springs (Chapter 10).
3   Wheel alignment incorrect (Chapter 10).
4   Front brakes dragging (Chapter 9).

59   Abnormal or excessive tire wear

1   Front wheel alignment incorrect (Chapter 10).
2   Sagging or broken springs (Chapter 10).
3   Tire out-of-balance (Chapter 10).
4   Worn strut or shock absorber (Chapter 10).
5   Overloaded vehicle.
6   Tires not rotated regularly.

60   Wheel makes a "thumping" noise

1   Blister or bump on tire (Chapter 1).
2   Improper strut or shock absorber action (Chapter 10).

61   Shimmy, shake or vibration

1   Tire or wheel out-of-balance or out-of-round (Chapter 10).
2   Loose or worn wheel bearings (Chapter 10).
3   Worn tie-rod ends (Chapter 10).
4   Worn balljoints (Chapter 10).
5   Excessive wheel runout (Chapter 10).
6   Blister or bump on tire (Chapter 1).

62   Hard steering

1   Lack of lubrication at balljoints, tie-rod ends and steering gear assembly (Chapter 10).
2   Front wheel alignment incorrect (Chapter 10).
3   Low tire pressure (Chapter 1).

63   Steering wheel does not return to center position correctly

1   Lack of lubrication at balljoints and tie-rod ends (Chapters 1 and 10).

2   Binding in steering column (Chapter 10).
3   Defective rack-and-pinion assembly (Chapter 10).
4   Front wheel alignment problem (Chapter 10).

64 Abnormal noise at the front end

1   Worn balljoints and tie-rod ends (Chapter 10).
2   Loose upper strut mount (Chapter 10).
3   Worn tie-rod ends (Chapter 10).
4   Loose stabilizer bar (Chapter 10).
5   Loose wheel lug nuts (Chapter 1).
6   Loose suspension bolts (Chapter 10).

65 Wander or poor steering stability

1   Mismatched or uneven tires (Chapter 10).
2   Worn balljoints or tie-rod ends (Chapters 1 and 10).
3   Broken or sagging springs (Chapter 10).
4   Front wheel alignment incorrect.
5   Worn steering gear clamp bushing (Chapter 10).
6   Wheel bearings worn (Chapter 10).

66 Erratic steering when braking

1   Wheel bearings worn (Chapter 10).
2   Broken or sagging springs (Chapter 10).
3   Leaking caliper (Chapter 9).
4   Warped brake discs (Chapter 9).
5   Worn steering gear clamp bushing (Chapter 10).
6   Wheel alignment incorrect.

67 Excessive pitching and/or rolling around corners or during braking

1   Loose stabilizer bar (Chapter 10).
2   Worn struts/shock absorbers or mounts (Chapter 10).
3   Broken or sagging springs (Chapter 10).
4   Overloaded vehicle.

68 Suspension bottoms

1   Overloaded vehicle.
2   Worn struts or shock absorbers (Chapter 10).
3   Incorrect, broken or sagging springs (Chapter 10).

69 Cupped tires

1   Front wheel alignment incorrect (Chapter 10).
2   Worn struts or shock absorbers (Chapter 10).
3   Hub bearings worn (Chapter 10).
4   Excessive tire or wheel runout (Chapter 10).
5   Worn balljoints (Chapter 10).

70 Excessive tire wear on outside edge

1   Inflation pressures incorrect (Chapter 1).
2   Excessive speed in turns.
3   Wheel alignment incorrect (excessive toe-in or positive camber). Have professionally aligned.
4   Suspension arm bent or twisted (Chapter 10).

71 Excessive tire wear on inside edge

1   Inflation pressures incorrect (Chapter 1).
2   Wheel alignment incorrect (toe-out or excessive negative camber). Have professionally aligned.
3   Loose or damaged steering components (Chapter 10).

72 Tire tread worn in one place

1   Tires out-of-balance.
2   Damaged or buckled wheel. Inspect and replace if necessary.
3   Defective tire (Chapter 1).

73 Excessive play or looseness in steering system

1   Hub bearings worn (Chapter 10).
2   Tie-rod end loose or worn (Chapter 10).
3   Steering gear loose (Chapter 10).

74 Rattling or clicking noise in steering gear

1   Steering gear mounting bolts loose (Chapter 10).
2   Steering gear defective (Chapter 10).

Chapter 1
Tune-up and routine maintenance

Contents

Specifications

Recommended lubricants and fluids

Note: *Listed here are manufacturer recommendations at the time this manual was written. Manufacturers occasionally upgrade their fluid and lubricant specifications, so refer to your vehicle owner's manual or check with your local auto parts store for current recommendations.*

Engine oil	API "certified for gasoline engines"
Viscosity	SAE 5W-30
Fuel	Unleaded gasoline, 87-octane minimum
Automatic transaxle fluid	
Models with a 2.4L, 3.0L engine, 6-speed (6T30/40/45, 6T70/75)	Dexron VI automatic transmission fluid
Models with a 3.4L V6 engine, 5-speed (AF33-5)	Saturn T-IV automatic transmission fluid
Models with a 3.6L V6 engine, 6-speed (6T70/75)	Dexron VI automatic transmission fluid
Transfer case	
Models with a 3.4L V6 engine (NVG 900)	GM VERSATRAK fluid
Models with a 2.4L, 3.0L, 3.6L engine (Getrag 760)	SAE 75W-90 synthetic gear oil
Differential fluid (rear)	SAE 75W-90 synthetic gear oil
Power steering fluid (models with a 3.6L V6 engine)	GM power steering fluid, or equivalent
Brake fluid	DOT 3 brake fluid
Engine coolant	50/50 mixture of DEX-COOL coolant and demineralized water
Parking brake mechanism grease	White lithium-based grease NLGI no. 2
Chassis lubrication grease	NLGI no. 2 GC or GC-LB chassis grease
Hood, door and trunk hinge lubricant	Lubriplate lubricant aerosol spray
Door hinge and check spring grease	NLGI no. 2 multi-purpose grease
Key lock cylinder lubricant	Graphite spray
Hood latch assembly lubricant	Lubriplate lubricant aerosol spray
Door latch lubricant	NLGI no. 2 multi-purpose grease or equivalent

Capacities*

Engine oil (including filter)	
2.4L four-cylinder engine	5.0 quarts
3.0L V6 engine	6.0 quarts
3.4L V6 engine	4.0 quarts
3.6L V6 engine	
2009 and earlier models	5.5 quarts
2013 and later models	6.0 quarts

All capacities are approximate. Add as necessary to bring to appropriate level

Capacities (continued)*

Automatic transaxle (drain and refill)

Models with a 2.4L four-cylinder engine, 6-speed (6T30/40/45)........	5.2 quarts
Models with a 3.0L V6 engine, 6-speed (6T70/75)	5.2 quarts
Models with a 3.4L V6 engine, 5-speed (AF33-5).............................	4.1 quarts
Models with a 3.6L V6 engine, 6-speed (6T70/75)	5.2 quarts

Note: *This is an initial-fill specification. Be sure to check the fluid level as described in Section 7.*

Differential (rear, AWD models)

2006 and earlier models	25.4 ounces
2007 and later models	16.9 ounces

Transfer case (AWD models)

Models with a 2.4L, 3.0L, 3.6L engine (Getrag 760)	1.0 quart
Models with a 3.4L V6 engine (NVG 900)	17 ounces

Cooling system

2.4L four-cylinder engine...	8.2 quarts
3.0L V6 engine	10.8 quarts
3.4L V6 engine	10.5 quarts
3.6L V6 engine	11.0 quarts

* All capacities are approximate. Add as necessary to bring to appropriate level

Brakes

Disc brake pad wear limit	3/32 inch
Drum brake shoe wear limit	1/16 inch
Parking brake shoe wear limit	1/16 inch

Ignition system

Spark plug type

2.4L four-cylinder engine...	AC Delco 41-108
3.0L V6 engine	AC Delco 41-109
3.4L V6 engine	AC Delco 41-101
3.6L V6 engine	
2009 and earlier models	41-990
2013 and later models	41-109

Spark plug gap

2.4L four-cylinder engine...	0.035 inch
3.0L V6 engine	0.043 inch
3.4L V6 engine	0.060 inch
3.6L V6 engine	0.040 inch

Firing order

2.4L four-cylinder engine...	1-3-4-2
3.0L, 3.4L, 3.6L V6 engines	1-2-3-4-5-6

Torque specifications   Ft-lbs (unless otherwise indicated)

Note: *One foot-pound (ft-lb) of torque is equivalent to 12 inch-pounds (in-lbs) of torque. Torque values below approximately 15 ft-lbs are expressed in inch-pounds, since most foot-pound torque wrenches are not accurate at these smaller values.*

Engine oil drain plug

2.4L four-cylinder, 3.4L V6	18
3.0L V6, 3.6L V6	15
Engine oil filter cap (2.4L four-cylinder engine)	18

Automatic transaxle drain plug

Models with a 2.4L, 3.0L engine, 6-speed (6T30/40/45, 6T70/75).....	106 in-lbs
Models with a 3.4L V6 engine, 5-speed (AF33-5)............................	29
Models with a 3.6L V6 engine, 6-speed (6T70/75)	106 in-lbs

Drivebelt tensioner bolt(s)

2.4L four-cylinder engine...	33
3.0L V6 engine	19
3.4L V6 engine	37
3.6L V6 engine	
2009 and earlier models (bracket bolts)	17
2013 and later models	18

Drivebelt idler pulley bolt

2.4L	not applicable
3.0L V6 engine	43
3.4L V6, 2009 and earlier 3.6L V6 engine......................................	37
2013 and later 3.6L V6 engine	43

Differential drain/fill plugs (rear, AWD models)

2006 and earlier models

Drain plug	22
Fill plug	26
2007 and later models (drain and fill)..	29

FRONT OF VEHICLE

Cylinder locations - four-cylinder engines

38027-2B-SPECS HAYNES

FRONT OF VEHICLE

Cylinder locations and coil terminal identification - 3.4L V6 engine

38027-2B-SPECS HAYNES

FRONT OF VEHICLE

Cylinder locations - 3.0L and 3.6L V6 engines

Torque specifications (continued)

Ft-lbs (unless otherwise indicated)

Transfer case
    2007 and earlier models (NVG 900)
        Drain plug 18
        Fill plug .... 132 in-lbs
    2008 and later models
        Models with a 3.4L V6 engine (NVG 900)
            Drain plug... 18
            Fill plug... 132 in-lbs
        Models with a 2.4L, 3.0L, 3.6L engine (Getrag 760) (drain and fill)... 29
Spark plugs
    3.4L V6 engine
        New (first install) 15
        Used .... 132 in-lbs
    2.4L and 3.0L engines.. 15
    3.6L V6 engine
        2009 and earlier models 15
        2013 and later models .... 156 in-lbs
Wheel lug nuts
    2009 and earlier models.. 100
    2010 and later models... 140

Engine compartment components

1	Brake fluid reservoir	5	Windshield washer fluid reservoir
2	Underhood fuse/relay box	6	Radiator hose
3	Coolant expansion tank	7	Engine oil dipstick
4	Battery (under cover)	8	Engine oil filler cap

9	Drivebelt
10	Air filter housing

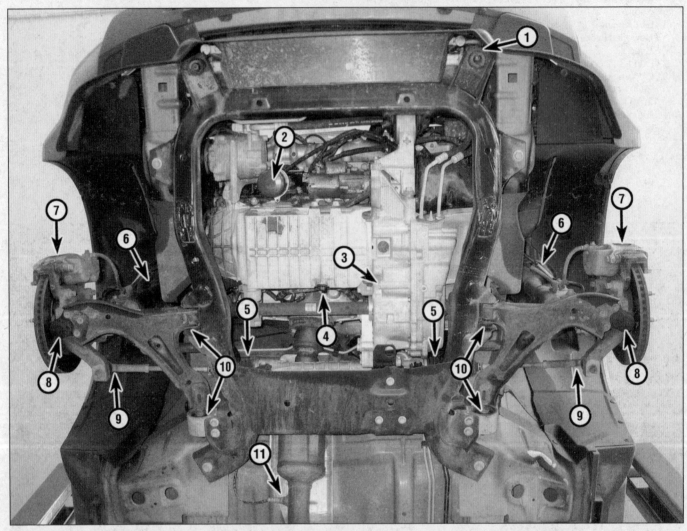

Typical engine compartment underside components (3.4L V6 model shown, 3.6L V6 similar)

1 Radiator drain (approximate location)	4 Engine oil drain plug	8 Lower balljoint
2 Engine oil filter	5 Steering gear boot	9 Tie-rod end
3 Automatic transmission fluid drain plug (AF33-5 transaxle)	6 Strut/coil spring assembly	10 Control arm bushing
	7 Brake caliper	11 Downstream oxygen sensor

Typical rear underside components

1	Trailing arm	5	Stabilizer bar bushing	8	Brake drum
2	Toe link	6	Coil spring	9	Muffler
3	Lower control arm	7	Shock absorber	10	Exhaust system hanger
4	Fuel tank				

1 Maintenance schedule

The following maintenance intervals are based on the assumption that the vehicle owner will be doing the maintenance or service work, as opposed to having a dealer service department do the work. These are the minimum maintenance intervals recommended by the factory for vehicles that are driven daily. If you wish to keep your vehicle in peak condition at all times, you may wish to perform some of these procedures even more often. Because frequent maintenance enhances the efficiency, performance and resale value of your vehicle, we encourage you to do

so. If you drive in dusty areas, tow a trailer, idle or drive at low speeds for extended periods or drive for short distances (less than four miles) in below freezing temperatures, shorter intervals are also recommended.

When the vehicle is new, follow the maintenance schedule to the letter, record the maintenance performed in your owners manual and keep all receipts to protect the new vehicle warranty. In many cases the initial maintenance check is done at no cost to the owner (check with your dealer service department for more information).

Every 250 miles or weekly, whichever comes first

Check the engine oil level (Section 4)
Check the coolant level (Section 4)
Check the windshield washer fluid level (Section 4)
Check the brake fluid level (Section 4)
Check the tires and tire pressures (Section 5)

Every 3,000 miles or 3 months, whichever comes first

All items listed above, plus . . .
Check the power steering fluid level (Section 6)
Change the engine oil and filter (Section 8)

Every 6,000 miles or 6 months, whichever comes first

All items listed above, plus . . .
Check the seat belts (Section 9)
Inspect the windshield wiper blades (Section 10)
Check and service the battery (Section 11)
Check the engine drivebelt (Section 12)
Inspect underhood hoses (Section 13)
Check the cooling system (Section 14)
Rotate the tires (Section 15)
Check the differential lubricant level (Section 16)

Every 15,000 miles or 12 months, whichever comes first

All items listed above, plus . . .
Check the driveaxle boots (Section 17)
Check the fuel system (Section 18)
Check the brake system (Section 19)*
Check the exhaust system (Section 20)
Check the transfer case lubricant level (AWD models) (Section 21)

Every 30,000 miles or 30 months, whichever comes first

All items listed above, plus . . .
Change the brake fluid (Section 22)
Replace the air filter (Section 23)
Replace the spark plugs (non-platinum or iridium type) (Section 24)
Check the steering and suspension components (Section 25)
Change the automatic transaxle fluid and filter (Section 26)**

Every 60,000 miles or 48 months, whichever comes first

Change the transfer case lubricant (Section 27)*
Change the differential lubricant (Section 28)*

Every 100,000 miles or 60 months, whichever comes first

Replace the spark plugs (platinum or iridium type) (Section 24)
Service the cooling system (drain, flush and refill) (Section 29)

* *This item is affected by "severe" operating conditions, as described below. If the vehicle is operated under severe conditions, perform all maintenance indicated with an asterisk (*) at half the indicated intervals. Severe conditions exist if you mainly operate the vehicle . . .*

in dusty areas
towing a trailer
idling for extended periods
driving at low speeds when outside temperatures remain below freezing and most trips are less than four miles long

** *Perform this procedure at half the recommended interval if operated under one or more of the following conditions:*

In heavy city traffic where the outside temperature regularly reaches 90-degrees F or higher in hilly or mountainous terrain
Frequent trailer towing
If the vehicle has been driven through deep water

2   Introduction

This Chapter is designed to help the home mechanic maintain the Chevrolet Equinox and Pontiac Torrent with the goals of maximum performance, economy, safety and reliability in mind.

Included is a master maintenance schedule, followed by procedures dealing specifically with each item on the schedule. Visual checks, adjustments, component replacement and other helpful items are included. Refer to the accompanying illustrations of the engine compartment and the underside of the vehicle for the locations of various components.

Servicing your vehicle in accordance with the mileage/time maintenance schedule and the step-by-step procedures will result in a planned maintenance program that should produce a long and reliable service life. Keep in mind that it's a comprehensive plan, so maintaining some items but not others at the specified intervals will produce the same results.

As you service your vehicle, you will discover that many of the procedures can - and should - be grouped together because of the nature of the particular procedure you're performing or because of the close proximity of two otherwise unrelated components to one another.

For example, if the vehicle is raised for chassis lubrication, you should inspect the exhaust, suspension, steering and fuel systems while you're under the vehicle. When you're rotating the tires, it makes good sense to check the brakes since the wheels are already removed. Finally, let's suppose you have to borrow or rent a torque wrench. Even if you only need it to tighten the spark plugs, you might as well check the torque of as many critical fasteners as time allows.

The first step in this maintenance program is to prepare you before the actual work begins. Read through all the procedures you're planning to do, then gather up all the parts and tools needed. If it looks like you might run into problems during a particular job, seek advice from a mechanic or an experienced do-it-yourselfer.

Owner's Manual and VECI label information

Your vehicle Owner's Manual was written for your year and model and contains very specific information on component locations, specifications, fuse ratings, part numbers, etc. The Owner's Manual is an important resource for the do-it-yourselfer to have; if one was not supplied with your vehicle, it can generally be ordered from a dealer parts department.

Among other important information, the Vehicle Emissions Control Information (VECI) label contains specifications and procedures for tune-up adjustments (if applicable) and, in some instances, spark plugs (see Chapter 6 for more information on the VECI label). The information on this label is the exact mainte-nance data recommended by the manufacturer. This data often varies by intended operating altitude, local emissions regulations, month of manufacture, etc.

This Chapter contains procedural details, safety information and more ambitious maintenance intervals than you might find in the manufacturer's literature. However, you may also find procedures and specifications in your Owner's Manual or VECI label that differ with what's printed here. In these cases, the Owner's Manual or VECI label can be considered correct, since it is specific to your particular vehicle.

3   Tune-up general information

The term tune-up is used in this manual to represent a combination of individual operations rather than one specific procedure that will maintain a gasoline engine in proper tune.

If, from the time the vehicle is new, the routine maintenance schedule is followed closely and frequent checks are made of fluid levels and high wear items, as suggested throughout this manual, the engine will be kept in relatively good running condition and the need for additional work will be minimized.

More likely than not, however, there may be times when the engine is running poorly due to lack of regular maintenance. This is even more likely if a used vehicle, which has not received regular and frequent maintenance checks, is purchased. In such cases, an engine tune-up will be needed outside of the regular routine maintenance intervals.

The first step in any tune-up or diagnostic procedure to help correct a poor running engine is a cylinder compression check. Compression check (see Chapter 2B) will help determine the condition of internal engine components and should be used as a guide for tune-up and repair procedures. If, for instance, the compression check indicates serious internal engine wear, a conventional tune-up won't improve the performance of the engine and would be a waste of time and money. Because of its importance, someone should do the compression check with the right equipment and the knowledge to use it properly.

The following procedures are those most often needed to bring a generally poor running engine back into a proper state of tune.

Minor tune-up
Check all engine-related fluids (Section 4)
Clean, inspect and test the battery (Section 11)
Check and adjust the drivebelt (Section 12)
Check all underhood hoses (Section 13)
Check the cooling system (Section 14)
Check the air filter (Section 23)

Major tune-up
All items listed under Minor tune-up, plus . . .
Replace the air filter (Section 23)

4.2 Engine oil dipstick (A) and oil filler cap (B). To prevent dirt from contaminating the engine, always make sure the area around the cap is clean before removing it (3.4L V6 engine shown)

Replace the spark plugs (Section 24)
Check the ignition system (Chapter 5)
Check the charging system (Chapter 5)

4   Fluid level checks (every 250 miles or weekly)

Note: *The following are fluid level checks to be done on a 250 mile or weekly basis. Additional fluid level checks can be found in specific maintenance procedures that follow. Regardless of intervals, be alert to fluid leaks under the vehicle, which would indicate a fault to be corrected immediately.*

1   Fluids are an essential part of the lubrication, cooling, brake and windshield washer systems. Because the fluids gradually become depleted and/or contaminated during normal operation of the vehicle, they must be periodically replenished. See *Recommended lubricants and fluids* in this Chapter's Specifications before adding fluid to any of the following components. **Note:** *The vehicle must be on level ground when fluid levels are checked.*

Engine oil
Refer to illustrations 4.2 and 4.4

2   The engine oil level is checked with a dipstick that extends through a tube and into the oil pan at the bottom of the engine **(see illustration)**.

3   The oil level should be checked before the vehicle has been driven, or about 5 minutes after the engine has been shut off. If the oil is checked immediately after driving the vehicle, some of the oil will remain in the upper engine components, resulting in an inaccurate reading on the dipstick.

4   Pull the dipstick out of the tube and wipe all the oil from the end with a clean rag or paper towel. Insert the clean dipstick all the way back into the tube, then pull it out again.

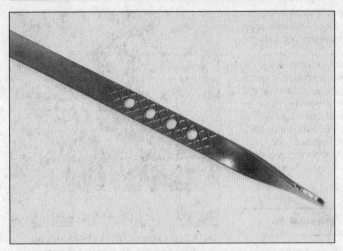

4.4 The oil level must be maintained within the cross-hatched zone at all times - it takes about one quart of oil to raise the level from the bottom of the zone to the top of the zone

4.8 The coolant expansion tank is located on the left side of the engine compartment

Note the oil at the end of the dipstick. Add oil as necessary to keep the level within the cross-hatched zone on the dipstick **(see illustration)**.

5   Do not overfill the engine by adding too much oil since this may result in oil-fouled spark plugs, oil leaks or oil seal failures.

6   Oil is added to the engine after unscrewing a cap from the valve cover. A funnel will help to reduce spills.

7   Checking the oil level is an important preventive maintenance step. A consistently low oil level indicates oil leakage through damaged seals, defective gaskets or past worn rings or valve guides. If the oil looks milky or has water droplets in it, the cylinder head gasket(s) may be blown or the head(s) or block may be cracked. The engine should be checked immediately. The condition of the oil should also be checked. Whenever you check the oil level, slide your thumb and index finger up the dipstick before wiping off the oil. If you see small dirt or metal particles clinging to the

dipstick, the oil should be changed (see Section 8).

Engine coolant

Refer to illustrations 4.8 and 4.9

Warning 1: *Do not allow antifreeze to come in contact with your skin or painted surfaces of the vehicle. Rinse off spills immediately with plenty of water. Antifreeze is highly toxic if ingested. Never leave antifreeze lying around in an open container or in puddles on the floor; children and pets are attracted by its sweet smell and may drink it. Check with local authorities on disposing of used antifreeze. Many communities have collection centers that will see that antifreeze is disposed of safely.*

Warning 2: *Never remove the expansion tank cap when the engine is warm.*

Caution: *Never mix green-colored ethylene glycol antifreeze and orange-colored "DEX-COOL" silicate-free coolant because doing so*

will destroy the efficiency of the "DEX-COOL" coolant which is designed to last for 100,000 miles or five years.

Note: *Non-toxic antifreeze is now manufactured and available at auto parts stores, but even this type should be disposed of properly.*

8   All vehicles covered by this manual are equipped with a coolant expansion tank, located at the left side of the engine compartment **(see illustration)** and connected by hoses to the cooling system.

9   The coolant level in the tank should be checked regularly. When the engine is cold, the coolant level should be at or slightly above the COLD FILL mark **(see illustration)**. If it isn't, add coolant to the tank.

Warning: *Never unscrew the expansion tank cap when the engine is warm. Only remove the cap when the engine and cooling system is cool.*

10   To add coolant, slowly twist open the cap; if you hear a hissing sound, STOP and wait until the hissing stops. Add a 50/50 mixture of

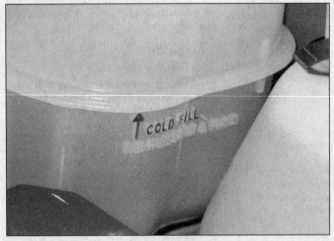

4.9 When the engine is cold, the coolant level must be up to the COLD FILL mark

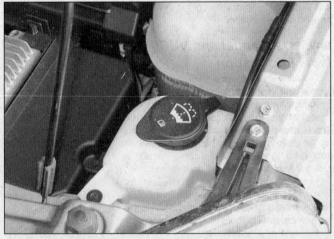

4.14 The windshield washer fluid tank is located at the left front corner of the engine compartment

4.19 The brake fluid reservoir is located at the left side of the firewall

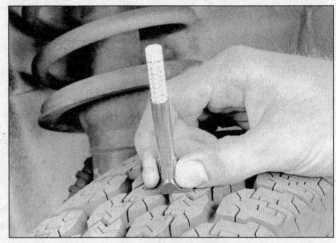

5.2 Use a tire tread depth indicator to monitor tire wear - they are available at auto parts stores and service stations and cost very little

"DEX-COOL" coolant and water until the level is up to the COLD FILL mark (see the **Caution** at the beginning of this Section).

11   Drive the vehicle and recheck the coolant level. If only a small amount of coolant is required to bring the system up to the proper level, water can be used. However, repeated additions of water will dilute the antifreeze and water solution. In order to maintain the proper ratio of antifreeze and water, always top up the coolant level with the correct mixture. An empty plastic milk jug or bleach bottle makes an excellent container for mixing coolant. Do not use rust inhibitors or additives.

12   If the coolant level drops consistently, there may be a leak in the system. Inspect the radiator, hoses, filler cap, drain plugs and water pump (see Section 14). If no leaks are noted, have the pressure cap tested by a service station.

13   Check the condition of the coolant as well. It should be relatively clear. If it is brown or rust colored, the system should be drained, flushed and refilled. Even if the coolant appears to be normal, the corrosion inhibitors wear out, so it must be replaced at the specified intervals. If the system is filled with standard green coolant/water, it must be flushed and replaced more frequently than if the original "DEX-COOL" coolant is retained.

Windshield washer fluid

Refer to illustration 4.14

14   Fluid for the windshield washer system is located in a plastic reservoir in the left side of the engine compartment **(see illustration)**.

15   In milder climates, plain water can be used in the reservoir, but it should be kept no more than 2/3 full to allow for expansion if the water freezes. In colder climates, use windshield washer system antifreeze, available at any auto parts store, to lower the freezing point of the fluid. Mix the antifreeze with water in accordance with the manufacturer's direc-

tions on the container. **Caution:** *Don't use cooling system antifreeze - it will damage the vehicle's paint.*

16   To help prevent icing in cold weather, warm the windshield with the defroster before using the washer.

Battery electrolyte

17   These vehicles are equipped with a battery which is permanently sealed (except for vent holes) and has no filler caps. Water doesn't have to be added to these batteries at any time. If a maintenance-type battery is installed, the caps on the top of the battery should be removed periodically to check for a low electrolyte level. If the level is low, add only distilled water until the level is above the plates.

Brake fluid

Refer to illustration 4.19

18   The brake master cylinder is mounted on the upper left of the engine compartment firewall.

19   The translucent plastic reservoir allows the fluid inside to be checked without removing the cap **(see illustration)**. Be sure to wipe the top of the reservoir cap with a clean rag to prevent contamination of the brake system before removing it.

20   When adding fluid, pour it carefully into the reservoir to avoid spilling it on surrounding painted surfaces. Be sure the specified fluid is used, since mixing different types of brake fluid can cause damage to the system. See *Recommended lubricants and fluids* in this Chapter's Specifications or your owner's manual. **Warning:** *Brake fluid can harm your eyes and damage painted surfaces, so use extreme caution when handling or pouring it. Do not use brake fluid that has been standing open or is more than one year old. Brake fluid absorbs moisture from the air. Moisture in the system can cause a danger-*

ous loss of brake performance.

21   At this time, the fluid and master cylinder can be inspected for contamination. The system should be drained and refilled if deposits, dirt particles or water droplets are seen in the fluid.

22   After filling the reservoir to the proper level, make sure the cap is installed tight to prevent fluid leakage.

23   The brake fluid level in the master cylinder will drop slightly as the pads at the front wheels wear down during normal operation. If the master cylinder requires repeated additions to keep it at the proper level, it's an indication of leakage in the brake system, which should be corrected immediately. Check all brake lines and connections (see Section 19 for more information).

24   If, upon checking the master cylinder fluid level, you discover the reservoir empty or nearly empty, the brake system should be bled and thoroughly inspected (see Chapter 9).

5   Tire and tire pressure checks (every 250 miles or weekly)

Refer to illustrations 5.2, 5.3, 5.4a, 5.4b and 5.8

1   Periodic inspection of the tires may spare you the inconvenience of being stranded with a flat tire. It can also provide you with vital information regarding possible problems in the steering and suspension systems before major damage occurs.

2   The original tires on this vehicle are equipped with 1/2-inch wide wear bands that will appear when tread depth reaches 1/16-inch, at which point the tires can be considered worn out. Tread wear can be monitored with a simple, inexpensive device known as a tread depth indicator **(see illustration)**.

UNDERINFLATION

CUPPING

OVERINFLATION

Cupping may be caused by:
- Underinflation and/or mechanical irregularities such as out-of-balance condition of wheel and/or tire, and bent or damaged wheel.
- Loose or worn steering tie-rod or steering idler arm.
- Loose, damaged or worn front suspension parts.

INCORRECT TOE-IN
OR EXTREME CAMBER

FEATHERING DUE
TO MISALIGNMENT

5.3 This chart will help you determine the condition of the tires and the probable cause(s) of abnormal wear

3   Note any abnormal tread wear **(see illustration)**. Tread pattern irregularities such as cupping, flat spots and more wear on one side than the other are indications of front end alignment and/or balance problems. If any of these conditions are noted, take the vehicle to a tire shop or service station to correct the problem.

4   Look closely for cuts, punctures and embedded nails or tacks. Sometimes a tire will hold air pressure for a short time or leak down very slowly after a nail has embedded itself in the tread. If a slow leak persists, check the valve stem core to make sure it's tight **(see illustration)**. Examine the tread for an object that may have embedded itself in the tire or for a plug that may have begun to leak (radial tire punctures are repaired with a plug that's installed in a puncture). If a puncture is suspected, it can be easily verified by spraying a solution of soapy water onto the puncture area **(see illustration)**. The soapy solution will bubble if there's a leak. Unless the puncture is unusually large, a tire shop or service station can usually repair the tire.

5   Carefully inspect the inner sidewall of each tire for evidence of brake fluid leakage. If you see any, inspect the brakes immediately.

6   Correct air pressure adds miles to the lifespan of the tires, improves mileage and enhances overall ride quality. Tire pressure cannot be accurately estimated by looking at a tire, especially if it's a radial. A tire pressure gauge is essential. Keep an accurate gauge in the vehicle. The pressure gauges attached to the nozzles of air hoses at gas stations are often inaccurate.

7   Always check tire pressure when the tires are cold. Cold, in this case, means the vehicle has not been driven over a mile in the three hours preceding a tire pressure check. A pressure rise of four to eight pounds is not uncommon once the tires are warm.

8   Unscrew the valve cap protruding from the wheel or hubcap and push the gauge

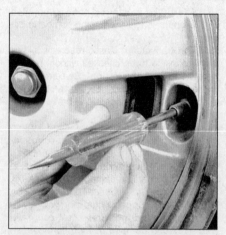

5.4a If a tire loses air on a steady basis, check the valve stem core first to make sure it's snug (special inexpensive wrenches are commonly available at auto parts stores)

5.4b If the valve stem core is tight, raise the corner of the vehicle with the low tire and spray a soapy water solution onto the tread as the tire is turned slowly - leaks will cause small bubbles to appear

5.8 To extend the life of the tires, check the air pressure at least once a week with an accurate gauge (don't forget the spare!)

6.7 The power steering fluid dipstick has marks on it so the fluid can be checked hot or cold

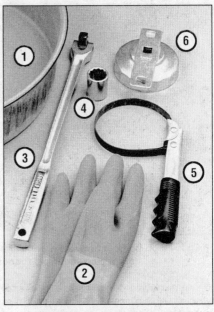

8.3 These tools are required when changing the engine oil and filter

1 **Drain pan** - It should be fairly shallow in depth, but wide to prevent spills
2 **Rubber gloves** - When removing the drain plug and filter, you will get oil on your hands (the gloves will prevent burns)
3 **Breaker bar** - Sometimes the oil drain plug is tight, and a long breaker bar is needed to loosen it
4 **Socket** - To be used with the breaker bar or a ratchet (must be the correct size to fit the drain plug - six-point preferred)
5 **Filter wrench** - This is a metal band-type wrench, which requires clearance around the filter to be effective
6 **Filter wrench** - This type fits on the bottom of the filter and can be turned with a ratchet or breaker bar (different-size wrenches are available for different types of filters)

firmly onto the valve stem **(see illustration)**. Note the reading on the gauge and compare the figure to the recommended tire pressure shown on the placard on the driver's side door pillar. Be sure to reinstall the valve cap to keep dirt and moisture out of the valve stem mechanism. Check all four tires and, if necessary, add enough air to bring them up to the recommended pressure.

9   Don't forget to keep the spare tire inflated to the specified pressure (see your owner's manual or the tire sidewall).

6   Power steering fluid level check (every 3000 miles or 3 months)

Models with 2.4L four-cylinder or 3.4L V6 engine

1   These models are equipped with Electric Power Steering (EPS). There is no power steering fluid to check.

Models with 3.0L or 3.6L V6 engine

Refer to illustration 6.7

2   The power steering system on these models relies on fluid, which may, over a period of time, require replenishing.
3   The fluid reservoir for the power steering pump is located on the pump body at the front (right side) of the engine, near the firewall.
4   For the check, the front wheels should be pointed straight ahead and the engine should be off.
5   Use a clean rag to wipe off the reservoir cap and the area around the cap. This will help prevent any foreign matter from entering the reservoir during the check.
6   Twist off the cap and check the temperature of the fluid at the end of the dipstick with your finger.
7   Wipe off the fluid with a clean rag, reinstall the dipstick, then withdraw it and read the

fluid level **(see illustration)**. The fluid should be at the proper level, depending on whether it was checked hot or cold.
8   If additional fluid is required, pour the specified type directly into the reservoir, using a funnel to prevent spills.
9   If the reservoir requires frequent fluid additions, all power steering hoses, hose connections, steering gear and the power steering pump should be carefully checked for leaks.

7   Automatic transaxle fluid level check

Checking the automatic transaxle fluid level is not a routine maintenance procedure on these vehicles. The only reason the fluid level would drop would be due to a leak.

Checking the fluid level is, of course, necessary when the transaxle fluid is drained and refilled. The fluid level checking procedure is included in Section 26.

8   Engine oil and filter change (every 3000 miles or 3 months)

Refer to illustrations 8.3, 8.9, 8.14 and 8.18
Note: *These vehicles are equipped with an oil life indicator system that illuminates a light on the instrument panel when the system deems it necessary to change the oil. A number of factors are taken into consideration to determine when the oil should be considered worn out. Generally, this system will allow the vehicle to accumulate more miles between oil changes than the traditional 3000 mile interval, but we believe that frequent oil changes are cheap insurance and will prolong engine life. If you do decide not to change your oil every 3000 miles and rely on the oil life indicator instead, make sure you don't exceed 10,000 miles before the oil is changed, regardless of what the oil life indicator shows.*

1   Frequent oil changes are the most important preventive maintenance procedures that can be done by the home mechanic. As engine oil ages, it becomes diluted and contaminated, which leads to premature engine wear.
2   Although some sources recommend oil filter changes every other oil change, we feel that the minimal cost of an oil filter and the relative ease with which it is installed dictate that a new filter be installed every time the oil is changed.
3   Gather together all necessary tools and materials before beginning this procedure **(see illustration)**.
4   You should have plenty of clean rags and newspapers handy to mop up any spills. Access to the underside of the vehicle may be improved if the vehicle can be lifted on a hoist, driven onto ramps or supported by jackstands. **Warning:** *Do not work under a vehicle which is supported only by a jack.*

8.9 Use a proper size box-end wrench or socket to remove the oil drain plug and avoid rounding it off

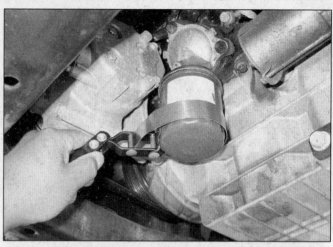

8.14 Since the oil filter is on very tight, you'll need a special wrench for removal - DO NOT use the wrench to tighten the new filter

5   If this is your first oil change, familiarize yourself with the locations of the oil drain plug and the oil filter.

6   Warm the engine to normal operating temperature. If the new oil or any tools are needed, use this warm-up time to gather everything necessary for the job. The correct type of oil for your application can be found in *Recommended lubricants and fluids* in this Chapter's Specifications.

7   With the engine oil warm (warm engine oil will drain better and more built-up sludge will be removed with it), raise and support the vehicle. Make sure it's safely supported!

8   Remove the under-vehicle splash shield. Move all necessary tools, rags and newspapers under the vehicle. Set the drain pan under the drain plug. Keep in mind that the oil will initially flow from the pan with some force; position the pan accordingly.

9   Being careful not to touch any of the hot exhaust components, use a wrench to remove the drain plug near the bottom of the oil pan **(see illustration)**. Depending on how hot the oil is, you may want to wear gloves while unscrewing the plug the final few turns.

10   Allow the oil to drain into the pan. It may be necessary to move the pan as the oil flow slows to a trickle.

11   After all the oil has drained, wipe off the drain plug with a clean rag. Small metal particles may cling to the plug and would immediately contaminate the new oil.

12   Clean the area around the drain plug opening and reinstall the plug. Tighten the plug securely with the wrench. If a torque wrench is available, use it to tighten the plug to the torque listed in this Chapter's Specifications.

V6 models

13   Move the drain pan into position under the oil filter.

14   Use the oil filter wrench to loosen the oil filter **(see illustration)**.

15   Completely unscrew the old filter. Be careful; it's full of oil. Empty the oil inside the filter into the drain pan.

16   Compare the old filter with the new one to make sure they're the same type.

17   Use a clean rag to remove all oil, dirt and sludge from the area where the oil filter mounts to the engine. Check the old filter to make sure the rubber gasket isn't stuck to the engine. If the gasket is stuck to the engine, remove it.

18   Apply a light coat of clean oil to the rubber gasket on the new oil filter **(see illustration)**.

19   Attach the new filter to the engine, following the tightening directions printed on the filter canister or packing box. Most filter manufacturers recommend against using a filter wrench due to the possibility of overtightening and damage to the seal.

Four-cylinder models

Refer to illustrations 8.20, 8.21 and 8.23

20   Move the drain pan into position under the oil filter. The canister type oil filter is located on the left, front side of the engine and is accessible from the top of the engine **(see illustration)**.

21   Unscrew the oil filter cap using a large

socket (not an open end wrench) and withdraw it together with the element **(see illustration)**.

22   Use a clean rag to remove all oil, dirt and sludge from the oil filter housing cap.

23   Install new O-ring seals in the grooves on the retaining cap and filter **(see illustration)**, then install the new element in the cap and insert them both into the filter housing. Screw on the cap and tighten it to the torque listed in this Chapter's Specifications.

All models

24   Remove all tools, rags, etc. from under the vehicle, being careful not to spill the oil in the drain pan, then lower the vehicle.

25   Move to the engine compartment and locate the oil filler cap.

26   Refer to the engine oil capacity in this Chapter's Specifications and add the proper amount of fresh oil into the engine. Wait a few minutes to allow the oil to drain into the pan, then check the level on the oil dipstick (see Section 4 if necessary). If the oil level is above the upper mark, start the engine and allow the new oil to circulate.

27   Run the engine for only about a minute and then shut it off. Immediately look under the vehicle and check for leaks at the oil pan drain plug and around the oil filter.

28   Wait about five minutes. With the new oil circulated and the filter now completely full, recheck the level on the dipstick and add more oil as necessary.

29   During the first few trips after an oil change, make it a point to check frequently for leaks and proper oil level.

30   The old oil drained from the engine cannot be reused in its present state and should be disposed of. Check with your local auto parts store, disposal facility or environmental agency to see if they will accept the oil for recycling. After the oil has cooled it can be drained into a container (capped plastic jugs, topped bottles, milk cartons, etc.) for transport to one of these disposal sites. Don't dispose of the oil by pouring it on the ground or down a drain!

8.18 Lubricate the oil filter gasket with clean engine oil before installing the filter on the engine

8.20 Unscrew the cap to access the oil filter

8.21 Remove the filter cap and cartridge assembly

Oil Life Monitor

31   The Oil Life Monitor is a function of the PCM that tracks engine operating temperature and rpm. If the PCM determines that your engine's oil has been used long enough, an indicator that shows "Change Engine Oil Soon" will light on the instrument panel.

32   When you change your engine oil and filter, whether you change it at the interval recommended in this Chapter or only when the light comes on, you will have to reset the system to make the indicator go out.

33   To reset, switch the ignition key to Run (engine not running) and depress/let up the accelerator pedal quickly three times (within five seconds). The light should flash to let you know the system is being reset. If the light does not flash, the procedure needs to be repeated.

9   Seat belt check (every 6000 miles or 6 months)

1   Check seat belts, buckles, latch plates, and guide loops for obvious damage and signs of wear.

2   Where the seat belt receptacle bolts to the floor of the vehicle, check that the bolts are secure.

3   See if the seat belt reminder light comes on when the key is turned to the Run or Start position. A chime should also sound.

10   Wiper blade inspection and replacement (every 6000 miles or 6 months)

Refer to illustrations 10.4a and 10.4b

1   The windshield wiper and blade assembly should be inspected periodically for damage, loose components and cracked or worn blade elements.

2   Road film can build up on the wiper blades and affect their efficiency, so they should be washed regularly with a mild detergent solution.

3   If the wiper blade elements are cracked, worn or warped, or no longer clean adequately, they should be replaced with new ones.

4   Lift the arm assembly away from the glass for clearance, depress the release lever, then slide the wiper blade assembly out of the hook in the end of the arm **(see illustrations)**.

5   Attach the new wiper to the arm. Connection can be confirmed by an audible click.

8.23 Install new O-rings on the filter and cap

11   Battery check, maintenance and charging (every 6000 miles or 6 months)

Refer to illustrations 11.1, 11.5, 11.7a, 11.7b and 11.7c

Warning: *Certain precautions must be followed when checking and servicing the battery. Hydrogen gas, which is highly flammable,*

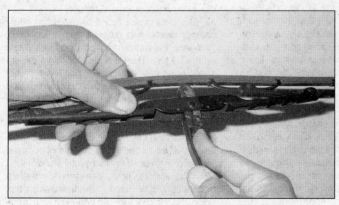

10.4a To release the blade holder, depress the release tab . . .

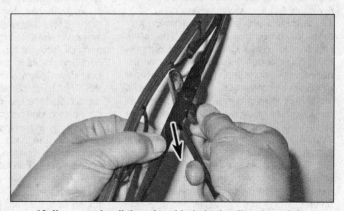

10.4b . . . and pull the wiper blade in the direction of the windshield to separate it from the arm

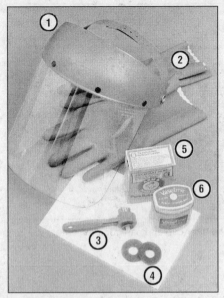

11.1 Tools and materials required for battery maintenance

1 **Face shield/safety goggles** - *When removing corrosion with a brush, the acidic particles can easily fly up into your eyes*
2 **Rubber gloves** - *Another safety item to consider when servicing the battery - remember that's acid inside the battery!*
3 **Battery terminal/cable cleaner** - *This wire brush cleaning tool will remove all traces of corrosion from the battery posts and cable clamps*
4 **Treated felt washers** - *Placing one of these on each post, under the cable clamps, will help prevent corrosion*
5 **Baking soda** - *A solution of baking soda and water can be used to neutralize corrosion*
6 **Petroleum jelly** - *A layer of this on the battery posts will help prevent corrosion*

is always present in the battery cells, so keep lighted tobacco and all other open flames and sparks away from the battery. The electrolyte inside the battery is actually dilute sulfuric acid, which will cause injury if splashed on your skin or in your eyes. It will also ruin clothes and painted surfaces. When removing the battery cables, always detach the negative cable first and hook it up last!

1   A routine preventive maintenance program for the battery in your vehicle is the only way to ensure quick and reliable starts. But before performing any battery maintenance, make sure that you have the proper equipment necessary to work safely around the battery **(see illustration)**. **Note:** *Some of the covered models have an auxiliary battery in addition to the standard battery. All of the following care and maintenance should be applied to both batteries.*

2   There are also several precautions that should be taken whenever battery maintenance is performed. Before servicing the battery, always turn the engine and all accessories off and disconnect the cable from the negative terminal of the battery.

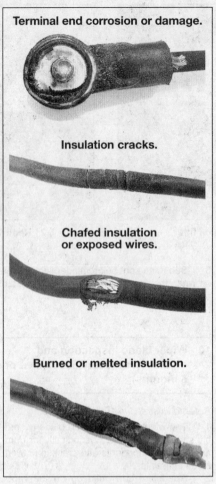

11.5 Typical battery cable problems

3   The battery produces hydrogen gas, which is both flammable and explosive. Never create a spark, smoke or light a match around the battery. Always charge the battery in a ventilated area.
4   Electrolyte contains poisonous and corrosive sulfuric acid. Do not allow it to get in your eyes, on your skin or on your clothes. Never ingest it. Wear protective safety glasses when working near the battery. Keep children away from the battery.
5   Note the external condition of the battery. If the positive terminal and cable clamp on your vehicle's battery is equipped with a rubber protector, make sure that it's not torn or damaged. It should completely cover the terminal. Look for any corroded or loose connections, cracks in the case or cover or loose hold-down clamps. Also check the entire length of each cable for cracks and frayed conductors **(see illustration)**.
6   If corrosion, which looks like white, fluffy deposits is evident, particularly around the terminals, the battery should be removed for cleaning. Loosen the cable bolts with a wrench, being careful to remove the ground cable first, and slide them off the terminals. Then remove the hold-down clamp and lift the battery from the engine compartment.
7   Clean the cable ends thoroughly with a battery brush or a terminal cleaner and a solu-

tion of warm water and baking soda. Wash the terminals and the side of the battery case with the same solution but make sure that the solution doesn't get into the battery. When cleaning the cables, terminals and battery case, wear safety goggles and rubber gloves to prevent any solution from coming in contact with your eyes or hands. Wear old clothes too - even diluted, sulfuric acid splashed onto clothes will burn holes in them. If the terminals have been corroded, clean them up with a terminal cleaner **(see illustrations)**. Thoroughly wash all cleaned areas with plain water.
8   Make sure that the battery tray is in good condition and the hold-down clamp is tight. If the battery is removed from the tray, make sure no parts remain in the bottom of the tray when the battery is reinstalled. When reinstalling the hold-down clamp bolts, do not overtighten them.
9   Any metal parts of the vehicle damaged by corrosion should be covered with a zinc-based primer, then painted.
10   Information on removing and installing the battery can be found in Chapter 5. Information on jump starting can be found at the front of this manual. For more detailed battery checking procedures, refer to the *Haynes Automotive Electrical Manual*.

Charging

Warning: *When batteries are being charged, hydrogen gas, which is very explosive and flammable, is produced. Do not smoke or allow open flames near a charging or a recently charged battery. Wear eye protection when near the battery during charging. Also, make sure the charger is unplugged before connecting or disconnecting the battery from the charger.*
Note: *The manufacturer recommends the battery be removed from the vehicle for charging because the gas that escapes during this procedure can damage the paint. Fast charging with the battery cables connected can result in damage to the electrical system.*

11   Slow-rate charging is the best way to restore a battery that's discharged to the point where it will not start the engine. It's also a good way to maintain the battery charge in a vehicle that's only driven a few miles between starts. Maintaining the battery charge is particularly important in the winter when the battery must work harder to start the engine and electrical accessories that drain the battery are in greater use.
12   It's best to use a one or two-amp battery charger (sometimes called a "trickle" charger). They are the safest and put the least strain on the battery. They are also the least expensive. For a faster charge, you can use a higher amperage charger, but don't use one rated more than 1/10th the amp/hour rating of the battery. Rapid boost charges that claim to restore the power of the battery in one to two hours are hardest on the battery and can damage batteries not in good condition. This type of charging should only be used in emergency situations.
13   The average time necessary to charge a battery should be listed in the instructions that come with the charger. As a general rule, a trickle charger will charge a battery in 12 to 16 hours.
14   Remove all the cell caps (if equipped)

11.7a A tool like this one (available at auto parts stores) is used to clean the side-terminal type battery cable contact area

11.7b Use the brush side of the tool to finish the job

11.7c Regardless of the type of tool used on the battery and cables, a clean, shiny surface should be the result

and cover the holes with a clean cloth to prevent spattering electrolyte. Disconnect the negative battery cable and hook the battery charger cable clamps up to the battery posts (positive to positive, negative to negative), then plug in the charger. Make sure it is set at 12-volts if it has a selector switch.

15   If you're using a charger with a rate higher than two amps, check the battery regularly during charging to make sure it doesn't overheat. If you're using a trickle charger, you can safely let the battery charge overnight after you've checked it regularly for the first couple of hours.

16   If the battery has removable cell caps, measure the specific gravity with a hydrometer every hour during the last few hours of the charging cycle. Hydrometers are available inexpensively from auto parts stores - follow the instructions that come with the hydrometer. Consider the battery charged when there's no change in the specific gravity reading for two hours and the electrolyte in the cells is gassing (bubbling) freely. The specific gravity reading from each cell should be very close to the others. If not, the battery probably has a bad cell(s).

17   Some batteries with sealed tops have

built-in hydrometers on the top that indicate the state of charge by the color displayed in the hydrometer window. Normally, a bright-colored hydrometer indicates a full charge and a dark hydrometer indicates the battery still needs charging.

18   If the battery has a sealed top and no built-in hydrometer, you can hook up a digital voltmeter across the battery terminals to check the charge. A fully charged battery should read 12.5 volts or higher.

19   Further information on the battery and jump-starting can be found in Chapter 5 and at the front of this manual.

12   Drivebelt and tensioner check and replacement (every 6000 miles or 6 months)

Refer to illustrations 12.2, 12.4 and 12.7

1   A serpentine drivebelt is located at the front of the engine and plays an important role in the overall operation of the engine and its components. Due to its function and material

make up, the belt is prone to wear and should be periodically inspected. The serpentine belt drives the alternator, power steering pump, water pump and air conditioning compressor.

2   With the engine off, open the hood and use your fingers (and a flashlight, if necessary), to move along the belt checking for cracks and separation of the belt plies. Also check for fraying and glazing, which gives the belt a shiny appearance **(see illustration)**. Both sides of the belt must be inspected.

3   Check the ribs on the underside of the belt. They should all be the same depth, with none of the surface uneven.

4   The tension of the belt is maintained by a spring-loaded tensioner assembly and isn't adjustable. The belt should be replaced when the indexing arrow is lined up with the indexing mark on the tensioner assembly **(see illustration)**.

5   Remove the air filter housing (see Chapter 4).

6   On V6 models, support the engine with a floor jack and a block of wood, then remove the right-side engine mount and bracket (see Chapter 2A or 2B).

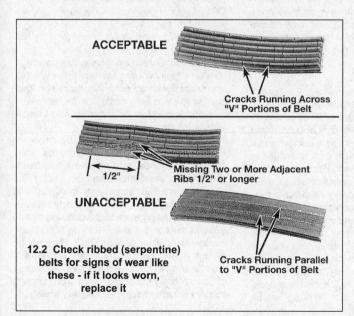

12.2 Check ribbed (serpentine) belts for signs of wear like these - if it looks worn, replace it

ACCEPTABLE

Cracks Running Across "V" Portions of Belt

1/2"

Missing Two or More Adjacent Ribs 1/2" or longer

UNACCEPTABLE

Cracks Running Parallel to "V" Portions of Belt

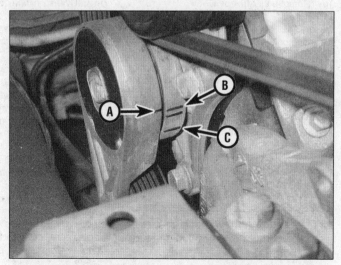

12.4 The indexing mark (A) on the drivebelt's tensioner must remain between the marks (B and C) on the tensioner assembly (C is the wear limit mark) (3.4L V6 engine shown - others similar)

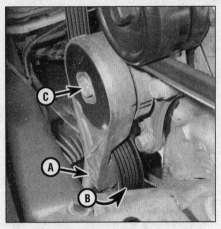

12.7 Insert a 3/8-inch drive tool into the square hole in the tensioner (A), then turn the tensioner counterclockwise (B, 2.4L four-cylinder and 3.4L V6) or clockwise (3.0L and 3.6L V6) for belt removal. C is the tensioner mounting bolt (3.4L V6 engine; on the 3.0L and 3.6L V6, the tensioner bracket must be unbolted from the engine)

7   Insert a 3/8-inch drive ratchet or breaker bar into the square hole in the tensioner, then rotate the tensioner counterclockwise (2.4L four-cylinder and 3.4L V6) or clockwise (3.0L and 3.6L V6) to release belt tension **(see illustration)**. **Note:** *If your ratchet or breaker bar is too thick to fit in the space between the tensioner and the chassis, special drivebelt tensioner tools are available at most auto parts stores.*
8   Noting how it's routed, remove the belt from the tensioner and auxiliary components, then slowly release the tensioner.
9   After verifying the new belt is the same length as the original belt, route the new belt over the various pulleys, again rotating the tensioner to allow the belt to be installed, then release the belt tensioner.

Tensioner replacement

10   Remove the drivebelt.
11   If you're working on a 2.4L four-cylinder or 3.4L V6 model, remove the tensioner mounting bolt **(see illustration 12.7)**.
12   If you're working on a 3.0L or 3.6L V6 model, remove the tensioner mounting bracket bolts and detach the tensioner/mounting bracket from the engine.
Note: *On 2013 and later 3.6L engines, it will be necessary to remove the idler pulley bolt and idler pulley to access the tensioner pulley bracket bolt.*
13   Installation is the reverse of the removal procedure. Tighten the mounting bolt(s) to the torque listed in this Chapter's Specifications.

13   Underhood hose check and replacement (every 6000 miles or 6 months)

Warning: *Replacement of air conditioning hoses must be left to a dealer service depart-*

ment or air conditioning shop that has the equipment to depressurize the system safely and recover the refrigerant. Never remove air conditioning components or hoses until the system has been depressurized.
1   High temperatures in the engine compartment can cause the deterioration of the rubber and plastic hoses used for engine, accessory and emission systems operation. Periodic inspection should be made for cracks, loose clamps, material hardening and leaks. Information specific to the cooling system hoses can be found in Section 14.
2   Some, but not all, hoses are secured to their fittings with clamps. Where clamps are used, check to be sure they haven't lost their tension, allowing the hose to leak. If clamps aren't used, make sure the hose has not expanded and/or hardened where it slips over the fitting, allowing it to leak.

Vacuum hoses

3   It's quite common for vacuum hoses, especially those in the emissions system, to be color-coded or identified by colored stripes molded into them. Various systems require hoses with different wall thickness, collapse resistance and temperature resistance. When replacing hoses, be sure the new ones are made of the same material.
4   Often the only effective way to check a hose is to remove it completely from the vehicle. If more than one hose is removed, be sure to label the hoses and fittings to ensure correct installation.
5   When checking vacuum hoses, be sure to include any plastic T-fittings in the check. Inspect the fittings for cracks and the hose where it fits over the fitting for distortion, which could cause leakage.
6   A small piece of vacuum hose (1/4-inch inside diameter) can be used as a stethoscope to detect vacuum leaks. Hold one end of the hose to your ear and probe around vacuum hoses and fittings, listening for the hissing sound characteristic of a vacuum leak. **Warning:** *When probing with the vacuum hose stethoscope, be very careful not to come into contact with moving engine components such as the drivebelt, cooling fan, etc.*

Fuel hose

Warning: *Gasoline is extremely flammable, so take extra precautions when you work on any part of the fuel system. Don't smoke or allow open flames or bare light bulbs near the work area, and don't work in a garage where a gas-type appliance (such as a water heater or clothes dryer) is present. Since gasoline is carcinogenic, wear fuel-resistant gloves when there's a possibility of being exposed to fuel, and, if you spill any fuel on your skin, rinse it off immediately with soap and water. Mop up any spills immediately and do not store fuel-soaked rags where they could ignite. When you perform any kind of work on the fuel system, wear safety glasses and have a Class B type fire extinguisher on hand. The fuel system is under pressure, so if any lines must be disconnected,*

the pressure in the system must be relieved first (see Chapter 4 for more information).
7   Check all rubber fuel lines for deterioration and chafing. Check especially for cracks in areas where the hose bends and just before fittings, such as where a hose attaches to the fuel filter and fuel injection unit.
8   High quality fuel line, specifically designed for high-pressure fuel injection applications, must be used for fuel line replacement. Never, under any circumstances, use regular fuel line, unreinforced vacuum line, clear plastic tubing or water hose for fuel lines.
9   Spring-type (pinch) clamps are commonly used on fuel lines. These clamps often lose their tension over a period of time, and can be sprung during removal. Replace all spring-type clamps with screw clamps whenever a hose is replaced.

Metal lines

10   Sections of metal line are routed along the frame, between the fuel tank and the engine. Check carefully to be sure the line has not been bent or crimped and no cracks have started in the line.
11   If a section of metal fuel line must be replaced, only seamless steel tubing should be used, since copper and aluminum tubing don't have the strength necessary to withstand normal engine vibration.
12   Check the metal brake lines where they enter the master cylinder and brake proportioning unit for cracks in the lines or loose fittings. Any sign of brake fluid leakage calls for an immediate and thorough inspection of the brake system.

14   Cooling system check (every 6000 miles or 6 months)

Refer to illustration 14.4
Caution: *Never mix green-colored ethylene glycol antifreeze and orange-colored "DEX-COOL" silicate-free coolant because doing so will destroy the efficiency of the "DEX-COOL" coolant which is designed to last for 100,000 miles or five years.*
1   Many major engine failures can be attributed to a faulty cooling system. If the vehicle is equipped with an automatic transaxle, the cooling system also cools the transmission fluid and thus plays an important role in prolonging transmission life.
2   The cooling system should be checked with the engine cold. Do this before the vehicle is driven for the day or after it has been shut off for at least three hours.
3   Remove the expansion tank cap by slowly unscrewing it. If you hear any hissing sounds (indicating there is still pressure in the system), wait until it stops. If there is no hissing sound, continue unscrewing it. Thoroughly clean the cap, inside and out, with clean water. Also clean the opening on the tank. All traces of corrosion should be removed. The coolant inside the tank should be relatively transpar-

Check for a chafed area that could fail prematurely.

Check for a soft area indicating the hose has deteriorated inside.

Overtightening the clamp on a hardened hose will damage the hose and cause a leak.

Check each hose for swelling and oil-soaked ends. Cracks and breaks can be located by squeezing the hose.

14.4 Hoses, like drivebelts, have a habit of failing at the worst possible time - to prevent the inconvenience of a blown radiator or heater hose, inspect them carefully as shown here

ent. If it is rust colored, the system should be drained and refilled (see Section 29). If the coolant level is not up to the COLD FILL mark, add additional antifreeze/coolant mixture (see Section 4).

4   Carefully check the large upper and lower radiator hoses along with any smaller diameter heater hoses that run from the engine to the firewall. Inspect each hose along its entire length, replacing any hose that is cracked, swollen or shows signs of deterioration. Cracks may become more apparent if the hose is squeezed **(see illustration)**.

5   Make sure all hose connections are tight. A leak in the cooling system will usually show up as white or rust-colored deposits on the areas adjoining the leak. If wire-type clamps are used at the ends of the hoses, it may be wise to replace them with more secure, screw-type clamps.

6   Use compressed air or a soft brush to remove bugs, leaves, etc. from the front of

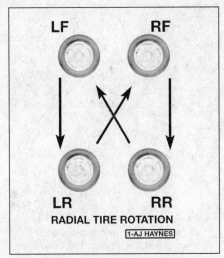

15.2 The recommended four-tire rotation pattern for these vehicles

the radiator or air conditioning condenser. Be careful not to damage the delicate cooling fins or cut yourself on them.

7   Every other inspection, or at the first indication of cooling system problems, have the cap and system pressure tested. If you don't have a pressure tester, most gas stations and repair shops will do this for a minimal charge.

15   Tire rotation (every 6000 miles or 6 months)

Refer to illustration 15.2

1   The tires should be rotated at the specified intervals and whenever uneven wear is noticed.

2   Tires must be rotated in the recommended pattern **(see illustration)**.

3   Refer to the information in *Jacking and towing* at the front of this manual for the proper procedures to follow when raising the vehicle and changing a tire. If the brakes are to be checked, don't apply the parking brake as stated. Make sure the tires are blocked to prevent the vehicle from rolling as it's raised.

4   Preferably, the entire vehicle should be raised at the same time. This can be done on a hoist or by jacking up each corner and then lowering the vehicle onto jackstands placed under the frame rails. Always use four jackstands and make sure the vehicle is safely supported.

5   After rotation, check and adjust the tire pressures as necessary. Tighten the lug nuts to the torque listed in this Chapter's Specifications.

16   Differential lubricant level check (AWD models) (every 6,000 miles or 6 months)

2005 and 2006 models

1   Before checking the fluid level, drive the vehicle on a flat surface (paved, preferably), in three tight circles at five mph. This will prime the

drive module and give a more accurate check.

2   Raise the vehicle and support it securely on jackstands.

3   Remove the plug from the filler hole in the rear of the differential drive module, just inboard of the left rear driveaxle.

4   The lubricant level should be measured from the bottom of the filler hole. The level should be below the fill-plug opening by 1/2 inch. If not, use a pump or squeeze bottle to add the recommended lubricant.

5   Install the plug and tighten it to the torque listed in this Chapter's Specifications.

2007 and later models

6   Raise the vehicle and support it securely on jackstands.

7   Remove the plug from the filler hole on the left side of the rear axle housing.

8   The lubricant level should be even with the bottom of the filler hole. If not, use a pump or squeeze bottle to add the recommended lubricant.

9   Install the plug and tighten it to the torque listed in this Chapter's Specifications.

17   Driveaxle boot check (every 15,000 miles or 12 months)

Refer to illustration 17.2

1   The driveaxle boots are very important because they prevent dirt, water and foreign material from entering and damaging the constant velocity (CV) joints. Oil and grease can cause the boot material to deteriorate prematurely, so it's a good idea to wash the boots with soap and water. Because it constantly pivots back and forth following the steering action of the front hub, the outer CV boot wears out sooner and should be inspected regularly.

2   Inspect the boots for tears and cracks as well as loose clamps **(see illustration)**. If there is any evidence of cracks or leaking lubricant, they must be replaced as described in Chapter 8.

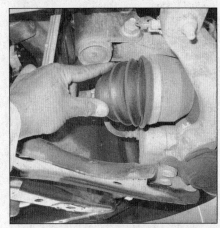

17.2 Inspect the inner and outer driveaxle boots for loose clamps, cracks or signs of leaking lubricant (AWD models have driveaxles at the rear, too)

19.7a With the wheel off, check the thickness of the inner pad through the inspection hole in the caliper (front caliper shown, rear caliper similar)

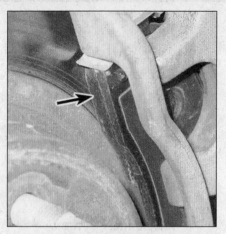

19.7b The outer pad is more easily checked at the edge of the caliper

18   Fuel system check (every 15,000 miles or 12 months)

Warning: *Gasoline is extremely flammable, so take extra precautions when you work on any part of the fuel system. Don't smoke or allow open flames or bare light bulbs near the work area, and don't work in a garage where a gas-type appliance (such as a water heater or clothes dryer) is present. Since gasoline is carcinogenic, wear fuel-resistant gloves when there's a possibility of being exposed to fuel, and, if you spill any fuel on your skin, rinse it off immediately with soap and water. Mop up any spills immediately and do not store fuel-soaked rags where they could ignite. When you perform any kind of work on the fuel system, wear safety glasses and have a Class B type fire extinguisher on hand. The fuel system is under constant pressure, so, before any lines are disconnected, the fuel system pressure must be relieved (see Chapter 4).*

1   If you smell gasoline while driving or after the vehicle has been sitting in the sun, inspect the fuel system immediately.

2   Remove the gas filler cap and inspect if for damage and corrosion. The gasket should have an unbroken sealing imprint. If the gasket is damaged or corroded, install a new cap.

3   Inspect the fuel feed and return lines for cracks. Make sure that the connections between the fuel lines and the fuel injection system are tight. **Warning:** *Your vehicle is fuel injected, so you must relieve the fuel system pressure before servicing fuel system components. The fuel system pressure relief procedure is outlined in Chapter 4.*

4   Since some components of the fuel system - the fuel tank and part of the fuel feed and return lines, for example - are underneath the vehicle, they can be inspected more easily with the vehicle raised on a hoist. If that's not possible, raise the vehicle and support it on jackstands.

5   With the vehicle raised and safely supported, inspect the gas tank and filler neck for punctures, cracks and other damage. The connection between the filler neck and the tank is particularly critical. Sometimes a rubber filler neck will leak because of loose clamps or deteriorated rubber. Inspect all fuel tank mounting brackets and straps to be sure that the tank is securely attached to the vehicle. **Warning:** *Do not, under any circumstances, try to repair a fuel tank (except rubber components). A welding torch or any open flame can easily cause fuel vapors inside the tank to explode.*

6   Carefully check all rubber hoses and metal lines leading away from the fuel tank. Check for loose connections, deteriorated hoses, crimped lines and other damage. Repair or replace damaged sections as necessary (see Chapter 4).

7   The evaporative emissions control system can also be a source of fuel odors. The function of the system is to store fuel vapors from the fuel tank in a charcoal canister until they can be routed to the intake manifold where they mix with incoming air before being burned in the combustion chambers.

8   The most common symptom of a faulty evaporative emissions system is a strong odor of fuel coming from the area of the charcoal canister. If a fuel odor has been detected, and you have already checked the areas described above, check the charcoal canister, located near the fuel tank, and the hoses connected to it (see Chapter 6).

19   Brake system check (every 15,000 miles or 12 months)

Warning: *The dust created by the brake system is harmful to your health. Never blow it out with compressed air and don't inhale any of it. An approved filtering mask should be worn when working on the brakes. Do not, under any circumstances, use petroleum-based solvents to clean brake parts. Use brake system cleaner only!*

Note: *For detailed photographs of the brake system, refer to Chapter 9.*

1   In addition to the specified intervals, the brakes should be inspected every time the wheels are removed or whenever a defect is suspected.

2   Any of the following symptoms could indicate a potential brake system defect: The vehicle pulls to one side when the brake pedal is depressed; the brakes make squealing or dragging noises when applied; brake pedal travel is excessive; the pedal pulsates; or brake fluid leaks, usually onto the inside of the tire or wheel.

3   Loosen the wheel lug nuts.

4   Raise the vehicle and place it securely on jackstands.

5   Remove the wheels (see *Jacking and towing* at the front of this book, or your owner's manual, if necessary).

Disc brakes

Refer to illustrations 19.7a, 19.7b, 19.9 and 19.11

6   There are two pads (an outer and an inner) in each caliper. The pads are visible with the wheels removed.

7   Check the pad thickness by looking at each end of the caliper and through the inspection window in the caliper body **(see illustrations)**. If the lining material is less than the thickness listed in this Chapter's Specifications, replace the pads. **Note:** *Keep in mind that the lining material is riveted or bonded to a metal backing plate and the metal portion is not included in this measurement.*

8   If it is difficult to determine the exact thickness of the remaining pad material by the above method, or if you are at all concerned about the condition of the pads, remove the caliper(s), then remove the pads from the calipers for further inspection (see Chapter 9).

9   Once the pads are removed from the calipers, clean them with brake cleaner and re-measure them with a ruler or a vernier caliper **(see illustration)**.

10   Measure the disc thickness with a micrometer to make sure that it still has service life remaining. If any disc is thinner than the specified minimum thickness, replace it (see Chapter 9). Even if the disc has service life remaining, check its condition. Look for scoring, gouging and burned spots. If these conditions exist, remove the disc and have it resurfaced (see Chapter 9).

11   Before installing the wheels, check all brake lines and hoses for damage, wear, deformation, cracks, corrosion, leakage, bends and twists, particularly in the vicinity of the rubber hoses at the calipers **(see illustration)**. Check the clamps for tightness and the connections for leakage. Make sure that all hoses and lines are clear of sharp edges, moving parts and the exhaust system. If any of the above conditions are noted, repair, reroute or replace the lines and/or fittings as necessary (see Chapter 9).

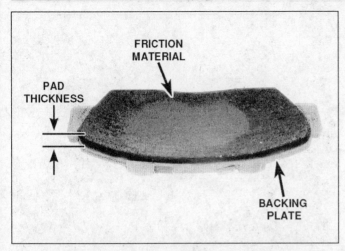

19.9 If a more precise measurement of pad thickness is necessary, remove the pads and measure the remaining friction material

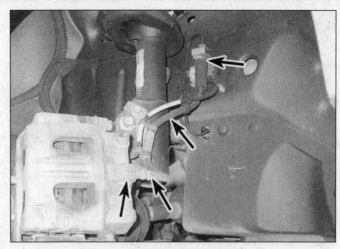

19.11 Check along the brake hoses and at each fitting for deterioration, cracks and leakage

Drum brakes

Refer to illustrations 19.15 and 19.17

12   With the rear wheels removed, make sure the parking brake is off and tap on the outside of the brake drums with a mallet to loosen them.

13   Remove the brake drums. If they won't come off, refer to Chapter 9.

14   Carefully clean the brake assemblies with brake system cleaner. **Warning:** *Don't blow the dust out with compressed air and don't inhale any of it (it's harmful to your health).*

15   Note the thickness of the lining material on both front and rear shoes **(see illustration).** Compare the measurement with the limit given in this Chapter's Specifications; if any lining is thinner than specified, then all of the brake shoes must be replaced (see Chapter 9). They should also be replaced if they're cracked, glazed with shiny areas or soaked with brake fluid.

16   Make sure that all of the brake springs are connected and in good condition.

17   Check the wheel cylinders for fluid leakage. Carefully pry back the wheel cylinder rubber cups with your fingers or a small screwdriver **(see illustration).** Any leakage is an indication that the wheel cylinders should be replaced at once (see Chapter 9). Also check all hoses and connections for signs of leakage.

18   Wipe the inside of the brake drums with a clean rag and brake cleaner or alcohol. Again, be careful not to breathe the dangerous brake dust.

19   Check the inside of the drum for score marks, deep scratches and hard spots that appear as small discolored areas. If imperfections can't be removed with fine emery cloth, the drum must be taken to an automotive machine shop for machining.

20   If your inspection reveals that all parts are in good condition, install the drums, install the wheels and lower the vehicle.

Brake booster check

21   Sit in the driver's seat and perform the following sequence of tests.

22   With the brake fully depressed, start the engine - the pedal should move down a little when the engine starts.

23   With the engine running, depress the brake pedal several times - the travel distance should not change.

24   Depress the brake, stop the engine and hold the pedal in for about 30 seconds - the pedal should neither sink nor rise.

25   Restart the engine, run it for about a minute and turn it off. Then firmly depress the brake several times - the pedal travel should decrease with each application.

26   If your brakes do not operate as

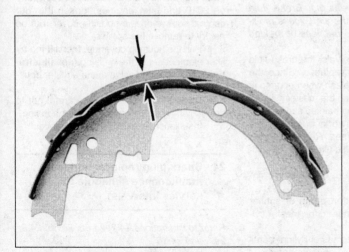

19.15 If the lining is bonded to the brake shoe, measure its thickness from the outer surface to the steel shoe, as shown here; if the lining is riveted to the shoe, measure from the lining outer surface to the rivet head

19.17 Carefully peel back the wheel cylinder boot and check for leaking fluid, indicating that the cylinder must be rebuilt

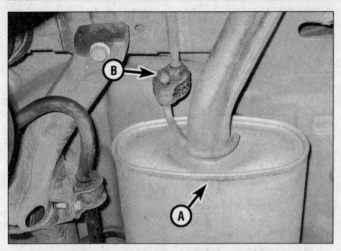

20.2a Inspect the muffler (A) and all hangers (B) for signs of deterioration

20.2b Inspect all flanged joints for signs of exhaust gas leakage

described, the brake booster has failed. Refer to Chapter 9 for the replacement procedure.

Parking brake

27   One method of checking the parking brake is to park the vehicle on a steep hill with the parking brake set and the transmission in Neutral (be sure to stay in the vehicle for this check). If the parking brake cannot prevent the vehicle from rolling, it's in need of attention (see Chapter 9).

20   Exhaust system check (every 15,000 miles or 12 months)

Refer to illustrations 20.2a and 20.2b

1   With the engine cold (at least three hours after the vehicle has been driven), check the complete exhaust system from the manifold to the end of the tailpipe. Be careful around the catalytic converter, which may be hot even after three hours. The inspection should be done with the vehicle on a hoist to permit unrestricted access. If a hoist isn't available, raise the vehicle and support it securely on jackstands.
2   Check the exhaust pipes and connections for signs of leakage and/or corrosion indicating a potential failure. Make sure that all brackets and hangers are in good condition and tight **(see illustrations)**.
3   Inspect the underside of the body for holes, corrosion, open seams, etc. which may allow exhaust gasses to enter the passenger compartment. Seal all body openings with silicone sealant or body putty.
4   Rattles and other noises can often be traced to the exhaust system, especially the hangers, mounts and heat shields. Try to move the pipes, mufflers and catalytic converter. If the components can come in contact with the body or suspension parts, secure the exhaust system with new brackets and hangers.

21   Transfer case lubricant level check (AWD models) (every 15,000 miles or 12 months)

1   Raise the vehicle and support it securely on jackstands.
2   The transfer case lubricant level is checked by removing the fill plug located at the side of the case.
3   The lubricant level should be just at the bottom of the hole. If not, add the appropriate lubricant through the opening.

22   Brake fluid change (every 30,000 miles or 24 months)

Warning: *Brake fluid can harm your eyes and damage painted surfaces, so use extreme caution when handling or pouring it. Do not use brake fluid that has been standing open or is more than one year old. Brake fluid absorbs moisture from the air. Excess moisture can cause a dangerous loss of braking effectiveness.*

1   At the specified intervals, the brake fluid should be drained and replaced. Since the brake fluid may drip or splash when pouring it, place plenty of rags around the master cylinder to protect any surrounding painted surfaces.
2   Before beginning work, purchase the specified brake fluid (see *Recommended lubricants and fluids* in this Chapter's Specifications).
3   Remove the cap from the master cylinder reservoir.
4   Using a hand suction pump or similar device, withdraw the fluid from the master cylinder reservoir.
5   Add new fluid to the master cylinder until it rises to the line indicated on the reservoir.
6   Bleed the brake system as described in Chapter 9 at all four brakes until new and uncontaminated fluid is expelled from the bleeder screw. Be sure to maintain the fluid

level in the master cylinder as you perform the bleeding process. If you allow the master cylinder to run dry, air will enter the system.
7   Refill the master cylinder with fluid and check the operation of the brakes. The pedal should feel solid when depressed, with no sponginess. **Warning:** *Do not operate the vehicle if you are in doubt about the effectiveness of the brake system.*

23   Air filter replacement (every 30,000 miles or 24 months)

Refer to illustrations 23.2 and 23.3

1   At the specified intervals, the air filter element should be replaced with a new one.
2   On all models, the air filter is housed in a black plastic box mounted on the right side of the engine compartment. Remove the intake tube, then on 2009 and earlier models disengage the retaining clips **(see illustration)** and on 2010 and later engines, loosen the four captive screws on the housing cover (but do not try to remove the screws).
3   Pull the housing cover up, then lift the air filter element out of the housing **(see illustration)**. Wipe out the inside of the air filter housing with a clean rag.
4   While the cover is off, be careful not to drop anything down into the air filter housing.
5   Installation is the reverse of removal.

24   Spark plug replacement (see maintenance schedule for service intervals)

Refer to illustrations 24.2, 24.5a, 24.5b, 24.8, 24.9 and 24.10

1   The spark plugs are threaded into the top (3.6L V6) or the side (3.4L V6) of each cylinder head. If you're working on a 3.6L V6, you'll have to remove the air intake duct for access.

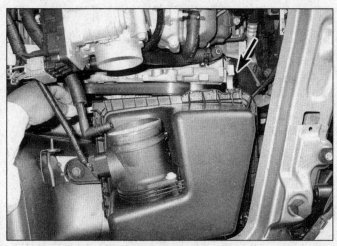

23.2 Disengage the retaining clamps . . .

23.3 . . . then lift the cover up and remove the filter element

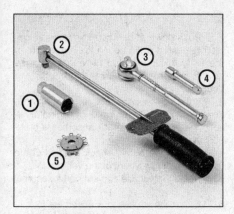

24.2 Tools required for changing spark plugs

1 *Spark plug socket - This will have special padding inside to protect the spark plug's porcelain insulator*
2 *Torque wrench - Although not mandatory, using this tool is the best way to ensure the plugs are tightened properly*
3 *Ratchet - Standard hand tool to fit the spark plug socket*
4 *Extension - Depending on model and accessories, you may need special extensions and universal joints to reach one or more of the plugs*
5 *Spark plug gap gauge - These come in a variety of styles. Make sure the gap for your engine is included*

2   In most cases, the tools necessary for spark plug replacement include a spark plug socket which fits onto a ratchet (spark plug sockets are padded inside to prevent damage to the porcelain insulators on the new plugs), various extensions and a gap gauge to check and adjust the gaps on the new plugs **(see illustration)**. A torque wrench should be used to tighten the new plugs.
3   The best approach when replacing the spark plugs is to purchase the new ones in advance, adjust them to the proper gap and

24.5a When checking the spark plug gap, slide the thin side between the electrodes and turn it until the gauge just fills the gap, then read the thickness on the gauge - do not force the tool into the gap or use the tapered portion to widen a gap

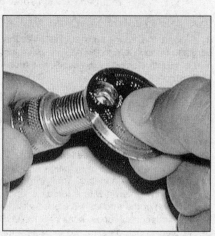

24.5b To change the gap, bend the side electrode only, using the adjuster hole in the tool, and be very careful not to crack or chip the porcelain insulator surrounding the center electrode

replace them one at a time. When buying the new spark plugs, be sure to obtain the correct plug type for your particular engine. This information can be found in your owner's manual and in this Chapter's Specifications.
4   Allow the engine to cool completely before attempting to remove any of the plugs. While you're waiting for the engine to cool, check the new plugs for defects and adjust the gaps.
5   The gap is checked by inserting the proper-thickness gauge between the electrodes at the tip of the plug **(see illustration)**. The gap between the electrodes should be the same as the one specified on the Emissions Control Information label or in this Chapter's Specifications. The gauge should just slide between the electrodes with a slight amount of drag. If the gap is incorrect, use the adjuster on the gauge body to bend the

curved side electrode slightly until the proper gap is obtained **(see illustration)**. If the side electrode is not exactly over the center electrode, bend it with the adjuster until it is. Check for cracks in the porcelain insulator (if any are found, the plug should not be used). **Note:** *Be careful not to scrape the thin platinum or iridium coating from the electrodes, as this would dramatically shorten the life of the plugs.*
6   If you're working on a 3.6L V6 engine, remove the ignition coils (see Chapter 5). If you're working on a 3.4L V6 engine, remove the spark plug wires by gently twisting the boots and pulling them straight off the plugs.
7   If compressed air is available, use it to blow any dirt or foreign material away from around the spark plugs. The idea here is to eliminate the possibility of debris falling into the cylinder as the spark plug is removed.

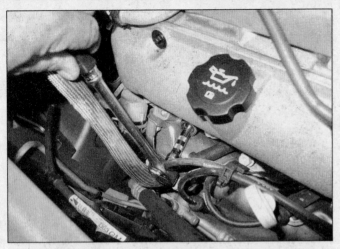

24.8 Use a socket and extension to unscrew the spark plugs - various length extensions and perhaps a flex-joint may be required to reach some plugs

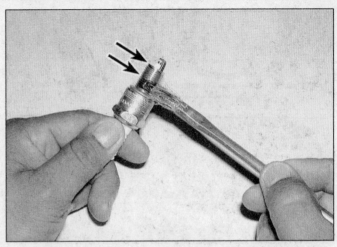

24.9 Apply a thin coat of anti-seize compound to the spark plug threads, being careful not to get any near the lower threads

8   Place the spark plug socket over the plug and remove it from the engine by turning it in a counterclockwise direction **(see illustration)**.

9   Compare the spark plug with the chart on the inside back cover of this manual to get an indication of the general running condition of the engine. Before installing the new plugs, it is a good idea to apply a thin coat of anti-seize compound to the threads **(see illustration)**.

10   Thread one of the new plugs into the hole until you can no longer turn it with your fingers, then tighten it with a torque wrench (if available) or the ratchet. It's a good idea to slip a short length of rubber hose over the end of the plug to use as a tool to thread it into place **(see illustration)**. The hose will grip the plug well enough to turn it, but will start to slip if the plug begins to cross-thread in the hole - this will prevent damaged threads and the accompanying repair costs.

11   Reinstall the ignition coil or reattach the spark plug wire, as applicable.

12   Repeat the procedure for the remaining spark plugs.

---

25   Suspension and steering check (every 30,000 miles or 30 months)

Note: *The steering linkage and suspension components should be checked periodically. Worn or damaged suspension and steering linkage components can result in excessive and abnormal tire wear, poor ride quality and vehicle handling and reduced fuel economy. For detailed illustrations of the steering and suspension components, refer to Chapter 10.*

Strut and shock absorber check

Refer to illustration 25.6

1   Park the vehicle on level ground, turn the engine off and set the parking brake. Check the tire pressures.

2   Push down at one corner of the vehicle, then release it while noting the movement of the body. It should stop moving and come to rest in a level position within one or two bounces.

3   If the vehicle continues to move up-and-down or if it fails to return to its original position, a worn or weak shock absorber is probably the reason.

4   Repeat the above check at each of the three remaining corners of the vehicle.

5   Raise the vehicle and support it securely on jackstands.

6   Check the struts/shock absorbers for evidence of fluid leakage **(see illustration)**. A light film of fluid is no cause for concern. Make sure that any fluid noted is from the struts/shocks and not from some other source. If leakage is noted, replace the struts/shocks as a set.

7   Check the struts/shocks to be sure that they are securely mounted and undamaged. Check the upper mounts for damage and wear. If damage or wear is noted, replace the

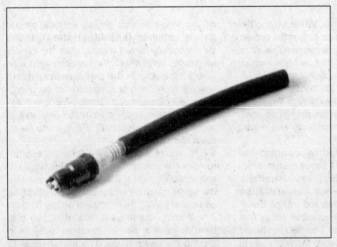

24.10 A length of snug-fitting rubber hose will save time and prevent damaged threads when installing the spark plugs

25.6 Check for signs of fluid leakage at this point on struts/shock absorbers (front strut shown)

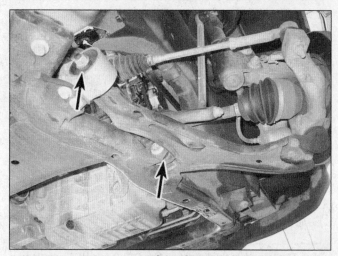

25.9a Examine the mounting points for the control arms on the front suspension

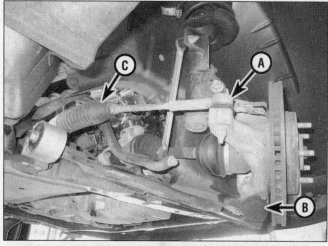

25.9b Inspect the tie-rod ends (A), the balljoints (B) and the steering gear boots (C)

25.11 With the steering wheel in the locked position and the vehicle raised, grasp the front tire as shown and try to move it back-and-forth - if any play is noted, check the steering gear mounts and tie-rod ends for looseness

26.5a Transaxle drain plug - five-speed transaxle/3.4L V6 engine

26.5b Transaxle drain plug - six-speed transaxle

struts/shocks as a set (front or rear).

8   If the struts/shocks must be replaced, refer to Chapter 10 for the procedure.

Steering and suspension check

Refer to illustrations 25.9a, 25.9b and 25.11

9   Visually inspect the steering and suspension components (front and rear) for damage and distortion. Look for damaged seals, boots and bushings and leaks of any kind. Examine the bushings where the control arms meet the chassis **(see illustrations)**.

10   Clean the lower end of the steering knuckle. Have an assistant grasp the lower edge of the tire and move the wheel in-and-out while you look for movement at the steering knuckle-to-control arm balljoint. If there is any movement the balljoint must be replaced.

11   Grasp each front tire at the front and rear edges, push in at the front, pull out at the rear and feel for play in the steering system com-

ponents. If any freeplay is noted, check the tie-rod ends for looseness or wear **(see illustration)**.

12   Additional steering and suspension system information and illustrations can be found in Chapter 10.

26   Automatic transaxle fluid change (every 30,000 miles or 30 months)

Refer to illustrations 26.5a and 26.5b

1   At the specified intervals, the transmission fluid should be drained and replaced. Since the fluid will remain hot long after driving, perform this procedure only after the engine has cooled down completely.

2   Before beginning work, purchase the specified transmission fluid (see *Recommended lubricants and fluids* in this Chapter's Specifications). **Note:** *On models with a five-speed transaxle, a new sealing washer for the*

drain plug should be obtained.

3   Other tools necessary for this job include a floor jack, jackstands to support the vehicle in a raised position, a drain pan capable of holding at least eight quarts, newspapers and clean rags.

4   Raise the vehicle and support it securely on jackstands.

5   Place the drain pan underneath the transaxle drain plug. Remove the drain plug and allow the fluid to drain **(see illustrations)**.

6   Install the drain plug and tighten it to the torque listed in this Chapter's Specifications.

Five-speed transaxle

Refer to illustrations 26.8 and 26.9

7   Raise the rear of the vehicle and support it securely on jackstands so the vehicle is level.

26.8 The automatic transaxle fluid dipstick on the five-speed transaxle is secured by a bolt (front transaxle mount removed for clarity)

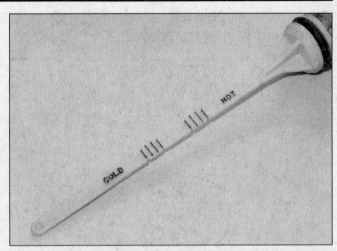

26.9 The transaxle fluid dipstick has COLD and HOT markings, but the final check of the level should always be made when the fluid is hot

26.21 Location of the automatic transaxle vent/fill cap

26.24 The automatic transaxle check plug is located near the driver's side driveaxle

8   These models have a short dipstick that's located behind the front transaxle mount, and it's bolted in place **(see illustration)**; it's easier to access it from below. Remove the bolt and take out the dipstick.

9   Add approximately 3.3 quarts of the specified type of automatic transmission fluid through the dipstick hole. Insert the dipstick and pull it out; the fluid level should be within the COLD range **(see illustration)**. If necessary, add fluid until the level is up to the COLD range.

10   Reinstall the dipstick (without the bolt) and start the engine. Allow the engine to run until the transmission fluid temperature is approximately 140 to 160-degrees F. With the brake pedal depressed, slowly shift through each gear range, then place the shifter in Park.

11   With the engine idling, pull out the dipstick, wipe it off, insert it and pull it out again, noting the fluid level. It should now be within the HOT range. If necessary, add fluid a little at a time until it is within the HOT range. **Caution:** *It is important to not overfill the transaxle.*

12   Turn off the engine, reinstall the dipstick and tighten the hold-down bolt securely. Lower the vehicle.

Six-speed transaxle with a dipstick

13   Lower the vehicle.

14   Remove the transaxle fluid dipstick located on the front side of the transaxle.

15   Add approximately 4-1/2 quarts of the specified type of automatic transmission fluid through the dipstick hole. Reinstall the dipstick.

16   Start the engine. With the brake pedal depressed, slowly shift through each gear range, then place the shifter in Park. Let the engine idle for one minute.

17   Using the Driver Information Center, display the transmission fluid temperature (TFT). If the fluid temperature is between 180 and 200-degrees F, proceed to the next Step. If the temperature is not up to 180-degrees F, drive the vehicle until it is. If it's above 200-degrees F, wait until the fluid has cooled back down and falls within the temperature range. **Note:** *Some scan tools are capable of displaying transmission fluid temperature.*

18   With the transaxle fluid at the proper temperature, remove the dipstick from the filler tube. Wipe the fluid from the dipstick with a clean rag, then reinsert it.

19   Pull the dipstick out again and note the fluid level; it should be in the crosshatched range. If additional fluid is required, add fluid a little at a time and keep checking the level until it's correct. It is important to not overfill the transmission.

Six-speed transaxle without a dipstick

Refer to illustrations 26.21 and 26.24

20   Raise the rear of the vehicle and support it securely on jackstands so the vehicle is level.

21   Remove the vent/fill cap (see illustration). Some models are equipped with a vent hose attached to the cap; if so, disconnect it.

22   Using a long funnel, add the specified amount of automatic transmission fluid through the vent/fill cap hole and check the fluid level by proceeding to the next Step.

23   Start the engine and allow it to idle. Depress the brake pedal and move the shifter into each gear position, holding it in each range for about three seconds, then returning the shifter back to the Park position.

24   With the engine running and the transaxle

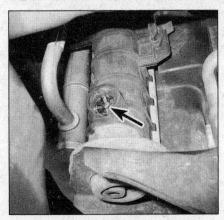

29.4 Use a screwdriver to open the radiator drain plug - it's located on the right side of the radiator

at normal operating temperature (having idled for 3 to 5 minutes), locate the check plug on the transaxle **(see illustration)**.

Caution: *The transmission fluid temperature must be 185 to 203-degrees F (85 to 95-degrees C) for an accurate check.*

25   Place a container under the check plug and remove it. Observe the fluid as it drips into the pan, indicating correct fluid level.

26   The fluid level should be at the bottom of the check hole. If fluid pours out excessively, the transaxle is overfilled. Double-check to make sure the vehicle is level. If no fluid drips from the check hole, add small amounts of fluid through the vent/fill cap at the top of the transaxle until it does, indicating the level is at the bottom of the check hole.

27   Install the check plug and tighten it securely. Also install the vent/fill cap.

27   Transfer case lubricant change (AWD models) (every 60,000 miles or 48 months)

1   This procedure should be performed after the vehicle has been driven so the lubricant will be warm and therefore will flow out of the transfer case more easily.

2   Raise the vehicle and support it securely on jackstands.

3   Remove the filler plug from the side of the case.

4   Remove the drain plug from the lower part of the case and allow the lubricant to drain completely.

5   After the case is completely drained, carefully clean and install the drain plug. Tighten the plug to the torque listed in this Chapter's Specifications.

6   Fill the case with the specified lubricant until it is at the proper level (see Section 21).

7   Install the filler plug and tighten it to the torque listed in this Chapter's Specifications.

8   Drive the vehicle for a short distance and recheck the lubricant level. In some instances a small amount of additional lubricant will have to be added.

28   Differential lubricant change (AWD models) (every 60,000 miles or 48 months)

1   This procedure should be performed after the vehicle has been driven, so the lubricant will be warm and therefore will flow out of the differential more easily.

2   Raise the rear of the vehicle and support it securely on jackstands.

3   Remove the check/fill plug, then remove the drain plug and drain the lubricant. **Note:** *On 2005 and 2006 models, the drain plug is located below the driveshaft at the front of the differential housing. On 2007 and later models, the drain plug is located on the underside of the differential housing.*

4   Reinstall the drain plug and tighten it securely.

5   Add new lubricant until it is up to the proper level (see Section 16). See *Recommended lubricants and fluids* for the specified lubricant type.

6   Reinstall the check/fill plug and tighten it to the torque listed in this Chapter's Specifications. **Note:** *It may take some driving with sharp right and left turns to distribute the fluid evenly. Recheck the level after about 50 miles of driving.*

29   Cooling system servicing (draining, flushing and refilling) (every 100,000 miles or 60 months)

Warning 1: *Wait until the engine is completely cool before beginning this procedure.*

Warning 2: *Do not allow engine coolant (antifreeze) to come in contact with your skin or painted surfaces of the vehicle. Rinse off spills immediately with plenty of water. Antifreeze is highly toxic if ingested. Never leave antifreeze lying around in an open container or in puddles on the floor; children and pets are attracted by its sweet smell and may drink it. Check with local authorities about disposing of used antifreeze. Many communities have collection centers which will see that antifreeze is disposed of safely.*

Caution: *Never mix green-colored ethylene glycol antifreeze and orange-colored "DEX-COOL" silicate-free coolant because doing so will destroy the efficiency of the "DEX-COOL" coolant, which is designed to last for 100,000 miles or five years.*

Note: *Non-toxic coolant is available at local auto parts stores. Although the coolant is non-toxic when fresh, proper disposal is still required.*

1   Periodically, the cooling system should be drained, flushed and refilled to replenish the antifreeze mixture and prevent formation of rust and corrosion, which can impair the performance of the cooling system and cause engine damage. When the cool-

ing system is serviced, all hoses and the expansion tank cap should be checked and replaced if necessary.

Draining

Refer to illustration 29.4

2   Apply the parking brake and block the wheels. If the vehicle has just been driven, wait several hours to allow the engine to cool down before beginning this procedure.

3   Once the engine is completely cool, remove the expansion tank cap.

4   Move a large container under the radiator drain to catch the coolant. Use a flat-blade screwdriver to open the radiator drain plug **(see illustration)**.

5   Remove the cap from the expansion tank.

6   While the coolant is draining, check the condition of the radiator hoses, heater hoses and clamps (refer to Section 14 if necessary).

Flushing

7   Fill the cooling system with clean water, following the *Refilling* procedure (see Step 13).

8   Start the engine and allow it to reach normal operating temperature, then rev up the engine a few times.

9   Turn the engine off and allow it to cool completely, then drain the system as described earlier.

10   Repeat Steps 6 through 8 until the water being drained is free of contaminants.

11   In severe cases of contamination or clogging of the radiator, remove the radiator (see Chapter 3) and have a radiator repair facility clean and repair it if necessary.

12   Many deposits can be removed by the chemical action of a cleaner available at auto parts stores. Follow the procedure outlined in the manufacturer's instructions. **Note:** *When the coolant is regularly drained and the system refilled with the correct antifreeze/water mixture, there should be no need to use chemical cleaners or descalers.*

Refilling

13   Close and tighten the radiator drain.

14   Place the heater temperature control in the maximum heat position.

15   Slowly add new coolant/water mixture to the expansion tank up to the COLD FILL mark.

16   Install the cap and run the engine in a well-ventilated area until the thermostat opens (coolant will begin flowing through the radiator and the upper radiator hose will become hot; the engine cooling fan will turn on). Allow the engine to run long enough for the cooling fan to cycle on and off twice.

17   Turn the engine off and let it cool. Add more coolant mixture, if necessary to bring the level to one-inch (3.4L V6) or 1/2-inch (3.6L V6) above the COLD FILL mark.

18   Start the engine, allow it to reach normal operating temperature and check for leaks.

30   Interior ventilation filter replacement (every 12,000 miles or 12 months)

2009 and earlier models

Refer to illustrations 30.2 and 30.3

Note: *On these models, the interior ventilation filter is housed inside the passenger's side cowl area.*

1   Remove the cowl cover (see Chapter 11).
2   Remove the filter cover (**see illustration**).
3   Pull the filter out of the case (**see illustration**).

4   Install the filter and the air inlet back into the cowl, and then reinstall the weatherstrip.

2010 and later models

Refer to illustrations 30.6 and 30.7

5   Remove the glove box (see Chapter 11).
6   Release the filter door clips and lower the door (**see illustration**).
7   Pull the filter out of the housing (**see illustration**).
8   Slide the filter back into the housing then close the door and reinstall the fasteners.
9   The remainder of installation is the reverse of removal.

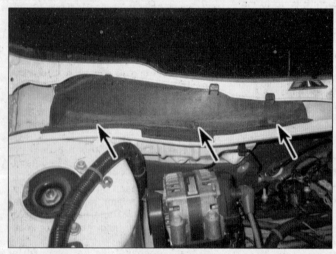

30.2 Remove the fasteners securing the filter cover and remove the cover

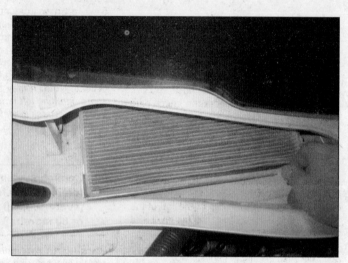

30.3 Pull the tab and lift the filter straight out

30.6 Release the retaining clips and open the filter door . . .

30.7 . . . then pull the filter out of the housing

Notes

Notes

Chapter 2 Part A
2.4L four-cylinder engine

Contents

Specifications

General

Displacement	146 cubic inches
Compression ratio	
2013 and earlier models	10:1
2014 and later models	11.16:1
Bore and stroke	3.4642 to 3.4649 x 3.861 inches
Firing order	1-3-4-2
Compression pressure	See Chapter 2D
Oil pressure	See Chapter 2D

FRONT OF VEHICLE ➊②③④

Cylinder locations

Timing chain tensioner

Timing chain tensioner compressed length	2.83 inches
Timing chain tensioner uncompressed length	3.35 inches

Hydraulic lash adjuster

Lash adjuster diameter	0.471 to 0.472 inches
Lash adjuster-to-bore clearance	0.0005 to 0.0020 inch

Camshafts

Lobe lift	Not available
Allowable lobe lift variation	0.005 inch
Endplay	0.0016 to 0.0057 inch
Journal diameter (all)	1.0604 to 1.0614 inches
Camshaft thrust clearance	0.8268 to 0.8252 inch

Torque specifications

Ft-lbs (unless otherwise indicated)

Note: *One foot-pound (ft-lb) of torque is equivalent to 12 inch-pounds (in-lbs) of torque. Torque values below approximately 15 ft-lbs are expressed in inch-pounds, because most foot-pound torque wrenches are not accurate at these smaller values.*

Camshaft position actuator solenoid bolts (intake and exhaust)	89 in-lbs
Camshaft position actuator bolts (intake and exhaust)*	
Step 1	22
Step 2	Tighten an additional 100-degrees
Camshaft bearing cap bolts	89 in-lbs
Crankshaft pulley bolt*	
Step 1	74
Step 2	Tighten an additional 125-degrees
Cylinder head bolts* (in sequence - **see illustration 12.17**)	
Bolts 1-10	
Step 1	22
Step 2	Tighten an additional 155-degrees
Bolts 11-14	26
Drivebelt tensioner bolt	See Chapter 1
Driveplate bolts	
Step 1	39
Step 2	Tighten an additional 25-degrees
Exhaust manifold-to-cylinder head nuts	
Step 1	124 in-lbs
Step 2	124 in-lbs
Exhaust manifold-to-cylinder head stud	132 in-lbs
Exhaust manifold heat shield bolts	89 in-lbs
Exhaust converter-to-manifold nuts	37
Engine front cover perimeter bolts	18
Engine front cover water pump bolt	18
Intake manifold bolts/nuts	89 in-lbs
Oil pump cover-to-engine front cover screws	53 in-lbs
Oil pump pressure relief valve plug	30
Oil pan-to-crankcase reinforcement bolts	18
Oil pan baffle bolts	124 in-lbs
Balance shaft chain tensioner	89 in-lbs
Balance shaft chain guides	
Adjustable balance shaft chain guide bolts	89 in-lbs
Fixed balance shaft chain guide bolts	106 in-lbs
Balance shaft carrier-to-block bolts	89 in-lbs
Timing chain tensioner	55
Timing chain guides	
Adjustable timing chain guide bolts	89 in-lbs
Fixed timing chain guide bolts	106 in-lbs
Timing chain upper guide bolts	89 in-lbs
Timing chain oiling nozzle bolt	89 in-lbs
Timing chain guide access hole plug	55
Valve cover bolts	89 in-lbs
Valve cover ground strap bolt	89 in-lbs
Water pump bolts	18
Water pump sprocket bolts	89 in-lbs
Water jacket drain bolt	15

* Bolt(s) must be replaced.

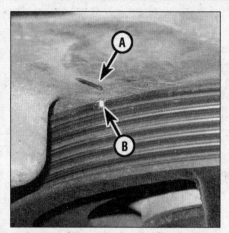

3.6 Timing marks - align the pointer on the engine front cover (A) with the notch in the crankshaft pulley (B)

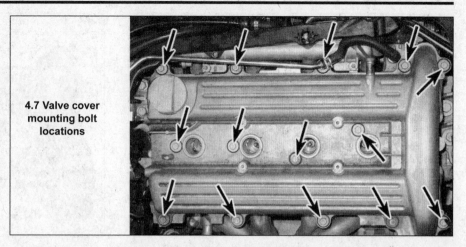

4.7 Valve cover mounting bolt locations

1 General information

1   This Part of Chapter 2 is devoted to in-vehicle repair procedures for the 2.4L DOHC (Double Overhead Camshaft) four-cylinder engines, as well as removal of the balance shafts. Information concerning engine removal and installation and engine overhaul can be found in Chapter 2D.
2   These engines are equipped with a single timing chain to drive the camshafts. The balance shaft chain drives the two balance shafts and the water pump sprocket. The balance shaft chain is driven by a sprocket on the crankshaft, directly behind the camshaft timing chain sprocket.
3   The Specifications included in this Part of Chapter 2 apply only to the procedures contained in this Part. Specifications related to engine removal and installation or overhaul can be found in Chapter 2D.

2 Repair operations possible with the engine in the vehicle

1   Many major repair operations can be accomplished without removing the engine from the vehicle.
2   Clean the engine compartment and the exterior of the engine with some type of degreaser before any work is done. It will make the job easier and help keep dirt out of the internal areas of the engine.
3   Depending on the components involved, it may be helpful to remove the hood to improve access to the engine as repairs are performed (refer to Chapter 11 if necessary). Cover the fenders to prevent damage to the paint. Special pads are available, but an old bedspread or blanket will also work.
4   If vacuum, exhaust, oil or coolant leaks develop, indicating a need for gasket or seal replacement, the repairs can generally be made with the engine in the vehicle. The intake and exhaust manifold gaskets, oil pan gasket,

crankshaft oil seals and cylinder head gasket are all accessible with the engine in place.
5   Exterior engine components, such as the intake and exhaust manifolds, the oil pan, the oil pump, the water pump, the starter motor, the alternator and the fuel system components can be removed for repair with the engine in place.
6   Since the cylinder head can be removed without pulling the engine, camshaft and valve component servicing can also be accomplished with the engine in the vehicle. Replacement of the timing chain, balance shaft chain and sprockets is also possible with the engine in the vehicle. Balance shaft removal, however, will require removal of the engine.
7   In extreme cases caused by a lack of necessary equipment, repair or replacement of piston rings, pistons, connecting rods and rod bearings is possible with the engine in the vehicle. However, this practice is not recommended because of the cleaning and preparation work that must be done to the components involved.

3 Top Dead Center (TDC) for number one piston - locating

1   Top Dead Center (TDC) is the highest point in the cylinder that each piston reaches as it travels up-and-down during crankshaft rotation. Each piston reaches TDC on the compression stroke and again on the exhaust stroke, but TDC generally refers to piston position on the compression stroke.
2   Positioning the piston(s) at TDC is an essential part of certain other repair procedures discussed in this manual.
3   Before beginning this procedure, place the transaxle in Neutral and apply the parking brake or block the rear wheels. Remove the spark plugs (see Chapter 1).
4   Install a compression pressure gauge in the number one spark plug hole (see Chapter 2D). It should be a gauge with a screw-in fitting and a hose at least six inches long.
5   Remove the inner fender splash shield (see Chapter 11). Rotate the crankshaft using a socket and large breaker bar while observing for pressure on the compression gauge.

The moment the gauge shows pressure indicates that the number one cylinder has begun the compression stroke.
6   Continue turning the crankshaft until the TDC notch in the crankshaft damper is aligned with the mark on the timing chain cover **(see illustration)**. At this point, the number one cylinder is at TDC on the compression stroke. If the marks are aligned but there was no compression, the piston was on the exhaust stroke. Continue rotating the crankshaft 360-degrees (1-turn).
7   After the number one piston has been positioned at TDC on the compression stroke, TDC for any of the remaining pistons can be located by turning the crankshaft and following the firing order. Divide the crankshaft pulley into two equal sections with chalk marks at each point, each indicating 180-degrees of crankshaft rotation. Rotating the engine past TDC no. 1 to the next mark will place the engine at TDC for cylinder no. 3.

4 Valve cover - removal and installation

Removal

Refer to illustration 4.7

1   Disconnect the cable from the negative battery terminal (see Chapter 5).
2   Remove the engine oil filler cap, then pull up on the engine cover to disconnect the ballstuds from the grommets on the back side of the cover, and remove the cover.
3   Remove the individual coils over each spark plug (see Chapter 1).
4   Detach the PCV hose from the intake manifold, but do not remove it from the valve cover.
5   Remove the camshaft housing cover fasteners or plate from the front of the valve cover.
6   Remove both engine lifting-eye bracket bolts and brackets from each end of the cylinder head.
7   Remove the valve cover bolts **(see illustration)** then lift off the valve cover. Tap gently with a soft-face hammer, if necessary, to break the gasket seal.

4.9 Install the gasket into the grooved recess in the valve cover

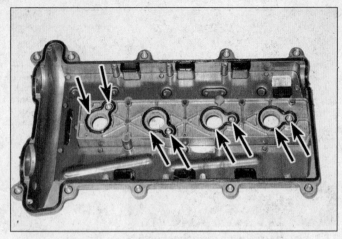

4.10 Change all the spark plug seals and O-rings in the valve cover

Installation

Refer to illustrations 4.9 and 4.10

8   Clean the gasket surfaces on the intake manifold, cylinder head and valve cover. Use a shop rag, lacquer thinner or acetone to wipe off all residue and gasket material from the sealing surfaces.
9   Insert a new valve cover gasket into the grooved recess in the valve cover. Make sure the gasket is positioned properly in the groove **(see illustration)**.
10   Install new O-rings and spark plug seals in the valve cover **(see illustration)**.
11   The remainder of installation is the reverse of removal. Tighten the valve cover bolts evenly, starting with the center bolts and working outward, to the torque listed in this Chapter's Specifications.
12   Reconnect the battery (see Chapter 5).

5   Intake manifold - removal and installation

Warning: *Wait until the engine is completely cool before beginning this procedure.*

Removal

Refer to illustrations 5.9 and 5.10

1   Disconnect the cable from the negative battery terminal (see Chapter 5).
2   Remove the engine oil filler cap.
3   Pull up on the engine cover to disconnect the ballstuds from the grommets on the back side of the cover, and remove the cover.
4   Detach the EVAP purge valve tube from the fitting on the manifold (see Chapter 6).
5   Remove the throttle body (see Chapter 4).
6   Remove the cover from the high-pressure fuel pump.
7   Remove the fuel rail harness connector bracket mounting bolt and connector, then remove the manifold insulator fasteners and insulator.
8   Disconnect any electrical connectors that would interfere with manifold removal. Open the wiring harness clips and detach all wiring harnesses from the manifold.
9   Remove the intake manifold insulator fasteners and insulator **(see illustration)**.
10   Remove the intake manifold mounting

fasteners **(see illustration)**.
11   Remove the intake manifold from the cylinder head.

Installation

Refer to illustration 5.15

12   Install new gaskets, if necessary.
Note: *The intake manifold gaskets do not have to be replaced unless they are not in perfect condition. Make sure the mating surfaces of the manifold and cylinder head are clean.*
13   Place the manifold against the cylinder head, aligning the bolt holes.
14   Install all the fasteners and tighten them only finger-tight.
15   Tighten the bolts in sequence **(see illustration)** to the torque listed in this Chapter's Specifications.
16   The remainder of installation is the reverse of removal.
17   Reconnect the battery (see Chapter 5).
18   Run the engine and check for vacuum and fuel leaks.

5.9 Remove the insulator fasteners and insulator

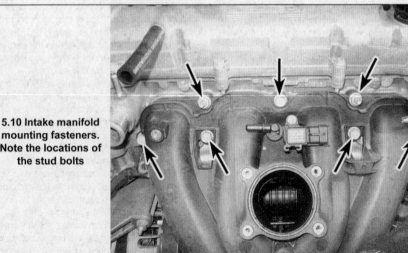

5.10 Intake manifold mounting fasteners. Note the locations of the stud bolts

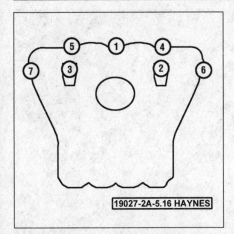

5.15 Intake manifold bolt tightening sequence

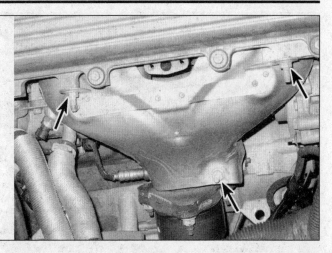

6.3 Location of the heat shield mounting bolts (typical)

6 Exhaust manifold - removal and installation

Warning: *Wait until the engine is completely cool before starting this procedure.*

Removal

Refer to illustrations 6.3, 6.5 and 6.7

1 Disconnect the cable from the negative battery terminal (see Chapter 5).
2 Remove the secondary air injection check valve assembly (see Chapter 6).
3 Remove the exhaust manifold heat shield **(see illustration)**.
4 Raise the vehicle and support it securely on jackstands. If equipped, remove the engine block heater.
Note: *If equipped with an engine block heater, drain the cooling system before removing it (see Chapter 1).*
5 Detach the exhaust pipe from the manifold **(see illustration)**.
6 Follow the lead from the oxygen sensor up to its electrical connector, then unplug the connector. Also detach the lead from its

retaining clip.
7 Remove the exhaust manifold mounting nuts and remove the manifold from the cylinder head **(see illustration)**.

Installation

8 Using a scraper, thoroughly clean the mating surfaces on the cylinder head, manifold and exhaust pipe. Remove any residue with brake system cleaner.
9 Check that the mating surfaces are perfectly flat and not damaged in any way. A warped or damaged manifold may require machining or, if severe enough, replacement. Install the new gasket to the cylinder head studs and place the manifold on the cylinder head. Tighten the nuts evenly, working from the center outwards, to the torque listed in this Chapter's Specifications.
10 Connect the exhaust pipe to the manifold and tighten the nuts evenly to the torque listed in this Chapter's Specifications.
11 The remainder of installation is the reverse of removal.
12 Reconnect the battery (see Chapter 5).
13 Refill the cooling system, if it was drained (see Chapter 1).

14 Run the engine and check for exhaust leaks.

7 Engine front cover - removal and installation

Removal

Refer to illustrations 7.7a and 7.7b

1 Disconnect the cable from the negative battery terminal (see Chapter 5).
2 Remove the air filter housing (see Chapter 4).
3 Drain the engine oil and remove the drivebelt (see Chapter 1).
4 Remove the drivebelt tensioner from the front cover (see Chapter 1).
5 Remove the crankshaft pulley (see Section 10). Support the engine with an engine hoist or engine support fixture and remove the right engine mount (see Section 17).
6 Loosen the right front wheel lug nuts, then raise the front of the vehicle and support it securely on jackstands. Remove the right front wheel.
7 Loosen the engine cover fasteners grad-

6.5 Soak the exhaust pipe retaining nuts with penetrating oil, then remove them (viewed from above)

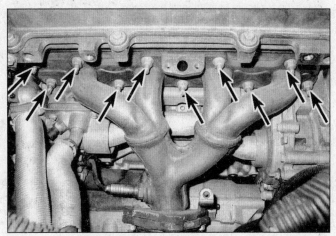

6.7 Exhaust manifold mounting nut locations

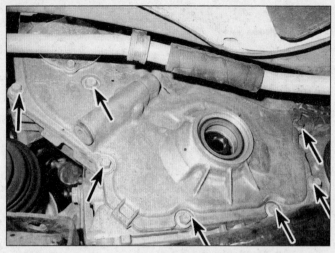

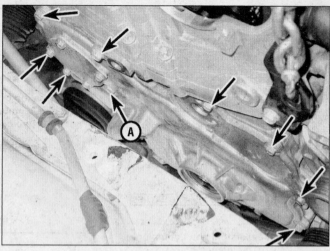

7.7a The engine cover mounting bolts can be accessed from below . . .

7.7b . . . and from above the engine compartment - don't forget the front cover/water pump bolt (A)

ually and evenly, then remove the fasteners **(see illustrations)**.

Note: *Draw a sketch of the engine cover and cover fasteners. Identify the location of all bolts for installation in their original locations.*

8   Remove the water pump bolt from the engine front cover.

9   Remove the front cover.

10   If the engine cover-to-block gasket is in good condition, leave it in place.

Installation

11   Inspect and clean all sealing surfaces of the engine front cover and the block.

Caution: *Be very careful when scraping on aluminum engine parts. Aluminum is soft and gouges easily. Severely gouged parts may require replacement.*

12   If necessary, replace the crankshaft front oil seal in the front cover (see Section 10).

13   Install a new gasket if necessary.

Caution: The front cover gasket is reusable if it is in good condition.

14   Install the front cover and fasteners. Make sure the hub on the oil pump inner rotor is aligned with the flats on the crankshaft and the fasteners are in their original locations. Tighten the fasteners by hand until the cover is contacting the block around its entire edge.

15   Install the long water pump bolt.

16   Tighten the bolts a little at a time to the torque listed in this Chapter's Specifications.

17   Install the drivebelt and tensioner (see Chapter 1). Tighten the drivebelt tensioner to the torque listed in the Chapter 1 Specifications.

18   Install the crankshaft pulley (see Section 10).

19   The remainder of installation is the reverse of removal.

20   Fill the crankcase with the recommended oil (see Chapter 1).

21   Reconnect the battery (see Chapter 5).

22   Start the engine and check for leaks. Check all fluid levels, adding as necessary (see Chapter 1).

8   Timing chain and sprockets - removal, inspection and installation

Removal

Refer to illustrations 8.8, 8.9, 8.10, 8.11a, 8.11b, 8.12, 8.14 and 8.15

Caution: *The timing system is complex. Severe engine damage will occur if you make any mistakes. Do not attempt this procedure*

unless you are highly experienced with this type of repair. If you are at all unsure of your abilities, consult an expert. Double-check all your work and be sure everything is correct before you attempt to start the engine.

1   Disconnect the cable from the negative battery terminal (see Chapter 5). Remove the spark plugs (see Chapter 1).

2   Remove the valve cover (see Section 4).

3   Drain the engine oil (see Chapter 1).

4   Remove the drivebelt (see Chapter 1).

5   Remove the drivebelt tensioner from the front cover (see Chapter 1).

6   Remove the engine front cover (see Section 7).

7   Remove the sparks plugs (see Chapter 1). Using a socket and breaker bar on the crankshaft pulley center bolt, rotate the engine (clockwise) so that the No. 1 piston is at TDC on the exhaust stroke, with the keyway on the crankshaft at 12 o'clock position, the mark on the intake camshaft sprocket at the 5 o'clock position and the mark on the exhaust camshaft sprocket at the 7 o'clock position on 2010 models. On 2011 and later models, the No. 1 piston is at TDC on the exhaust stroke, with the mark on the crankshaft key in the twelve o'clock position, the mark on the intake camshaft sprocket at the 10 o'clock position and the mark on the exhaust camshaft sprocket at the 7 o'clock position **(see illustrations 11.34a or 11.34b)**.

Note: *The marks are adjacent to the diamond-shaped or triangular-shaped holes on the sprockets/actuators.*

8   Remove the timing chain tensioner **(see illustration)**.

9   Remove the upper timing chain guide **(see illustration)**.

10   Use a wrench on the exhaust cam hex to hold the camshaft, and remove the exhaust camshaft sprocket bolt **(see illustration)**. Discard the bolt and install a new one on reassembly.

11   Remove the adjustable timing chain guide **(see illustrations)**.

12   Unscrew the access plug and remove

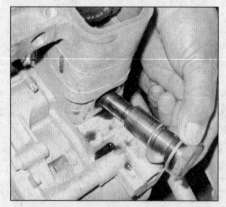

8.8 Remove the timing chain tensioner from the cylinder head

8.9 Upper timing chain guide mounting bolts

8.10 Use a wrench on the hex on the camshaft to prevent the camshaft from turning while loosening the sprocket bolt

8.11a Remove the adjustable timing chain guide mounting bolt . . .

8.11b . . . then lift the guide out through the top of the cylinder head

8.12 Access plug location

8.14 Use a wrench on the hex on the camshaft when loosening the camshaft sprocket bolt

the fixed timing chain guide upper mounting bolt **(see illustration)**.

13   Remove the fixed timing chain guide lower mounting bolt and lift out the guide.

14   Remove the intake camshaft sprocket bolt **(see illustration)**. Discard the bolt and install a new bolt on reassembly.

15   Remove the timing chain through the top of the cylinder head **(see illustration)**.

Caution: *Once you have removed the timing chain, do NOT turn either camshaft or the crankshaft. Doing so will damage the pistons and/or the valves.*

16   Remove the outer friction washer from the crankshaft sprocket. Verify that the timing mark on the crankshaft gear is at the 5 o'clock position and the keyway is at the 12 o'clock position, then remove the gear.

17   Remove the inner friction washer from the end of the crankshaft.

Caution: *The lower timing chain crankshaft gear may be equipped with a second spacing washer installed in front of crankshaft gear that should be removed. The washer has a dot mark and may be mistaken for the correct crankshaft timing mark.*

Inspection

18   Clean all parts with solvent and dry with compressed air, if available.

19   Inspect the chain tensioner for excessive wear or other damage. Drain all the oil out of the chain tensioner if it is to be reused.

20   Inspect the timing chain guides for deep grooves, excessive wear, or other damage.

21   Inspect the timing chain for excessive wear or damage.

22   Inspect the crankshaft and camshaft sprockets for chipped or broken teeth, excessive wear, or damage.

23   Replace any component that is in questionable condition.

Installation

Refer to illustrations 8.24, 8.26a, 8.26b, 8.30, 8.34a, 8.34b, 8.34c, 8.34d and 8.34e

Caution: *Before starting the engine, carefully rotate the crankshaft by hand through at least two full revolutions (use a socket and breaker bar on the crankshaft pulley center bolt). If you feel any resistance, STOP! There is something wrong most likely, valves are contacting*

the pistons. You must find the problem before proceeding. Check your work and see if any updated repair information is available.

24   Before installing the timing chain, make sure the timing mark (round dot) on the crank-

8.15 Remove the timing chain and the intake camshaft sprocket through the top of the cylinder head

8.24 The round dot (alignment mark) on the sprocket should be in the 5 o'clock position. When installing the chain, the first silver plated link must be aligned with this dot

8.26a The different colored link must align with the INT on the intake camshaft, and the matching colored links with the EXH on the exhaust camshaft

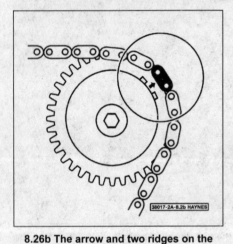

8.26b The arrow and two ridges on the camshaft sprockets align with the colored timing chain links

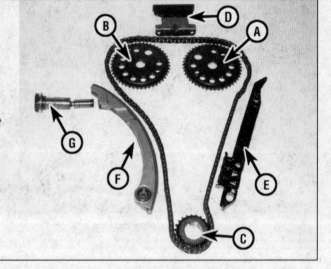

8.30 Timing chain component details

A   Intake camshaft sprocket
B   Exhaust camshaft sprocket
C   Crankshaft sprocket
D   Upper timing chain guide
E   Fixed timing chain guide
F   Adjustable timing chain guide
G   Timing chain tensioner

shaft sprocket is pointing to the 5 o'clock position **(see illustration)**.

25   Install the intake camshaft sprocket onto the camshaft, using a new bolt. Tighten the intake camshaft sprocket bolt lightly, finger-tight at this time.

Caution: *Do not turn the camshaft more than 1/2 turn to avoid any valve/piston contact. The camshafts should be positioned correctly before the timing chain is installed.*

26   Install the timing chain by lowering it from the top through the opening. The chain has three colored links, two are matching and one is a different color. The different-colored timing chain link must be aligned with the arrow and two ridges on the intake camshaft sprocket **(see illustrations)**.

Note: *The different colored link will be installed at the intake camshaft sprocket (front) while the other matching links will be installed at the crankshaft sprocket and the exhaust camshaft sprocket (rear).*

27   Drape the timing chain over the crankshaft sprocket and engage the plated link (matching color) on the chain with the crank-

shaft sprocket timing mark located in the 5 o'clock position **(see illustration 8.24)**.

28   Install the adjustable timing chain guide. Install the bolts and tighten them to the torque listed in this Chapter's Specifications.

29   Install the exhaust camshaft sprocket onto the camshaft, installing a new bolt. Be sure the plated link (matching colored) on the chain is aligned with the EXH designation and the mark(s) on the sprocket (which should be pointing to the 10 o'clock position) **(see illustration 8.26a)**. Tighten the exhaust camshaft sprocket bolt lightly, finger tight at this time.

30   Install the fixed timing chain guide **(see illustration)**. Tighten the bolts to the torque listed in this Chapter's Specifications.

31   Install the upper timing chain guide **(see illustration 8.9)**. Tighten the bolts to the torque listed in this Chapter's Specifications.

32   Hold the intake camshaft with a wrench on the camshaft's hex to prevent it from turning, then tighten the intake camshaft bolt to the torque listed in this Chapter's Specifications.

33   Hold the exhaust camshaft with a wrench on the camshaft's hex to prevent it from turning, then tighten the intake camshaft bolt to the torque listed in this Chapter's Specifications.

34   Install the timing chain tensioner. The timing chain tensioner must be installed in its compressed state. Follow the steps below to correctly compress the tensioner **(see illustrations)**.

Caution: *Do not install a tensioner in its released state. Damage to the tensioner and timing chain will occur.*

• *Disassemble the tensioner and drain all the oil. Inspect the tensioner body, the piston and all components for scoring or damage. If necessary, replace the tensioner with a new one.*

• *Install the tensioner piston into the vise with the flats seated in the jaws of the vise.*

• *Install the ratchet cylinder into the piston, aligning the groove with the locating pin.*

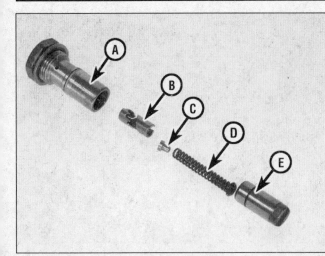

8.34a Timing tensioner details

A Timing chain tensioner body
B Ratchet cylinder
C Spring adjuster
D Spring
E Piston

8.34b Install the piston with the flats locked into the jaws of the vise

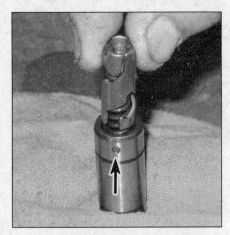

8.34c Align the groove in the ratchet cylinder with the pin in the piston

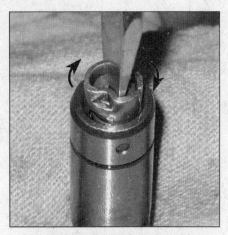

8.34d Using a flat-bladed screwdriver, drive the ratchet cylinder down to the bottom and rotate it clockwise to lock it into place

8.34e The tensioner should measure the correct length in its compressed state or it must be replaced with a new one

• *Drive the ratchet cylinder into the piston with a flat-bladed screwdriver. Rotate the ratchet cylinder clockwise when it reaches the bottom. The ratchet cylinder should be locked into position. The tensioner must measure 2.83 inches (72 mm) from end-to-end.*

35   Install the timing chain oiling nozzle. Tighten the bolt to the torque listed in this Chapter's Specifications. Using a tool with a plastic or rubber tip, strike a sharp blow to the tensioner at the chain end; this should release the tensioner so it can apply force against the chain.

36   Apply a small amount of RTV sealant to the threads and install the timing chain guide access plug. Tighten the bolt to the torque listed in this Chapter's Specifications.

37   Install the valve cover (see Section 4).

38   Install the engine front cover (see Section 7).

39   The remainder of installation is the reverse of removal. Install a new oil filter and refill the crankcase with oil (see Chapter 1).

40   Reconnect the battery (see Chapter 1).

41   Run the engine and check for leaks.

9   Balance shaft chain and balance shafts - removal, inspection and installation

Note: *This procedure covers removal of the balance shaft chain and balance shafts, but take note that the shafts themselves can only be removed from the engine block after the engine has been removed from the vehicle. If there is a problem with the balance shafts that does warrant their removal, the engine would have to be removed anyway, since replacement of the balance shaft bushings is a job that must be left to an automotive machine shop. If you're just removing or replacing the chain, ignore the steps that don't apply.*

Removal

Refer to illustrations 9.5, 9.6 and 9.9

1   Disconnect the cable from the negative battery terminal (see Chapter 5).

2   Drain the engine oil (see Chapter 1).

3   Position the engine at Top Dead Center (TDC) for cylinder number 1 (see Section 3).

4   Remove the timing chain, timing chain guides and sprockets (see Section 8).

5   Remove the balance shaft chain tensioner **(see illustration)**.

6   Remove the adjustable balance shaft

9.5 Location of the balance shaft chain tensioner mounting bolts

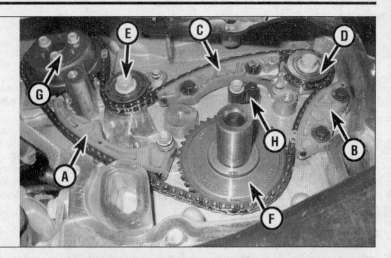

9.6 Balance shaft chain and guide details

A   Adjustable balance shaft chain guide
B   Small balance shaft chain guide
C   Upper balance shaft chain guide
D   Intake side (front) balance shaft sprocket
E   Exhaust side (rear) balance shaft sprocket
F   Crankshaft/balance shaft sprocket
G   Water pump sprocket
H   Timing chain oiling nozzle

9.9 Balance shaft sprocket/chain alignment marks (A) and retainer bolts (B)

9.18 The timing mark (round dot) on the crankshaft sprocket should point to the 6 o'clock position (approximately)

9.20 With the arrow on the intake side balance shaft sprocket pointing up (and aligned with the cutout on the balance shaft retainer, not visible in this photo, but similar to the one shown in illustration 9.21a), install a drill bit into the hole to lock the sprocket in place

chain guide **(see illustration)**.
7   Remove the small balance shaft chain guide **(see illustration 9.6)**.
8   Remove the upper balance shaft chain guide **(see illustration 9.6)**.

9   Remove the balance shaft drive chain **(see illustration)**.
Note: *To aid in removal, gather all the slack in the chain between the water pump sprocket and the crankshaft sprocket.*
10   If you're removing the balance shafts (engine removed from the vehicle), remove the balance shaft retainer bolts.
11   Remove the balance shafts from the engine block.
Caution: *Mark each balance shaft to insure correct reassembly. The balance shafts are not interchangeable. Do not install the balance shaft into the wrong bore or extreme engine vibration will occur.*

Inspection

12   Clean all parts with clean solvent and dry with compressed air, if available.
13   Inspect the chain tensioners for excessive wear or other damage.
14   Inspect the balance shaft chain guides for deep grooves, excessive wear, or other damage.
15   Inspect the balance shaft chain for excessive wear or damage.
16   Inspect the crankshaft and water pump sprockets for chipped or broken teeth, exces-

sive wear, or damage.
17   Replace any component that is damaged.

Installation

Refer to illustrations 9.18, 9.20, 9.21a, 9.21b, 9.26a, 9.26b and 9.28
18   Before installing the balance shaft chain, make sure the crankshaft timing mark (round dot) is pointing to the 6 o'clock position **(see illustration)**.
Caution: *Do not rotate the engine to find TDC number 1 after the timing chain has been removed unless the engine has been rotated accidentally. If the engine is not positioned at TDC number 1, the camshafts must be removed before rotating the engine to prevent damage to the valves (see Section 11).*
19   Install the balance shafts into the bores and tighten the balance shaft retainer bolts to the torque listed in this Chapter's Specifications.
20   Align the balance shaft sprockets before installing the balance shaft chain. Starting with the intake side balance shaft, place the alignment arrow pointing up, then temporarily install a drill bit into the alignment hole and the sprocket teeth to lock the balance shaft

9.21a Location of the alignment notch for the sprocket arrow (A) and the alignment hole (B) on the exhaust side balance shaft sprocket

9.21b Install the drill bit into the exhaust balance shaft retainer to lock it into position

9.26a Rotate the plunger 90-degrees, align the holes in the body and piston . . .

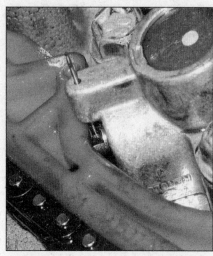

9.26b . . . then install a drill bit to retain the piston in the locked position

sprocket in place **(see illustration)**.

21   Position the exhaust side (rear) balance shaft sprocket with the arrow pointing down and aligned with the cutout in the retainer, then install a drill bit into the alignment hole to hold the sprocket **(see illustrations)**.

22   Install the balance shaft chain onto the balance shaft/crankshaft sprocket and the balance shafts. Align the colored links with the alignment marks on each sprocket. Position the different colored or anodized link onto the intake side balance shaft, aligning the mark with the colored link at approximately the 12 o'clock position **(see illustration 9.9)**.

Note: *The balance shaft chain has three colored links; two of them are the same color and one is different. The different colored link will be installed at the intake balance shaft sprocket (front) while the other colored links will be installed at the crankshaft sprocket and the exhaust balance shaft sprocket (rear).*

23   Working clockwise, position the second colored link on the crankshaft sprocket, aligning the mark on the sprocket with the colored link at the 6 o'clock position **(see illustra-**

tion 9.18).

24   Pass the chain over the water pump sprocket, under the exhaust balance shaft sprocket and into position. Align the third colored link on the exhaust balance shaft sprocket, aligning the mark on the sprocket with the colored link at the 6 o'clock position.

25   Install the balance shaft chain guides **(see illustration 9.6)**. Tighten the bolts to the torque listed in this Chapter's Specifications.

26   Reset the balance shaft chain tensioner: turn the tensioner plunger 90-degrees in the bore and compress the tensioner plunger **(see illustration)**; rotate the plunger back to the original position at 12 o'clock; and insert a paper clip through the hole in the body into the plunger **(see illustration)**.

27   Install the balance shaft chain tensioner and tighten the bolts to the torque listed in this Chapter's Specifications.

28   Remove the drill bit to release the plunger **(see illustration)**.

29   Recheck all the balance shaft chain timing marks.

30   Install the timing chain (see Section 8) and all components removed previously.

31   Reconnect the battery (see Chapter 5).

32   Install a new oil filter and refill the crankcase with oil (see Chapter 1).

33   Run the engine and check for leaks.

9.28 After the tensioner is installed and the bolts tightened, remove the drill bit

10.4 Remove the splash shield mounting fasteners, then remove the splash shield

10.5 A large pin spanner can be used to prevent the pulley from rotating while the bolt is loosened

10 Crankshaft pulley and front oil seal - removal and installation

Removal

Refer to illustrations 10.4, 10.5 and 10.7

1   Disconnect the cable from the negative battery terminal (see Chapter 5).
2   Remove the drivebelt (see Chapter 1).
3   Raise the vehicle and support it securely on jackstands.
4   Remove the splash shield from below the engine compartment **(see illustration)**.
5   Use a breaker bar and socket to remove the crankshaft pulley center bolt **(see illustration)**. Discard the bolt and obtain a new one for installation.
Note: *It will be necessary to lock the pulley in position using a strap wrench or a large pin spanner. Wrap a length of old drivebelt around the pulley for a good grip if you are using a strap wrench.*
6   Slide the puller off the nose of the crankshaft. If the pulley is stuck, use a puller that

bolts to the three threaded holes in the pulley hub. Additionally, a spacer, such as a deep socket that just fits into the hole in the pulley and bears on the crankshaft, will be required to avoid damage to the crankshaft.
7   Use a seal puller to remove the crankshaft front oil seal **(see illustration)**. A screwdriver may be used instead, if the tip is wrapped with tape to avoid scratching the crankshaft.
8   Clean the seal bore and check it for nicks or gouges. Also examine the area of the hub that rides in the seal for signs of abnormal wear or scoring.

Installation

Refer to illustration 10.9

9   Coat the lip of the new seal with clean engine oil and drive it into the bore with a seal driver or a socket slightly smaller in diameter than the seal **(see illustration)**. The open side of the seal faces into the engine.
10   Using clean engine oil, lubricate the sealing surface of the hub. Install the crankshaft pulley/damper with a special installation tool,

available at most auto parts stores. Do not use a hammer to install the pulley/damper. Install a new center bolt and tighten it to the torque listed in this Chapter's Specifications.
Note: *A new pulley bolt must be used.*
11   The remainder of installation is the reverse of removal.
12   Reconnect the battery (see Chapter 1).

11 Camshafts and hydraulic lash adjusters - removal, inspection and installation

Removal

Note: *This is a difficult procedure, involving special tools. Read through the entire Section and obtain the necessary tools before beginning the procedure. New camshaft sprocket bolts must be purchased ahead of time.*
Note: *These engines require three special tools - EN-48953 and two EN-48749 (or equivalent) - that secure the chain in relationship to the cylinder head on both the intake and exhaust sides. Once secured, the intake and exhaust camshaft actuators (hydraulic units attached to the front of each sprocket) can be marked in their relationship to the chain and removed. If the tools are not available, the timing chain will have to be removed (see Section 8).*
1   Relieve the fuel system pressure (see Chapter 4) then disconnect the cable from the negative battery terminal (see Chapter 5).
2   Remove the air filter housing (see Chapter 4).
3   Remove the ignition coils and spark plugs (see Chapter 1).
4   Remove the valve cover (see Section 4).
5   Remove the timing chain tensioner from the side of the cylinder head (see Section 8).
6   Rotate the crankshaft until special tool EN-48953 can be installed onto the camshaft actuators. Working through the holes in the tool, loosen each of the camshaft actuator bolts. Once the bolts are loosened, the tool can be removed, but the

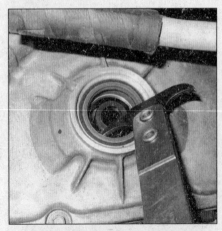

10.7 Use a seal puller to remove the old crankshaft seal, taking care not to damage the crankshaft or the seal bore in the cover

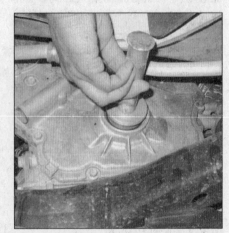

10.9 Driving the new front cover seal in with a seal driver

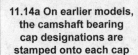

11.14a On earlier models, the camshaft bearing cap designations are stamped onto each cap

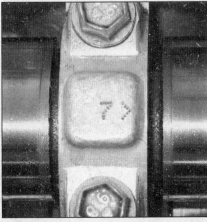

11.14b Note that the arrow on the cap faces the timing chain end of the engine

engine can't be rotated.

7 Mark the camshaft actuators to the chains to insure they are reinstalled in the exact locations.

8 Remove the upper chain guide two mounting bolts and remove the chain guide.

9 Remove the timing chain tensioner from the side of the cylinder head (see Section 8).

10 Install special tool EN-48749 from the top until it locks into the intake side of the chain, then install the second EN-48749 from the top into the exhaust side of the chain.

Note: *Make sure the tips of the special tools are fully engaged into the timing chain.*

11 The timing chain should be supported in position by the locking tool. Remove the exhaust or intake camshaft actuator bolt and slip the actuator down and away from the chain and the camshaft and discard the actuator bolt.

12 Remove the camshaft position actuator solenoid retaining bolt the pull the solenoid out of the cylinder head (see Chapter 6).

Note: *If both camshafts are being removed, remove both solenoids.*

Intake camshaft

Refer to illustrations 11.14a, 11.14b, 11.15, 11.16 and 11.17

13 Remove the three rear cylinder head cover plate bolts and remove the cover. Remove the high pressure fuel pump (see Chapter 4).

14 Each camshaft cap must be marked for location and direction. Earlier models may be marked from the factory (see illustrations); if they are not, they must be marked prior to disassembly. Working a little at time, loosen each bearing cap bolt slowly and evenly until the spring tension is no longer pushing on the camshaft. Lift the camshaft from the cylinder head, parallel to the head surface.

Caution: *The caps must be installed in their original locations. Keep all parts from each camshaft together; never mix parts from one camshaft with those for another.*

15 Remove the rocker arms (see illustration).

16 Place the rocker arms in a suitable container, in order, so they can be reinstalled in their original positions (see illustration).

17 Remove the hydraulic lash adjusters from their bores in the cylinder head (see illustration). Store these with their corresponding rocker arms so they can be reinstalled in their original locations.

Exhaust camshaft

18 Each camshaft cap must be marked for location and direction. Earlier models may be marked from the factory (see illustrations 11.14a and 11.14b), but if they are not, they must be marked prior to disassembly. Working a little at time, loosen each bearing cap bolt slowly and evenly. Lift the camshaft from the cylinder head, parallel to the head surface.

Caution: The caps must be installed in their original locations. Keep all parts from each camshaft together; never mix parts from one camshaft with those for another.

19 Mark the positions of the rocker arms so they can be reinstalled in their original locations, then remove the rocker arms.

20 Place the rocker arms in a suitable container so they can be separated and identified (see illustration 11.16).

11.15 Remove each rocker arm

11.16 Store the rocker arms in an organized manner so they can be returned to their original locations

11.17 Pull the lash adjusters from their bores in the head and store them along with their corresponding rocker arms

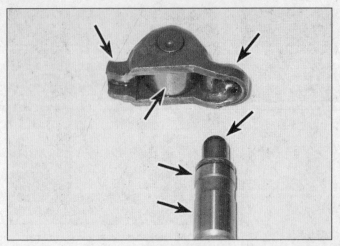

11.22 Check the rocker arms and lash adjusters for wear at the indicated points

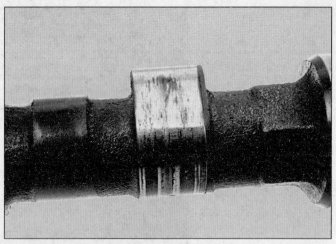

11.25 Check the cam lobes for pitting, excessive wear, and scoring. If scoring is excessive, as shown here, replace the camshaft

21   Lift the hydraulic lash adjusters from their bores in the cylinder head. Identify and separate the adjusters so they can be reinstalled in their original locations **(see illustration 11.17)**.

Inspection

Refer to illustrations 11.22, 11.25, 11.26, 11.27 and 11.29

22   Check each hydraulic lash adjuster for excessive wear, scoring, pitting, or an out-of-round condition **(see illustration)**. Replace as necessary.

23   Measure the outside diameter at the top and bottom of each adjuster, then take a second set of measurements at a right angle to the first. If any measurement is significantly different from the others, the adjuster is tapered or out of round and must be replaced. If the necessary equipment is available, measure the diameter of the lash adjuster and the inside diameter of the corresponding cylinder head bore. Subtract the diameter of the lash adjuster from the bore diameter to obtain the

oil clearance. Compare the measurements obtained to those given in this Chapter's Specifications. If the adjusters or the cylinder head bores are excessively worn, new adjusters or a new cylinder head, or both, may be required. If the valve train is noisy, particularly if the noise persists after a cold start, you can suspect a faulty lash adjuster.

24   Inspect the rocker arms for signs of wear or damage. The areas of wear are the tip that contacts the valve stem, the socket that contacts the lash adjuster and the roller that contacts the camshaft **(see illustration 11.22)**.

25   Examine the camshaft lobes for scoring, pitting, galling (wear due to rubbing), and evidence of overheating (blue, discolored areas). Look for flaking of the hardened surface layer of each lobe **(see illustration)**. If any such wear is evident, replace the camshaft.

26   Measure the lobe height of each cam lobe on the intake camshaft, and record your measurements **(see illustration)**. Compare the measurements for excessive variation; if the

lobe heights vary more than 0.005 inch (0.125 mm), replace the camshaft. Compare the lobe height measurements on the exhaust camshaft and follow the same procedure. Do not compare intake camshaft lobe heights with exhaust camshaft lobe heights, as they are different. Only compare intake lobes with intake lobes and exhaust lobes with exhaust lobes.

27   Inspect the camshaft bearing journals and the cylinder head bearing surfaces for pitting or excessive wear. If any such wear is evident, replace the component concerned. Using a micrometer, measure the diameter of each camshaft bearing journal at several points **(see illustration)**. If the diameter of any journal is less than specified, replace the camshaft.

28   To check the bearing journal oil clearance, remove the rocker arms and hydraulic lash adjusters (if not already done), use a suitable solvent and a clean lint-free rag to clean all bearing surfaces, then install the camshafts and bearing caps with a piece of

11.26 Measure each camshaft lobe height with a micrometer

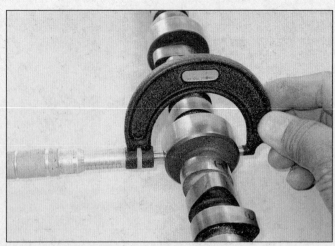

11.27 Measure each journal diameter with a micrometer. If any journal is less than the specified minimum, replace the camshaft

11.28 Lay a strip of Plastigage on each camshaft journal, in line with the camshaft

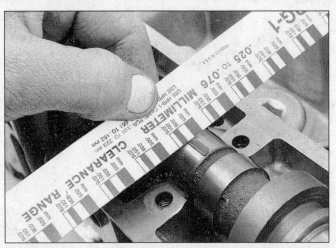

11.29 Compare the width of the crushed Plastigage to the scale on the package to determine the journal oil clearance

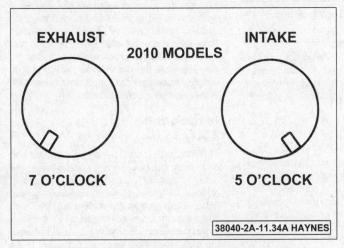

11.34a Camshaft notch positioning 2010 models

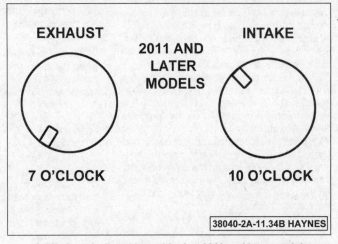

11.34b Camshaft notch positioning 2011 and later models

Plastigage across each journal **(see illustration)**. Tighten the bearing cap bolts to the specified torque. Don't rotate the camshafts.

29   Remove the bearing caps and measure the width of the flattened Plastigage with the Plastigage scale **(see illustration)**. Scrape off the Plastigage with your fingernail or the edge of a credit card. Don't scratch or nick the journals or bearing caps.

30   If the oil clearance of any bearing is worn beyond the service limit, install a new camshaft and repeat the check. If the clearance is still excessive, replace the cylinder head.

31   To check camshaft endplay, remove the hydraulic lash adjusters, clean the bearing surfaces carefully, and install the camshafts and bearing caps. Tighten the bearing cap bolts to the specified torque, then measure the endplay using a dial indicator mounted on the cylinder head so that its tip bears on the camshaft end.

32   Lightly but firmly tap the camshaft fully toward the gauge, zero the gauge, then tap the camshaft fully away from the gauge and note the gauge reading. If the measured endplay is

at or beyond the specified service limit, install a new camshaft thrust cap and repeat the check. If the clearance is still excessive, the camshaft or the cylinder head must be replaced.

Installation

Refer to illustrations 11.34a, 11.34b and 11.40

33   Lubricate the rocker arms and hydraulic lash adjusters with engine assembly lubricant or fresh engine oil. Install the adjusters into their original bores, then install the rocker arms in their correct locations.

34   Lubricate the camshafts with camshaft installation lubricant and install them in their correct locations. Position the camshafts with the notches in the ends of the shafts as shown **(see illustrations)**. Make sure the No. 1 piston is at TDC on the exhaust stroke, with the keyway on the crankshaft at 12 o'clock position, the mark on the intake camshaft sprocket at the 5 o'clock position and the mark on the exhaust camshaft sprocket at the 7 o'clock position on 2010 models. On 2011 and later models, the No. 1 piston is at TDC on the

exhaust stroke, with the mark on the crankshaft key in the twelve o'clock position, the mark on the intake camshaft sprocket at the 10 o'clock position and the mark on the exhaust camshaft sprocket at the 7 o'clock position

35   Turn the oil seals for the camshaft actuators on the camshaft(s) so that the split lines (where the two ends of the seal meet) of the seal are exactly opposite of each other.

36   Install the camshaft bearing caps in their correct locations except for the rear cap (No.6) on the intake camshaft. Install the cap bolts and tighten by hand until snug.

37   Tighten the bolts in four to five steps (no more than three turns at a time), starting with the center cap and working to the outside caps, to the torque listed in this Chapter's Specifications.

38   On the rear intake camshaft cap (No.6) apply a bead of liquid sealer approximately 1/8-inch wide to three areas of the sealing surface on the rear camshaft cap and install the cap. Install the cap bolts and tighten by hand until snug. This must be done within 20 minutes of applying the liquid sealer.

11.40 Camshaft and timing sprocket alignment details

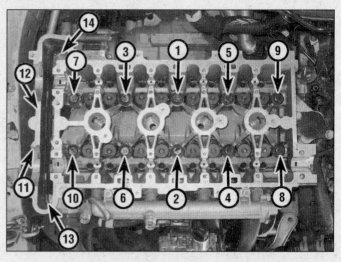

12.17 Cylinder head bolt tightening sequence

39   Tighten the rear intake camshaft cap (No.6) bolts to 44-inch lbs (5 Nm) then tighten them to the torque listed in this Chapter's Specifications, once the bolts are tightened loosen the bolts 120-degrees, then re-tighten the bolts to the this Chapter's Specifications.

40   Slide the camshaft sprocket actuators and timing chain along the guide pins toward the camshafts making sure the slots are aligned with the projections on the sprockets **(see illustration)**.

41   Install new camshaft actuator bolt(s) and tighten them hand tight then remove the special tool EN-48749 locking the chain from the intake side of the chain. Verify that the marks on the chain are aligned with marks made on the camshaft actuator.

Note: *If the marks made on the actuator are not aligned with the marks on the chain, use a wrench on the flats of the camshaft and rotate the actuator slightly until the marks are aligned. If the marks can't be aligned the timing chain cover and chain must be removed and re-installed (see Section 8).*

42   Remove the special tool EN-48749 locking the chain from the exhaust side.

43   Reset the timing chain tensioner (see Section 8) and install the tensioner.

Note: *If the tensioner is not reset, it will over extend causing premature wear on the chain and tensioner.*

44   Install special tool EN-48953 on to the camshaft actuators, working through the holes in the tool, tighten the camshaft sprockets to the torque listed in this Chapter's Specifications.

45   Release the tensioner, by turning the crankshaft balancer bolt counterclockwise using about 33 ft-lbs (45 Nm) of torque.

46   Remove the camshaft locking tool EN-48953 from the cylinder head.

47   Install the upper timing chain guide and fasteners (see Section 8).

48   The remainder of installation is the reverse of removal.

49   Reconnect the battery (see Chapter 5).

12   Cylinder head - removal and installation

Caution: *The engine must be completely cool when the head is removed. Failure to allow the engine to cool off could result in head warpage. New head bolts should be purchased ahead of time.*

Removal

1   Relieve the fuel system pressure (see Chapter 4), then disconnect the cable from the negative battery terminal (see Chapter 5).

2   Wait until the engine is completely cool, then drain the cooling system (see Chapter 1).

3   Remove the coolant recovery tank and upper radiator hose (see Chapter 3).

4   Remove the drivebelt and the drivebelt tensioner (see Chapter 1).

5   Remove the exhaust manifold (see Section 6).

6   Remove the intake manifold (see Section 5).

7   Remove the valve cover (see Section 4).

8   Remove the timing chain (see Section 8).

Note: *It is possible to remove the cylinder head without removing the timing chain cover and the timing chain completely by using special tools EN-48789, EN-48749 and EN-48953, but it is highly recommended to take the extra steps and remove the timing chain cover and timing chain.*

9   Label and disconnect the electrical connectors from the cylinder head that will interfere with removal. Use tape and mark each connector to insure correct reassembly.

10   Remove and discard the cylinder head bolts, following the reverse of the tightening sequence **(see illustration 12.17)**. Loosen the bolts in sequence 1/4-turn at a time. If the head is to be completely overhauled, refer to Section 11 for removal of the camshafts, rocker arms and hydraulic lash adjusters.

11   Use a prybar at the corners of the head-to-block mating surface to break the gasket seal. Do not pry between the cylinder head and engine block in the gasket sealing area.

12   Lift the cylinder head off the engine. If resistance is felt, place a wood block against the end and strike the wood block with a hammer. Store the cylinder head on wood blocks to prevent damage to the gasket sealing surfaces.

13   Remove the old cylinder head gasket. Before removing, note the correct orientation of the gasket for correct installation.

Installation

Refer to illustration 12.17

14   The mating surfaces of the cylinder head and block must be perfectly clean when the head is installed. Use a gasket scraper to remove all traces of carbon and old gasket material, then clean the mating surfaces with brake system cleaner. If there's oil on the mating surfaces when the cylinder head is installed, the gasket may not seal correctly and leaks may develop. When working on the engine block, cover the open areas of the engine with shop rags to keep debris out during repair and reassembly. Use a vacuum cleaner to remove any debris that falls into the cylinders.

15   Check the engine block and cylinder head mating surfaces for nicks, deep scratches and other damage. **Caution:** *Do not use a tap to chase the threads in the cylinder head bolt holes. Clean the holes with a nylon bristle brush to remove the dirt, corrosion and sealant that will affect torque readings.*

16   Install the new gasket, locating it on the dowels in the block.

17   Carefully position the cylinder head on the engine block without disturbing the gasket. Install new cylinder head bolts and, following the recommended sequence **(see illustration)**, tighten the bolts to the torque listed in this Chapter's Specifications. All the main cylinder head bolts (numbers 1 through 10) are tightened in the first Step and second Step. The four smaller bolts located on the front of the cylinder head are the only ones tightened in the third Step. Mark a stripe on each of the main cylinder head bolts to help keep track of the bolts that have been tightened the additional 155-degrees.

Note: *The method used for the head bolt tightening procedure is referred to as a torque-angle method. A special torque angle gauge (available at most auto parts stores) is available to attach to a breaker bar and socket for better accuracy during the tightening procedure.*

18   Install the timing chain (see Section 8).
19   Install the exhaust manifold (see Section 6).
20   Install the intake manifold (see Section 5).
21   The remainder of installation is the reverse of removal.
22   Reconnect the battery (see Chapter 5).
23   Change the engine oil and filter and refill the cooling system (see Chapter 1), then start the engine and check carefully for oil and coolant leaks.

13   Oil pan - removal and installation

Removal

1   Drain the engine oil and remove the drivebelt (see Chapter 1).
2   Loosen the right-front wheel lug nuts, raise the front of the vehicle and support it securely on jackstands. Remove the right front wheel.
3   Remove the splash shield from below the right side of the engine compartment **(see illustration 10.4)**.
4   Remove the lower air conditioning compressor mounting bolt (see Chapter 3). Loosen, but don't remove, the other compressor mounting bolts.
5   Remove the dipstick and the dipstick tube (the tube is bolted to the intake manifold).
6   Install an engine support fixture of the type that straddles the top of the engine onto the fender edges and attaches to the top of the engine to take its weight. These can be rented at most rental yards. Bolt it securely to the engine and tighten it to lift the weight of the engine off of its mounts.
7   Remove the right engine mount (see Section 17).
8   Remove the secondary air injection pump mounting bolts (see Chapter 6) and separate the pump from the side of the oil pan, secur-

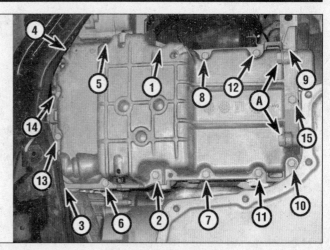

13.13 Oil pan bolt tightening sequence. Tighten the pan-to-transaxle bolts (A) until they're snug (but not too tight). Tighten the pan-to-block bolts in numerical order to the torque listed in this Chapter's Specifications, then tighten bolts (A) to the torque listed in this Chapter's Specifications

ing the pump out of the way.
9   Remove the four oil pan-to-transaxle bolts and the remaining oil pan bolts in the reverse of the tightening sequence **(see illustration 13.13)**.
10   Carefully remove the oil pan from the lower crankcase.
Caution: *If the oil pan is difficult to separate from the lower crankcase, use a rubber mallet or a block of wood and a hammer to jar it loose. If it's stubborn and still won't come off, pry carefully on casting protrusions (not the mating surfaces!).*

Installation

Refer to illustration 13.13

11   Using a gasket scraper, thoroughly clean all old gasket material from the lower crankcase and oil pan. Remove residue and oil film with a solvent such as brake system cleaner.
12   Apply a 2 mm bead of RTV sealant to the perimeter of the oil pan, inboard of the bolt holes, and around the oil suction port. Allow the sealant to set-up before installing the oil pan to the engine (install the pan within the time given by the sealant manufacturer).
13   Install the oil pan and bolts. Tighten the bolts in sequence **(see illustration)** to the torque listed in this Chapter's Specifications.
14   The remainder of installation is the

reverse of removal. Tighten the wheel lug nuts to the torque listed in the Chapter 1 Specifications.
15   Refill the engine with oil and install a new oil filter (see Chapter 1), then run the engine and check for leaks.

14   Oil pump - removal, inspection and installation

Removal

Refer to illustrations 14.5a and 14.5b

1   Drain the engine oil (see Chapter 1).
2   Remove the drivebelt (see Chapter 1).
3   Loosen the right-front wheel lug nuts, raise the front of the vehicle and support it securely on jackstands. Remove the right front wheel.
4   Remove the engine front cover (see Section 7).
5   Working on the backside of the engine cover, loosen the oil pump cover screws a little at a time until they're all loose **(see illustrations)**. When all of the screws are loose, remove the cover, and remove the oil pump gears.

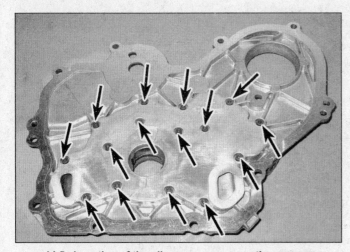

14.5a Location of the oil pump cover mounting screws

14.5b Lift the oil pump cover from the oil pump assembly

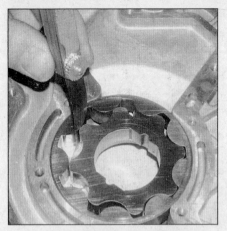

14.8a Using a feeler gauge to check the inner-to-outer rotor tip clearance . . .

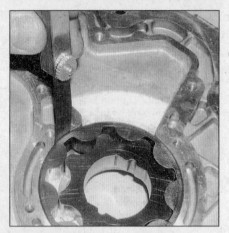

14.8b . . . and the outer rotor-to-housing clearance

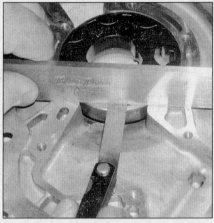

14.8c Use a straightedge and a feeler gauge to check the rotor-to-cover clearance

14.10 Oil pressure relief valve component details

1   *Oil pressure relief valve plug*
2   *Sealing washer*
3   *Spring*
4   *Piston*

Inspection

Refer to illustrations 14.8a, 14.8b, 14.8c and 14.10

6   Note any identification marks on the rotors and withdraw the rotors from the pump body. If no marks can be seen, use a permanent marker and make your own to ensure that they will be installed correctly.

7   Thoroughly clean and dry the components.

8   Inspect the rotors for obvious wear or damage. If either rotor, the pump body or the cover is scored or damaged, the complete oil pump assembly must be replaced. Also check the inner-to-outer rotor tip clearance, the outer rotor-to-housing clearance, and the rotor-to-cover side clearance **(see illustrations)**.

9   If the oil pump components are in acceptable condition, dip the rotors in clean engine oil and install them into the pump body with any identification marks positioned as noted during disassembly. If they are not, the front cover must be replaced as a unit.

10   Remove the oil pressure relief valve components from the front cover. Thoroughly

clean and dry the components. Inspect the components for obvious wear or damage. Install them in the correct order **(see illustration)**.

Installation

11   Install the rotors into the housing with the hub of the inner rotor facing the engine front cover. The inner rotor hub must be installed correctly or the engine front cover/oil pump cover will not fasten properly.

12   Install the oil pump cover and screws and tighten by hand until snug. Then tighten the screws gradually and evenly to the torque listed in this Chapter's Specifications. Install the oil pressure relief valve components.

13   Install the engine front cover (see Section 7).

14   Fill the engine with fresh engine oil (see Chapter 1). Install a new oil filter.

15   Start the engine and check for leaks.

16   Run the engine and make sure oil pressure comes up to normal quickly. If it doesn't, stop the engine and find out the cause. Severe engine damage can result from running an engine with insufficient oil pressure!

15   Driveplate - removal and installation

Removal

1   Raise the vehicle and support it securely on jackstands, then remove the transaxle (see Chapter 7A). If it's leaking, now would be a very good time to replace the front pump seal/O-ring.

2   Use a center punch or paint to make alignment marks on the driveplate and crankshaft to ensure correct alignment during reinstallation.

3   Remove the bolts that secure the driveplate to the crankshaft. If the crankshaft turns, wedge a screwdriver in the ring gear teeth.

4   Remove the driveplate from the crankshaft. If there is a spacer between the crankshaft and the driveplate, note which way it is installed.

Installation

5   Inspect the surface of the driveplate for cracks.

6   Clean and inspect the mating surfaces of the driveplate and the crankshaft. If the crankshaft rear seal is leaking, replace it before reinstalling the driveplate (see Section 16).

7   Position the driveplate against the crankshaft, installing the spacer if one was present originally. Align the mating marks made during removal. Note that some engines have an alignment dowel or staggered bolt holes to ensure correct installation. Before installing the bolts, apply thread locking compound to the threads.

8   Wedge a screwdriver in the ring gear teeth to keep the driveplate from turning and tighten the bolts to the torque listed in this Chapter's Specifications. Work up to the final torque in three or four steps.

9   The remainder of installation is the reverse of removal.

16 Rear main oil seal - replacement

1   The one-piece rear main oil seal is pressed into the engine block and the crankcase reinforcement section. Remove the transaxle (see Chapter 7A) and the driveplate (see Section 15).

2   Pry out the old seal with a special seal removal tool or a flat-blade screwdriver.

Caution: *To prevent an oil leak after the new seal is installed, be very careful not to scratch or otherwise damage the crankshaft sealing surface or the bore in the engine block.*

3   Clean the crankshaft and seal bore in the block thoroughly and de-grease these areas by wiping them with a rag soaked in lacquer thinner or acetone. Lubricate the lip of the new seal and the outer diameter of the crankshaft with engine oil.

4   Position the new seal onto the crankshaft. Make sure the edges of the new oil seal are not rolled over. Use a special rear main oil seal installation tool or a socket with the exact diameter of the seal to drive the seal in place. Make sure the seal is not off-set; it must be flush along the entire circumference of the engine block and the crankcase reinforcement section.

Note: *When installing the new seal, if so marked, the words* THIS SIDE OUT *on the seal must face out, toward the rear of the engine.*

5   The remainder of installation is the reverse of removal.

17 Powertrain mounts - check and replacement

Check

1   Engine mounts seldom require attention, but broken or deteriorated mounts should be replaced immediately or the added strain placed on the driveline components may

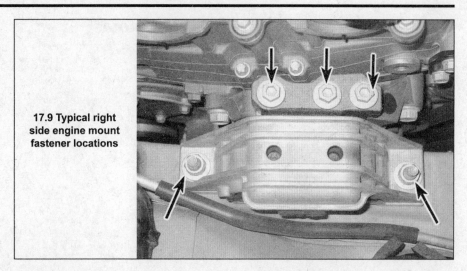

17.9 Typical right side engine mount fastener locations

cause damage or wear.

2   During the check, the engine must be raised slightly to remove the weight from the mounts.

3   Raise the vehicle and support it securely on jackstands, then position a jack under the engine oil pan. Place a large block of wood between the jack head and the oil pan, then carefully raise the engine just enough to take the weight off the mounts.

Warning: DO NOT place any part of your body under the engine when it's supported only by a jack!

4   Check the mounts to see if the rubber is cracked, hardened or separated from the bushing in the center of the mount.

5   Check for relative movement between the mount and the engine or frame (use a large screwdriver or prybar to attempt to move the mounts). If movement is noted, lower the engine and tighten the mount fasteners.

Replacement

Caution: *Do not disconnect more than one mount at a time unless the engine will be removed from the vehicle.*

6   Disconnect the cable from the negative terminal of the battery (see Chapter 5).

Right side mount

Refer to illustration 17.9

7   Remove the air filter housing (see Chapter 4).

8   Place a large block of wood between the jack head and the oil pan, then carefully raise the engine just enough to take the weight off the mounts.

9   Remove the engine mount-to-body bolts, then the mount-to-engine bracket bolts, and detach the mount from the engine bracket **(see illustration)**.

10   Installation is the reverse of removal. Use thread-locking compound on the mount bolts and tighten them securely.

11   Reconnect the battery (see Chapter 5).

All other mounts

12   To replace the transaxle mounts, see Chapter 7A.

Notes

Chapter 2 Part B
3.4L V6 engine

Contents

Specifications

General

Displacement	204 cubic inches
Bore and stroke	3.62 x 3.31 inches
Cylinder numbers (drivebelt end-to-transaxle end)	
Front bank (radiator side)	2-4-6
Rear bank	1-3-5
Firing order	1-2-3-4-5-6
Compression	See Chapter 2D
Oil pressure	See Chapter 2D

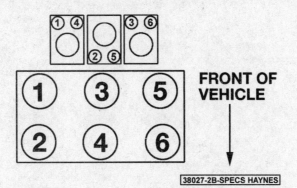

38027-2B-SPECS HAYNES

Cylinder location and coil terminal identification diagram

Torque specifications

Ft-lbs (unless otherwise indicated)

Note: *One foot-pound (ft-lb) of torque is equivalent to 12 inch-pounds (in-lbs) of torque. Torque values below approximately 15 foot-pounds are expressed in inch-pounds, because most foot-pound torque wrenches are not accurate at these smaller values.*

Camshaft sprocket/variable camshaft timing sprocket bolt(s)	
2005 and 2006 (single bolt)	103
2007 and later (3 bolts)	144 in-lbs
Catalytic converter-to-exhaust manifold flange nuts	23
Crankshaft pulley bolt	
2005	
Step 1	52
Step 2	Tighten an additional 70 degrees
2006 and later*	
Step 1	92
Step 2	Tighten an additional 130 degrees
Cylinder head bolts* (in sequence - **see illustration 8.22**)	
Step 1	44
Step 2	Tighten an additional 95 degrees
Drivebelt idler pulley (right and left side) bolts	37
Driveplate-to-crankshaft bolts	52
Engine mount bolts	
Right engine mount bolts	37
Front engine mount bolts	
Mount-to-transmission bolts	37
Mount-to-frame bracket bolt	81
Exhaust manifold fasteners	144 in-lbs
Exhaust heat shield bolts	89 in-lbs
Intake manifold (lower) bolts	
Bolts 1 to 4	
Step 1	62 in-lbs
Step 2	115 in-lbs
Bolts 5 to 8	
Step 1	115 in-lbs
Step 2	18
Intake manifold (upper) bolts	18
Oil pan bolts/nuts	
To block	18
Side bolts	37
Oil pump mounting bolt	30
Rocker arm bolts	
2005	31
2006 and later	24
Timing chain cover bolts	
Small bolts	20
Medium and large bolts	41
Timing chain tensioner plate bolts	15
Transaxle brace	37
Transaxle-to-oil pan brace bolts	37
Valve cover-to-cylinder head bolts	89 in-lbs
Valve lifter guide bolts	89 in-lbs

*Bolt(s) must be replaced with **new** bolts

4.3 Pull the PCV tube out of the valve cover

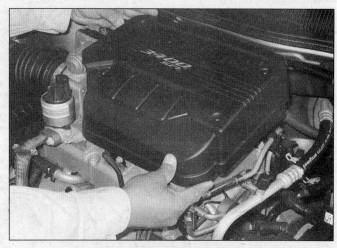

4.5 Pull the engine cover off of its mounting studs

1 General information

This Part of Chapter 2 is devoted to in-vehicle repair procedures for the 3.4L V6 engine. This engine utilizes a cast-iron block with six cylinders arranged in a "V" shape at a 60-degree angle between the two banks. The overhead valve aluminum cylinder heads are equipped with replaceable valve guides and seats. Hydraulic lifters actuate the valves through tubular pushrods.

The engine is easily identified by looking for the designations printed directly on the engine cover.

Information concerning engine removal and installation and overhaul can be found in Part D of this Chapter. The following repair procedures are based on the assumption that the engine is installed in the vehicle. If the engine has been removed from the vehicle and mounted on a stand, many of the Steps outlined in this Part of Chapter 2 will not apply.

2 Repair operations possible with the engine in the vehicle

Many major repair operations can be accomplished without removing the engine from the vehicle.

Clean the engine compartment and the exterior of the engine with some type of degreaser before any work is done. It'll make the job easier and help keep dirt out of the internal areas of the engine.

Depending on the components involved, it may be helpful to remove the hood to improve access to the engine as repairs are performed (refer to Chapter 11 if necessary). Cover the fenders to prevent damage to the paint. Special pads are available, but an old bedspread or blanket will also work.

If vacuum, exhaust, oil or coolant leaks develop, indicating a need for gasket or seal replacement, the repairs can generally be done with the engine in the vehicle. The intake and exhaust manifold gaskets, timing chain cover gasket, oil pan gasket, crankshaft oil seals and cylinder head gaskets are all accessible with the engine in place.

Exterior engine components, such as the intake and exhaust manifolds, the oil pan (and the oil pump), the water pump, the starter motor, the alternator and the fuel system components can be removed for repair with the engine in place.

Since the cylinder heads can be removed without pulling the engine, valve component servicing can also be accomplished with the engine in the vehicle. Replacement of the timing chain and sprockets is also possible with the engine in the vehicle, although camshaft removal cannot be performed with the engine in the chassis (see Part D of this Chapter).

In extreme cases caused by a lack of necessary equipment, repair or replacement of piston rings, pistons, connecting rods and rod bearings is possible with the engine in the vehicle. However, this practice is not recommended because of the cleaning and preparation work that must be done to the components involved.

3 Top Dead Center (TDC) - locating

Refer to Chapter 2, Part C for this procedure.

4 Valve covers - removal and installation

Removal

Refer to illustrations 4.3, 4.5 and 4.13

1   Disconnect the cable from the negative terminal of the battery (see Chapter 5).

2   Disconnect the spark plug wires from the side of the engine on which you're working (see Chapter 1). Label the wires to ensure correct installation and move them out of the way.

Front valve cover

3   Remove the PCV tube from the PCV valve and the valve cover **(see illustration)**.

4   Remove the spark plug wire bracket.

Rear valve cover

5   Remove the engine cover by pulling it from its mounting studs **(see illustration)**.

6   Refer to Chapter 4 and remove the air filter housing and the intake duct.

7   Remove the drivebelt (see Chapter 1).

8   Remove the drivebelt tensioner.

9   Refer to Chapter 5 and remove the alternator.

10   Remove the alternator mounting bracket.

11   Refer to Chapter 5 and remove the ignition coil and its mounting bracket. You can leave the spark plug wires connected to it as you move it out of the way.

12   Detach the PCV tube from the valve cover.

Both valve covers

13   Unscrew the valve cover mounting bolts **(see illustration)**.

4.13 Valve cover mounting bolt locations

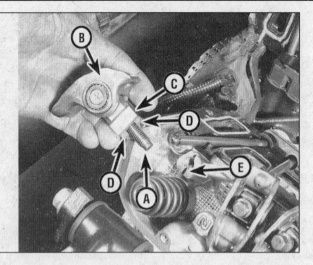

5.3 Rocker arm details - the rocker arms are kept as an assembly by a small sleeve between the bolt and the pedestal - note the projections on the pedestal; they fit into grooves in the head

A   *Rocker arm bolt*
B   *Rocker arm*
C   *Rocker arm pedestal*
D   *Pedestal projections*
E   *Grooves in the head*

14   Detach the valve cover. **Note:** *If the cover sticks to the cylinder head, use a block of wood and a hammer to dislodge it. If the cover still won't come loose, pry on it carefully, but don't distort the sealing flange. The valve cover gasket should stay with the cylinder head when the valve cover is lifted off.*

15   Carefully cut the gasket material from the lower intake manifold gasket at the cylinder head. **Caution:** *The lower intake manifold gasket can be damaged if this isn't done, which could cause an engine oil leak.*

Installation

16   The mating surfaces of each cylinder head and valve cover must be perfectly clean when the covers are installed. Use a gasket scraper to remove all traces of sealant or old gasket material, then clean the mating surfaces with brake system cleaner (if there's sealant or oil on the mating surfaces when the cover is installed, oil leaks may develop). The valve covers are made of aluminum, so be extra careful not to nick or gouge the mating

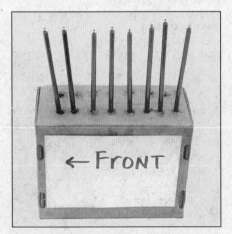

5.4 A perforated cardboard box can be used to store the pushrods to ensure they are reinstalled in their original locations - note the label indicating the front end of the engine

surfaces with the scraper. Take special care to clean the points where the intake manifold contacts the cylinder heads.

17   Clean the mounting bolt threads with a die if necessary to remove any corrosion and restore damaged threads. Use a tap to clean the threaded holes in the heads.

18   Apply a dab of RTV sealant to the two joints where the intake manifold and cylinder head meet.

19   Place the new gasket in position in the groove of the valve cover, then install it. Tighten the bolts in several steps to the torque listed in this Chapter's Specifications.

20   The remainder of installation is the reverse of removal.

21   Start the engine and check carefully for oil leaks at the joints.

5   Rocker arms and pushrods - removal, inspection and installation

Removal

Refer to illustrations 5.3 and 5.4

1   Disconnect the cable from the negative terminal of the battery (see Chapter 5).

2   Remove the valve cover(s) (see Section 4).

3   Beginning at the drivebelt end of one cylinder head, remove the rocker arm mounting bolts one at a time and detach the rocker arms, pivot balls and pedestals **(see illustration)**. Store each set of rocker arm components separately in a marked plastic bag to ensure they're reinstalled in their original locations. **Note:** *The rocker arms have the pedestal mount captured on the rocker arm bolt by a metal sleeve inside. The components can be separated if necessary by tapping the bolt out of the pedestal, but normally all components for a particular valve will stay as an assembly.*

4   Remove the pushrods and store them separately to make sure they don't get mixed up during installation **(see illustration)**. **Note:**

Intake and exhaust pushrods are different lengths. Intake pushrods are approximately 5-3/4 inches long, while exhausts are 6 inches long. They may also have color codes to easily tell them apart.

Inspection

5   Inspect each rocker arm for wear, cracks and other damage, especially where the push-rods and valve stems make contact.

6   Make sure the rollers operate freely as well.

7   Make sure the hole at the pushrod end of each rocker arm is open.

8   Inspect the pushrods for cracks and excessive wear at the ends. Roll each push-rod across a piece of plate glass to see if it's bent (if it wobbles, it's bent).

Installation

9   Lubricate the lower end of each pushrod with clean engine oil or moly-base grease and install them in their original locations. Make sure each pushrod seats completely in the lifter socket.

10   Apply moly-base grease to the ends of the valve stems and the upper ends of the pushrods.

11   Install the rocker arm assemblies and tighten them to the torque listed in this Chapter's Specifications. As the bolts are tightened, make sure the pushrods engage properly in the rocker arms and that the projections on the bottom of the pedestals fit into the grooves on the head before tightening the bolts **(see illustration 5.3)**.

12   Install the valve covers. Start and run the engine, then check for oil leaks and unusual sounds coming from the valve cover area.

6   Intake manifold - removal and installation

1   Disconnect the cable from the negative terminal of the battery (see Chapter 5).

2   Remove the engine cover.

Upper intake manifold

Refer to illustration 6.10

3   Disconnect the brake booster vacuum tube from the intake manifold.

4   Disconnect the front spark plug wires from the spark plugs and the bracket.

5   Refer to Chapter 5 and remove the ignition coil along with its bracket (you can leave the spark plug wires attached). Position it out of the way.

6   Remove the air intake duct (see Chapter 4).

7   Unbolt the heater outlet tube from the top of the engine and move it to the side. Don't disconnect the heater hoses from it.

8   Remove the EGR tube and the PCV tube.

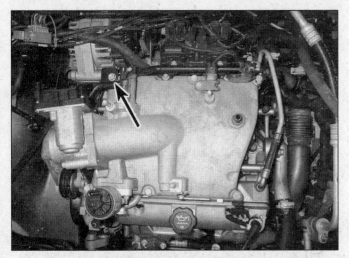

6.10 **Loosen the upper alternator mounting bolt and slide it out far enough to remove this bracket**

6.25 **Pry the manifold loose at a casting boss - don't pry between the gasket surfaces!**

9   Unscrew the alternator mounting bolt that's closest to the intake manifold. Allow it to remain in place but pulled out about an inch.

10   Remove the alternator bracket into which the mounting bolt was threaded **(see illustration)**.

11   Remove the spark plug wire retainer from the top of the intake manifold. Disconnect the MAP sensor and any remaining interfering components.

12   Loosen the upper intake manifold bolts a little at a time, starting with the outer bolts and working towards the inner bolts, then remove the upper intake manifold with the throttle body attached.

13   Clean the mounting surfaces of the lower intake manifold and the upper intake manifold with brake system cleaner, removing all traces of the old gasket material or sealant.

14   Install the new gaskets over the lower intake manifold. Install the upper intake manifold onto the lower intake manifold and tighten the bolts a little at a time, starting with the inner bolts and working towards the outer bolts, to the torque listed in this Chapter's Specifications. The remainder of the installation is the reverse of removal.

Lower intake manifold

Refer to illustrations 6.25, 6.28, 6.29 and 6.31

Warning: *The engine must be completely cool before starting this procedure.*

15   Drain the cooling system (see Chapter 1) and relieve the fuel system pressure (see Chapter 4).

16   Remove the upper intake manifold.

17   Disconnect the heater core inlet hose from its tube at the engine.

18   Unbolt the tube, pull it out of the socket and remove it.

19   Remove the upper radiator bracket from the radiator.

20   Remove the radiator inlet hose.

21   Refer to Chapter 4 and remove the fuel rail.

22   Remove the valve covers (see Section 4).

23   Loosen the rocker arm bolts, rotate the rocker arms out of the way and remove the pushrods that go through the manifold gaskets (see Section 5).

24   Loosen the manifold mounting bolts/nuts in 1/4-turn increments until they can be removed by hand.

25   The manifold will probably be stuck to the cylinder heads and force may be required to break the gasket seal **(see illustration)**. **Caution:** *Don't pry between the manifold and the heads or damage to the gasket sealing surfaces may occur, leading to vacuum leaks.*

26   Lift the old gaskets off. Use a gasket scraper to remove all traces of sealant and old gasket material, then clean the mating surfaces with brake system cleaner. **Note:** *The mating surfaces of the cylinder heads, block, coolant crossover housing and manifold must be perfectly clean when the mani-* fold *is installed. Gasket removal solvents are available at most auto parts stores and may be helpful when removing old gasket material that's stuck to the heads and manifold (since the manifold and the coolant crossover housing is made of aluminum, aggressive scraping can cause damage). Be sure to follow the directions printed on the container. If there's old sealant or oil on the mating surfaces when the manifold is installed, oil or vacuum leaks may develop. Use a vacuum cleaner to remove any gasket material that falls into the intake ports or the lifter valley.*

27   Use a tap of the correct size to chase the threads in the bolt holes, if necessary, then use compressed air (if available) to remove the debris from the holes. **Warning:** *Wear safety glasses or a face shield to protect your eyes when using compressed air!*

28   Place the intake manifold gaskets in position on the heads **(see illustration)**. Install the pushrods and rocker arms (see Section 4).

29   Apply a 3/16-inch (5 mm) bead of RTV sealant to the front and rear ridges of the

6.28 **Install the intake gaskets against each cylinder head . . .**

6.29 . . . then apply a bead of sealant to the end ridges between the cylinder heads

6.31 Intake manifold TIGHTENING sequence - make sure the bolts in the center (1 through 4) are completely tightened before tightening the end bolts (5 through 8)

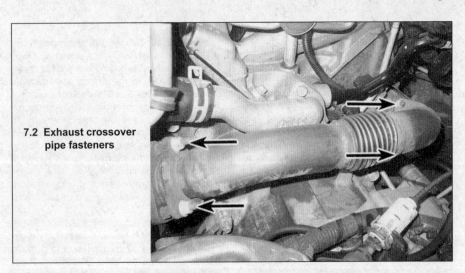

7.2 Exhaust crossover pipe fasteners

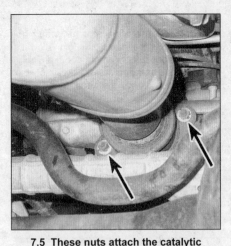

7.5 These nuts attach the catalytic converters and the rear portion of the exhaust system to the exhaust manifolds

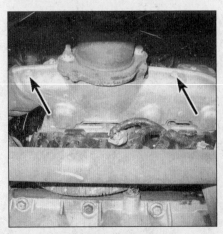

7.6 Exhaust manifold heat shield fasteners

engine block between the heads **(see illustration)**.

30   Carefully lower the manifold into place and install the mounting bolts/nuts finger tight. **Note:** *Coat the bolt threads with thread-locking liquid before installing them.*

31   Tighten the four vertical bolts (1 through 4) at the center of the manifold in the recommended tightening sequence **(see illustration)** to the torque listed in this Chapter's Specifications.

32   Tighten the four angled bolts (5 through 8) at the ends of the manifold in the recommended tightening sequence to the torque listed in this Chapter's Specifications.

33   Install the remaining components in the reverse order of removal.

34   Refill the cooling system (see Chapter 1). Start the engine and check for leaks.

7   Exhaust manifolds - removal and installation

Removal

Refer to illustrations 7.2, 7.5, 7.6 and 7.7

1   Disconnect the cable from the negative battery terminal (see Chapter 5).

2   Unbolt the crossover exhaust pipe that connects the two exhaust manifolds **(see**

illustration). It doesn't have to be completely removed from the vehicle.

Front manifold

3   Remove the EGR tube from the EGR valve and the exhaust manifold.

Rear manifold

4   Disconnect the oxygen sensor wiring.

5   Raise the vehicle and support it securely on jackstands. Disconnect the exhaust pipe/ catalytic converter from the exhaust manifold **(see illustration)**. Secure the exhaust pipe away from the manifold.

Both manifolds

6   Remove the exhaust manifold heat shield **(see illustration)**.

7   Remove the manifold mounting fasteners and remove the exhaust manifold **(see illustration)**. Remove the gaskets.

Installation (front or rear)

8   Clean the mating surfaces to remove all traces of old gasket material, then inspect the

7.7 Remove the six fasteners (arrows indicate the upper three) from the exhaust manifold

8.7 Remove the bolt holding the oil dipstick tube to the front cylinder head

manifold for distortion and cracks. Warpage can be checked with a precision straightedge held against the mating flange. If a feeler gauge thicker than 0.030-inch can be inserted between the straightedge and flange surface, take the manifold to an automotive machine shop for resurfacing.

9   Using a new gasket, place the manifold against the head and install the fasteners.

10   Starting in the middle and working out toward the ends, tighten the mounting fasteners a little at a time until all of them are at the torque listed in this Chapter's Specifications.

11   Install the remaining components in the reverse order of removal.

12   Start the engine and check for exhaust leaks between the manifold and cylinder head and between the manifold and exhaust pipe.

8   Cylinder heads - removal and installation

Warning: *The engine must be completely cool before starting this procedure.*

Removal

Refer to illustration 8.7

1   Disconnect the cable from the negative terminal of the battery (see Chapter 5).

2   Raise the vehicle and support it securely on jackstands.

3   Drain the cooling system and the engine oil (see Chapter 1). Disconnect the exhaust pipe(s) from the manifold(s), then lower the vehicle.

4   Remove the rocker arms and pushrods (see Section 5).

5   Refer to Section 6 and remove the lower intake manifold.

6   Remove the exhaust manifold (see Section 7).

7   Remove the dipstick tube if you're removing the front head **(see illustration)**.

8.17 Remove the old gasket and carefully scrape off all old gasket material and sealant

Rear cylinder head

8   Disconnect the heater outlet hose from the heater outlet tube. Also disconnect the heater core outlet hose from the heater outlet tube.

9   Disconnect the fuel tube retainer from the heater tube bracket, then unbolt both ends of the heater outlet tube from the top of the engine and remove it.

10   Remove the engine coolant temperature (ECT) sensor (see Chapter 6).

11   Remove the engine hoist bracket.

Both cylinder heads

12   Make sure that all interfering components are disconnected from the cylinder heads.

13   Remove the spark plugs (see Chapter 1).

14   Loosen each of the cylinder head bolts 1/4-turn at a time until they can be removed by hand - work from bolt-to-bolt in a pattern that's the reverse of the tightening sequence **(see illustration 8.22)**. Discard the bolts - new ones must be used during installation, but note which ones are studs and their locations.

15   Lift the head(s) off the engine. If resistance is felt, don't pry between the head and block, as damage to the mating surfaces will result. Recheck for head bolts that may have been overlooked, then use a hammer and block of wood to tap up on the head and break the gasket seal. Be careful because there are locating dowels in the block that position each head. As a last resort, pry each head up at the rear corner only and be careful not to damage anything. After removal, place the head on blocks of wood to prevent damage to the gasket surfaces.

Installation

Refer to illustrations 8.17, 8.20a, 8.20b and 8.22

16   The mating surfaces of each cylinder head and block must be perfectly clean when the head is installed.

17   Use a gasket scraper to remove all traces of carbon and old gasket material **(see illustration)**, then clean the mating surfaces with brake system cleaner. If there's oil on the mating surfaces when the head is installed,

8.20a Position the new gasket over the dowel pins . . .

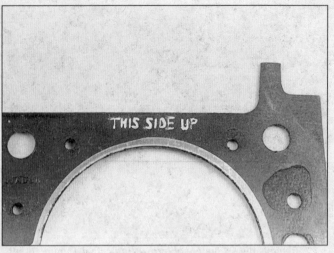

8.20b . . . with the correct side facing up

the gasket may not seal correctly and leaks may develop. When working on the block, it's a good idea to cover the lifter valley with shop rags to keep debris out of the engine. Use a shop rag or vacuum cleaner to remove any debris that falls into the cylinders.

18　Check the block and head mating surfaces for nicks, deep scratches and other damage. If damage is slight, it can be removed with a file; if it's excessive, machining may be the only alternative.

19　Use a tap of the correct size to chase the threads in the head bolt holes. Dirt, corrosion, sealant and damaged threads will affect torque readings.

20　Position the new gasket over the dowel pins in the block. Some gaskets are marked TOP or THIS SIDE UP to ensure correct installation **(see illustrations)**.

21　Carefully position the head on the block without disturbing the gasket.

22　Install the *new* cylinder head bolts. Tighten the bolts, using the recommended sequence **(see illustration)**, to the torque listed in this Chapter's Specifications. Then, using the same sequence, turn each bolt the amount of angle listed in this Chapter's Specifications.

23　The remainder of installation is the reverse of removal.

24　Refill the cooling system and change the engine oil and filter (see Chapter 1).

9　Crankshaft pulley - removal and installation

Refer to illustrations 9.5 and 9.6

1　Disconnect the cable from the negative terminal of the battery (see Chapter 5).

2　With the parking brake applied and the shift lever in Park, loosen the lug nuts from the right front wheel, then raise the front of the vehicle and support it securely on jackstands.

3　Remove the right front wheel and the right splash shield from the wheelwell (see Chapter 11).

4　Remove the drivebelt (see Chapter 1).

5　Remove the bolt from the front of the crankshaft **(see illustration)**. The bolt is normally very tight, so use a large breaker bar and a six-point socket to remove it. If you have a 2006 or later vehicle, obtain a new bolt, but save the old one for the initial installation of the pulley. **Note:** *Use a pin spanner to prevent the crankshaft from turning. If you don't have access to this tool, remove the starter (see Chapter 5) and position a large screwdriver in the ring gear teeth to keep the crankshaft from turning while an assistant removes the crankshaft pulley bolt.*

6　Use a puller that has three hooks that

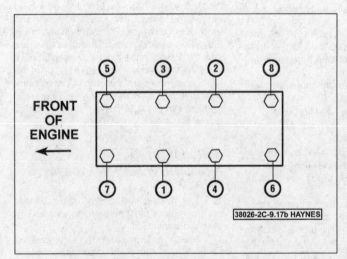

8.22 Cylinder head bolt TIGHTENING sequence

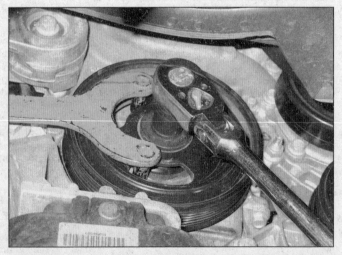

9.5 A pin spanner can be used to prevent the crankshaft from turning while loosening the pulley bolt

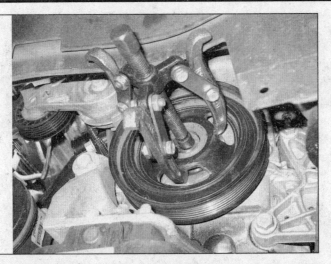

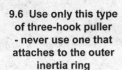
9.6 Use only this type of three-hook puller - never use one that attaches to the outer inertia ring

can grasp the inside diameter of the outer ring of the crankshaft pulley. Don't attach the puller's hooks to the outside of the pulley as this will damage it. Remove the crankshaft pulley/balancer from the crankshaft (see illustration). Additionally, a spacer, such as a socket that just fits into the hole in the pulley and bears on the nose of the crankshaft will be required to avoid damage to the crankshaft.

7   Apply a small amount of RTV sealant onto the crankshaft keyway.

8   Position the crankshaft pulley/balancer on the crankshaft and slide it on as far as it will go. Note that the slot (keyway) in the hub must be aligned with the Woodruff key in the end of the crankshaft.

9   Using a crankshaft balancer installation tool, available at most auto parts stores, press the crankshaft pulley/balancer onto the crankshaft. Note that the crankshaft bolt can also be used to press the crankshaft balancer into position, but when doing so, use a liberal amount of clean engine oil on the bolt threads to prevent galling.

10   If you're working on a 2005 model, install the old bolt and washer and tighten it to the torque listed in this Chapter's Specifications. If you're working on a 2006 or later model, purchase a new bolt. Install the old bolt and washer and tighten the bolt to the Step 1 torque listed in this Chapter's Specifications, then remove the bolt. Install the new bolt and washer, tightening it to the Step 1 torque listed in this Chapter's Specifications, followed by the Step 2 angle torque.

11   The remainder of installation is the reverse of removal.

10   Crankshaft front oil seal - removal and installation

Refer to illustrations 10.2 and 10.3

1   Remove the crankshaft pulley (see Section 9).

2   Note how the seal is installed - the new one must be installed to the same depth and facing the same way. Carefully pry the oil seal out of the cover with a seal puller or a large screwdriver (see illustration). Be very careful not to distort the cover or scratch the crankshaft!

3   Apply clean engine oil or multi-purpose grease to the outer edge of the new seal, then install it in the cover with the lip (spring side) facing IN. Drive the seal into place (see illustration) with a seal driver or a large socket and a hammer. Make sure the seal enters the bore squarely and stop when the front face is at the proper depth.

4   Check the surface on the pulley hub that the oil seal rides on. If the surface has been grooved from long-time contact with the seal, the pulley should be replaced with a new one.

5   Lubricate the pulley hub with clean engine oil and reinstall the crankshaft pulley. Use a vibration damper installation tool to press the pulley onto the crankshaft.

6   The remainder of installation is the reverse of removal.

11   Timing chain and sprockets - removal, inspection and installation

Warning: *The engine must be completely cool before starting this procedure.*

Removal

Refer to illustrations 11.8a, 11.8b, 11.16a, 11.16b and 11.18

1   Disconnect the cable from the negative terminal of the battery (see Chapter 5).

2   Refer to Chapter 4 and remove the air filter housing and the air intake duct.

3   Drain the coolant and engine oil (see Chapter 1).

4   Remove the drivebelt (see Chapter 1).

5   Unbolt the drivebelt tensioner (see Chapter 1).

6   Remove the thermostat and the water pump pulley (see Chapter 3).

7   Remove the crankshaft pulley (see Section 9).

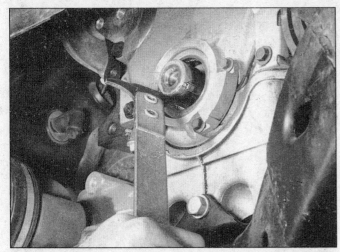
10.2 Carefully pry the old seal out of the timing chain cover - don't damage the crankshaft in the process

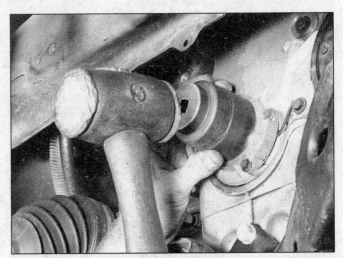
10.3 Drive the new seal into place with a seal driver or a large socket and hammer

11.8a Support the engine with an engine support fixture and chains attached to the front and rear engine lifting eyes

11.8b There should be lifting brackets on the engine; if they've been removed, bolt the chains directly to the cylinder heads

8   Attach an engine support fixture to the engine lifting eyes **(see illustrations)**, then remove the passenger side engine mount and the mount support bracket from the front of the timing chain cover (see Section 17). **Note:** *Engine support fixtures can usually be obtained at equipment rental yards, as well as some auto parts stores.*

9   Remove the lower drivebelt idler pulley.

10   Remove the oil pan (see Section 13).

11   Remove the left drivebelt idler pulley.

12   Remove the engine mount bracket from the top of the timing chain cover.

13   Refer to Chapter 3 and remove the water pump.

14   Remove the thermostat bypass hose adapter.

15   Detach the radiator outlet hose from the timing chain cover.

16   Remove the timing chain cover-to-engine block bolts **(see illustrations)**.

17   Separate the cover from the engine. If it's stuck, tap it with a soft-face hammer, but don't try to pry it off.

18   Temporarily install the crankshaft pulley bolt and turn the crankshaft with the bolt to align the timing marks on the crankshaft and camshaft sprockets. When aligned at TDC for number 1 piston, the crankshaft sprocket timing mark should align with the mark on the bottom of the chain dampener plate, and the small hole in the camshaft sprocket should be at the 6 o'clock position, aligned with the timing mark in the top of the chain dampener plate **(see illustration)**.

19   Remove the camshaft sprocket bolt(s). Do not turn the camshaft in the process (if you do, realign the timing marks before the bolt(s) are removed).

20   Use two large screwdrivers to carefully pry the camshaft sprocket off the camshaft dowel pin. Slip the timing chain and camshaft sprocket off the engine.

Inspection

Refer to illustration 11.22

21   The timing chain should be replaced with a new one if the engine has high mileage, the chain has visible damage, or total freeplay midway between the sprockets exceeds one inch. Failure to replace a worn timing chain may result in erratic engine performance, loss of power and decreased fuel mileage. Loose chains can jump timing. In the worst case, chain jumping or breakage will result in severe engine damage. Always replace the timing chain and sprockets in sets. If you intend to install a new timing chain, remove the crankshaft sprocket with a puller and install a new one. Be sure to align the key in the crankshaft with the keyway in the sprocket during installation.

22   Inspect the timing chain dampener plate for cracks and wear and replace it if necessary **(see illustration)**. The plate should be reinstalled before installing the new timing chain and sprockets.

23   Clean the timing chain and sprockets with solvent and dry them with compressed air (if available). **Warning:** *Wear eye protection when using compressed air.*

24   Inspect the components for wear and damage. Look for teeth that are deformed, chipped, pitted or cracked.

11.16a Timing chain cover bolt locations, upper . . .

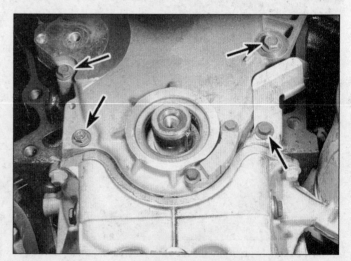

11.16b . . . and lower

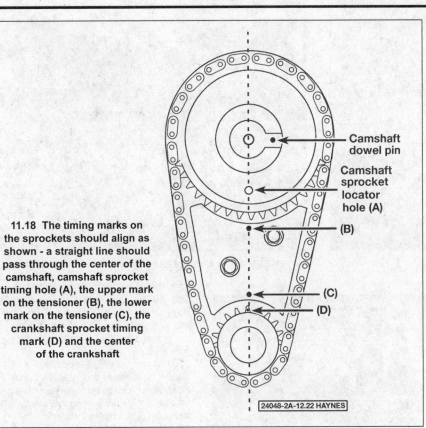

11.18 The timing marks on the sprockets should align as shown - a straight line should pass through the center of the camshaft, camshaft sprocket timing hole (A), the upper mark on the tensioner (B), the lower mark on the tensioner (C), the crankshaft sprocket timing mark (D) and the center of the crankshaft

Camshaft dowel pin

Camshaft sprocket locator hole (A)

(B)

(C)
(D)

24048-2A-12.22 HAYNES

11.22 The timing chain damper (guide) is retained by two bolts

Installation

25   Use a gasket scraper to remove all traces of old gasket material and sealant from the cover and engine block. Be careful not to nick or gouge it. Clean the gasket sealing surfaces with brake system cleaner.

26   Install the dampener plate, tightening the bolts to the torque listed in this Chapter's Specifications.

27   If the camshaft has turned at all since removal of the sprocket, turn the camshaft to position the dowel pin at 3 o'clock. Mesh the timing chain with the camshaft sprocket, then engage it with the crankshaft sprocket. The timing marks should be aligned as shown in **illustration 11.18**. **Note:** *If the crankshaft has been disturbed, turn it until the "O" stamped on the crankshaft sprocket is exactly at the top and is aligned with the mark on the dampener plate.*

28   Install the camshaft sprocket bolt(s) and make sure the dowel hole in the sprocket is aligned with the dowel pin in the camshaft. Verify that the top mark on the crankshaft sprocket and the bottom mark on the camshaft sprocket are precisely aligned with their respective marks on the damper plate. Tighten the bolt(s) to the torque listed in this Chapter's Specifications.

29   Lubricate the chain, the sprocket and the front of the crankshaft gear with clean engine oil.

30   Apply a thin layer of RTV sealant to both sides of the new gasket, then position the gasket on the engine block (the dowel pins should keep it in place). Apply an extra 1/4-inch bead of sealant to the bottom edges of

the gasket, where it meets the oil pan.

31   Attach the cover to the engine and install the bolts. Follow a criss-cross pattern when tightening the fasteners and work up to the torque listed in this Chapter's Specifications in three steps.

32   The remainder of installation is the reverse of removal.

33   Add oil and coolant, start the engine and check for leaks.

12   Valve lifters - removal, inspection and installation

1   A noisy valve lifter can be isolated when the engine is idling. Hold a mechanic's stethoscope or a length of hose near the location of

each valve while listening at the other end. Another method is to remove the valve cover and, with the engine idling, touch each of the valve spring retainers, one at a time. If a valve lifter is defective, it'll be evident from the shock felt at the retainer each time the valve seats.

2   The most likely causes of noisy valve lifters are dirt trapped inside the lifter and lack of oil flow, viscosity or pressure. Before condemning the lifters, check the oil for fuel contamination, correct level, cleanliness and correct viscosity.

Removal

Refer to illustrations 12.5, 12.6a, 12.6b and 12.7

3   Remove the valve cover(s) and lower intake manifold as described in Sections 4 and 6.

4   Remove the rocker arms and pushrods (see Section 5).

5   Remove the bolts holding the roller lifter guide to the block, and remove the two roller lifter guides **(see illustration)**. Mark the guides as to which side they came from.

6   There are several ways to extract the lifters from the bores. A special tool designed to grip and remove lifters is manufactured

12.5 Remove the bolts (A) and pull up the roller lifter guides (B)

12.6a A magnetic pick-up tool . . .

12.6b . . . or a scribe can be used to remove the lifters

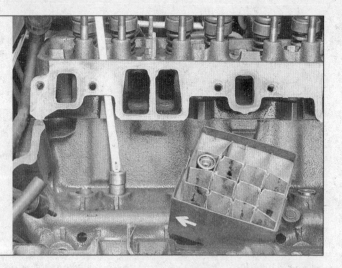

12.7 Store the lifters in order to ensure installation in their original locations

freely without excessive play **(see illustration)**. If the lifters walls are worn (not very likely), inspect the lifter bores in the block. If the pushrod seats are worn, inspect the push-rods also.

11   When reinstalling used lifters, make sure they're replaced in their original bores. Soak new lifters in oil to remove trapped air. Coat all lifters with moly-base grease or engine assembly lube prior to installation.

12   Install the lifters and lifter guides, tightening the bolts to the torque listed in this Chapter's Specifications.

13   The remainder of installation is the reverse of removal.

14   Run the engine and check for oil leaks.

13 Oil pan - removal and installation

Removal

Refer to illustrations 13.4 and 13.10

1   Disconnect the cable from the negative terminal of the battery (see Chapter 5).

by many tool companies and is widely available, but it may not be required in every case. On engines without a lot of varnish buildup, the lifters can often be removed with a small magnet or even with your fingers. A machinist's scribe with a bent end can be used to pull the lifters out by positioning the point under the retainer ring in the top of each lifter **(see illustrations)**. **Caution:** *Don't use pliers to remove the lifters unless you intend to replace them with new ones. The pliers may damage the precision machined and hardened lifters, rendering them useless.*

7   Before removing the lifters, arrange to store them in a clearly labeled box to ensure they're reinstalled in their original locations. Remove the lifters and store them where they won't get dirty **(see illustration)**.

Inspection and installation

Refer to illustrations 12.10a and 12.10b

8   Parts for valve lifters are not available separately. The work required to remove them from the engine again if cleaning is unsuccessful outweighs any potential savings from repairing them.

9   Clean the lifters thoroughly with solvent

and dry them thoroughly, without mixing them up.

10   Check each lifter wall and plunger seat for scuffing, score marks or uneven wear **(see illustration)**. Check the rollers carefully for wear or damage and make sure they turn

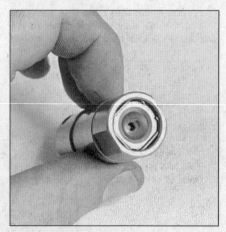

12.10a Check the pushrod seat in the top of each lifter for wear

12.10b The roller on the roller lifters must turn freely - check for wear and excessive play as well

13.4 Remove the transaxle-to-oil pan brace

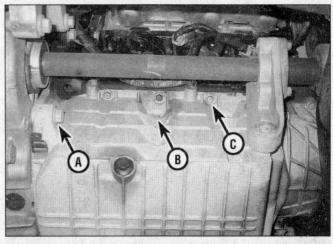

13.10 Oil pan fastener details

A   Oil pan-to-transaxle bolt
B   Oil pan side bolt
C   Oil pan flange bolt

2   Refer to **illustration 11.8a** and install an engine support fixture. Raise the vehicle and support it securely on jackstands.
3   Drain the engine oil and remove the oil filter (see Chapter 1).
4   Remove the transaxle brace from the oil pan **(see illustration)**.
5   Remove the starter (see Chapter 5).
6   Remove the air conditioning compressor mounting bolts, then position it to the side. **Warning:** *Don't disconnect the refrigerant lines.* On 2005 models, remove the compressor mounting bracket if necessary for clearance.
7   If you're working on an AWD vehicle, refer to Chapter 7B and remove the transfer case.
8   Remove the driveaxle and intermediate shaft if you're working on a FWD model (see Chapter 8).
9   Use the engine support fixture to remove the weight from the engine mounts, then remove the front engine mount. Raise the engine about two inches or as necessary.
10   Remove the rear oil pan-to-transaxle bolt **(see illustration)**.
11   Remove the oil pan side bolts, then the vertical mounting bolts.
12   Carefully separate the oil pan from the block. Don't pry between the block and the pan or damage to the sealing surfaces could occur and oil leaks may develop. Instead, tap the pan with a soft-face hammer to break the gasket seal.

Installation
Refer to illustration 13.14

13   Clean the pan with solvent and remove all old sealant and gasket material from the block and pan mating surfaces. Clean the mating surfaces with brake system cleaner and make sure the bolt holes in the block are clear.

14   Apply a bead of RTV sealant to the joints where the block meets the front cover, and a short bead (1/4-inch wide) to each side corner of the rear main cap where it meets the block **(see illustration)**, then install the new oil pan gasket.
15   Place the oil pan in position on the block and install the upper nuts/bolts.
16   After the pan-to-block fasteners are installed, tighten them to the torque listed in this Chapter's Specifications. Starting at the center, follow a criss-cross pattern and work up to the final torque in three steps.
17   After all the pan-to-block bolts have been tightened, install the oil pan side bolts and tighten them to torque listed in this Chapter's Specifications.
18   Install the oil pan-to-transaxle brace and tighten the bolts to the torque listed in this Chapter's Specifications.
19   The remainder of installation is the reverse of removal.
20   Refill the engine with oil, run it until normal operating temperature is reached and check for leaks.

14   Oil pump - removal and installation

Refer to illustration 14.2

1   Remove the oil pan (see Section 13).
2   Unbolt the oil pump and lower it from the engine **(see illustration)**. **Note:** *The oil pump driveshaft will come out with the pump as you lower it. It's a rod with a flat-sided portion at each end.*
3   If the pump is defective, replace it with a new one - don't reuse the original or attempt to rebuild it. Inspect the ends of the oil pump driveshaft and the plastic collar (if used) that retains the driveshaft to the oil pump.

13.14 Apply a bead of RTV sealant on each side of the rear main cap, where the pan gasket will meet it

14.2 Oil pump mounting bolt

15.2a Most driveplates have locating dowels - if the one you're working on doesn't have one, make some marks to ensure proper alignment on reassembly

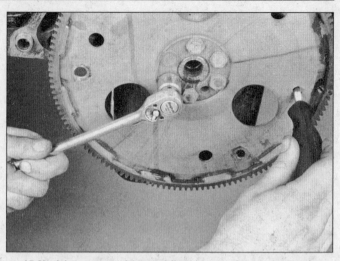

15.2b A large screwdriver wedged in one of the holes in the driveplate can be used to keep the driveplate from turning as the mounting bolts are removed

4   Prime the pump by pouring clean engine oil into the pick-up screen while turning the pump driveshaft.

5   To install the pump, turn the flat on the driveshaft so it mates with the slot in the oil pump shaft. Make sure that any plastic collar is fitted over the oil pump-to-oil pump driveshaft joint, then install the oil pump and driveshaft assembly into the block while engaging the upper end of the oil pump driveshaft into the oil pump drive.

6   Install the pump mounting bolt and tighten it to the torque listed in this Chapter's Specifications.

7   The remainder of installation is the reverse of removal.

15   Driveplate - removal and installation

Removal

Refer to illustrations 15.2a and 15.2b

1   Raise the vehicle and support it securely on jackstands, then refer to Chapter 7 and remove the transaxle.

2   Remove the bolts that secure the drive-plate to the crankshaft **(see illustration)**. If the crankshaft turns, insert a screwdriver through one of the holes in to driveplate **(see illustration)**. **Note:** *If there is a retaining ring between the bolts and the driveplate, note which side faces the driveplate when removing it.*

3   Remove the driveplate from the crank-shaft. **Caution:** *When removing a driveplate, wear gloves to protect your fingers - the edges of the ring gear teeth may be sharp.*

4   Clean the driveplate to remove grease and oil. Inspect the surface for cracks, and check for cracked and broken ring gear teeth. Lay the driveplate on a flat surface to check for warpage.

5   Clean and inspect the mating surfaces of the driveplate and the crankshaft. If the crankshaft rear seal is leaking, replace it before reinstalling the driveplate (see Section 17).

Installation

6   Position the driveplate against the crankshaft. Be sure to align the marks made during removal. Note that some engines have an alignment dowel or staggered bolt holes to ensure correct installation. Before installing the bolts, apply thread locking compound to the threads and place the retaining ring in position on the driveplate.

7   Wedge a screwdriver through the ring gear teeth to keep the driveplate from turning as you tighten the bolts to the torque listed in this Chapter's Specifications. If the front pump seal/O-ring is leaking, now would be a very good time to replace it.

8   The remainder of installation is the reverse of removal.

16   Rear main oil seal - replacement

Refer to illustration 16.5

1   Remove the transaxle (see Chapter 7).

2   Remove the driveplate (see Section 15).

3   Inspect the oil seal, as well as the oil pan and engine block surface for signs of leakage. Sometimes an oil pan gasket leak can appear to be a rear oil seal leak.

4   Carefully note how the old seal is installed. The replacement seal must be installed in the same way.

5   Pry the oil seal from the block with a screwdriver **(see illustration)**. Be careful not to nick or scratch the crankshaft or the seal bore. Thoroughly clean the seal bore in the block with a shop towel. Remove all traces of oil and dirt. **Note:** *There is a special tool EN-48672 that can be used to remove the seal without any possibility of damage to the*

crankshaft or seal bore. These tools are available at GM dealers and some automotive tool suppliers.

6   Obtain a special installation tool EN-48108 from one of the above-mentioned sources. This tool is required to properly install the new seal. **Caution:** *Don't remove the protector sleeve from the new seal. Don't use any lubricant on the new seal.*

7   Clean the crankshaft and the seal bore. Remove any nicks or burrs with fine crocus paper.

8   The installation tool should come with instructions. Install the tool to the crankshaft and tighten it. Place the new seal onto the tool and put the drive drum over it, then install the other tool pieces onto the shaft. **Note:** *The seal will fit onto the tool in only one way.*

9   Tighten the nut to push the new seal into place, stopping when the drum is against the engine block. Remove the tool and check the position of the new seal.

10   Install the driveplate (see Section 15).

11   Install the transaxle (see Chapter 7).

16.5 Carefully pry the old seal out

17.4 The right engine mount can be safely removed when the engine is supported either by an engine support fixture or a jack that's padded and placed under the oil pan

17.6 Remove this through-bolt to disconnect the front engine mount from the frame bracket - the rear mount is similar

17   Powertrain mounts - check and replacement

Refer to illustrations 17.4 and 17.6

1   Raise the vehicle and support it securely on jackstands.

2   Use a floor jack with a block of wood to solidly support the engine under the oil pan.

Right side mount

3   This mount is at the passenger side (drivebelt end) of the vehicle. Refer to Chapter 4 and remove the air filter housing.

4   Unbolt the mount from the engine bracket and from the vehicle frame, then remove it **(see illustration)**.

5   Installation is the reverse of removal. Tighten the bolts to the torque listed in this Chapter's Specifications.

Front and rear mounts

6   Remove the mount through-bolt **(see illustration)**.

7   Remove the bolts from the transmission, then remove the mount.

8   Installation is the reverse of removal. Tighten the bolts to the torque listed in this Chapter's Specifications.

Notes

Chapter 2 Part C
3.0L and 3.6L V6 engines

Contents

Specifications

General
Displacement
3.0L engines	183 cubic inches

3.6L engines
2009 and earlier models (LY7)	217 cubic inches
2013 and later models (LFX)	220 cubic inches

Bore and stroke
3.0L engines	3.50 x 3.16 inches

3.6L engines
2009 and earlier models (LY7)	3.70 x 3.37 inches
2013 and later models (LFX)	3.46 x 3.86 inches

Cylinder numbers (front to rear)
Left (front) side	2-4-6
Right (rear) side	1-3-5
Firing order	1-2-3-4-5-6

Cylinder location diagram

Camshaft
Journal diameters
Front journal	1.376 inches
Other journals	1.061 inches
Camshaft endplay	0.0018 to 0.0085 inch
Valve lift	0.425 inch

Lobe height
Intake	1.668 to 1.680 inches
Exhaust	1.670 to 1.682 inches

Torque specifications   Ft-lbs (unless otherwise indicated)

Note: *One foot-pound (ft-lb) of torque is equivalent to 12 inch-pounds (in-lbs) of torque. Torque values below approximately 15 ft-lbs are expressed in inch-pounds, because most foot-pound torque wrenches are not accurate at these smaller values.*

Camshaft bearing cap bolts	89 in-lbs
Camshaft (actuator) sprocket bolts (use NEW bolts)	43
Camshaft actuator oil control valve bolts	89 in-lbs
Camshaft position actuator solenoid valve bolts	89 in-lbs
Crankshaft pulley bolt (use NEW bolt)	
Step 1	74
Step 2	Tighten an additional 150 degrees

Torque specifications (continued)   **Ft-lbs (unless otherwise indicated)**

Note: *One foot-pound (ft-lb) of torque is equivalent to 12 inch-pounds (in-lbs) of torque. Torque values below approximately 15 ft-lbs are expressed in inch-pounds, because most foot-pound torque wrenches are not accurate at these smaller values.*

Cylinder head bolts (in sequence)
    8 mm bolts
        Step 1 .... 132 in-lbs
        Step 2 .... Tighten an additional 75 degrees
    11 mm bolts (use NEW bolts)
        Step 1 22
        Step 2 .... Tighten an additional 150 degrees
Driveplate bolts (use NEW bolts)
    Step 1 22
    Step 2 .... Tighten an additional 45 degrees
Engine mount bolt (3.6L models)................................... 37
Engine mount bracket-to-block bolts
    3.0L M8 bolt.. 16
    3.0L M10 bolt.. 43
Engine mount strut bracket-to-engine bolts (3.0L models)...................... 66
Engine mount bracket bolts (3.6L models)
    Upper right bolt.. 81
    Remaining bolt/nuts... 37
Engine mount strut insulator-to-strut bracket bolts (3.0L models)........... 43
Engine mount strut-to-insulator bolt
    3.0L 43
    3.6L 18
Engine mount strut bolt/nuts (3.6L models)
    Bracket bolts.. 18
    Bracket nuts 81
    Strut nuts 81
Engine mount strut-to-chassis bolt (3.0L models) 43
Front lower engine mount nut-to-frame (3.0L models) 55
Front upper engine mount nut-to-engine mount (3.0L models)............... 66
Rear upper engine mount nut-to-engine mount (3.0L models)............... 44
Exhaust manifold bolts 18
Exhaust manifold heat shield bolts .... 89 in-lbs
Exhaust manifold-to-catalytic converter nuts 33
Upper intake manifold
    3.0L models.. 18
    3.6L models.. 17
Lower intake manifold
    3.0L models.. 18
    3.6L models.. 17
Oil pan baffle bolts.. 89 in-lbs
Oil pan-to-block bolts
    All except two rear pan-to-rear main oil seal retainer bolts 18
    Two rear pan-to-rear main oil seal retainer bolts............................ 89 in-lbs
Oil pump mounting bolts... 18
Rear main oil seal housing bolts................................... 89 in-lbs
Timing chain cover bolts
    3.0L models and 2009 and earlier 3.6L models (illustration 12.47a)
      8 mm
        Step 1.. 168 in-lbs
        Step 2.. Tighten an additional 60 degrees
      12 mm
        3.0L models... 48
        3.6 models
          Step 1 48
          Step 2 .... Tighten an additional 60 degrees
    2013 and later 3.6L V6 models (illustration 12.47b)
        Step 1, bolts 1 through 23 15
        Step 2, bolts 1 through 23 15
        Step 3, bolts 1 through 23 .... Tighten an additional 60 degrees
        Step 4, bolt 24 .... 132 in-lbs
        Step 5, bolt 25 (not on all models)..................... 48
Timing chain guide assembly bolts................................ 17
Timing chain tensioner assembly bolts.......................... 17
Timing chain idler sprocket bolts 43
Valve cover bolts .... 89 in-lbs

1   General information

1   This Part of Chapter 2 is devoted to in-vehicle repair procedures for the 3.0L and 3.6L V6 engine. Information concerning engine removal, installation and overhaul can be found in Chapter 2D.

2   Since the repair procedures included in this Part are based on the assumption the engine is still in the vehicle, if they are being used during a complete engine overhaul (with the engine already removed from the vehicle and on a stand) many of the Steps included here will not apply.

3   The engine utilizes an aluminum block with cast-iron sleeves. The six cylinders are arranged in a "V" shape at a 60-degree angle between the two banks. The cylinder heads utilize a twin overhead camshaft arrangement with four valves per cylinder. Variable cam timing is used via adjustable camshaft drive sprockets controlled by the Powertrain Control Module. The engine uses aluminum cylinder heads with powdered metal guides and valve seats. Hydraulic lash adjusters negate the need for valve adjustments and roller rocker arms actuate the valves. The oil pump is mounted at the front of the engine behind the timing chain cover and is driven by the crankshaft.

2   Repair operations possible with the engine in the vehicle

1   Many major repair operations can be accomplished without removing the engine from the vehicle, but due to the vehicle's configuration, there are many jobs that can't. Where this is the case, a reference to Chapter 2D will be provided.

2   Clean the engine compartment and the exterior of the engine with some type of pressure washer before any work is done. A clean engine will make the job easier and will help keep dirt out of the internal areas of the engine.

3   If oil or coolant leaks develop, indicating a need for gasket or seal replacement, some repairs can be made with the engine in the vehicle. The intake and exhaust manifold gaskets, and the crankshaft oil seals are accessible with the engine in place.

4   Exterior engine components, such as the water pump, the starter motor, the alternator, the power steering pump and the fuel injection components, as well as the intake and exhaust manifolds, can be removed for repair with the engine in place.

5   Since the camshafts can be removed without removing the engine, valve lifter servicing and valve spring/valve stem oil seal replacement can also be accomplished with the engine in the vehicle.

6   Replacement of, repairs to or inspection of the timing chain cover, timing chain, oil pan and oil pump, as well as cylinder head removal and installation, will require engine removal.

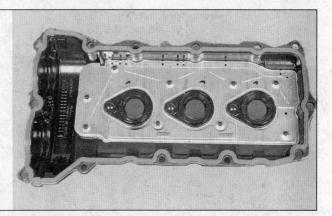

4.9 Make sure the gasket is seated in the valve cover groove

3   Top Dead Center (TDC) for number one piston - locating

1   Top Dead Center (TDC) is the highest point in the cylinder that each piston reaches as it travels up the cylinder bore. Each piston reaches TDC on the compression stroke and again on the exhaust stroke, but TDC generally refers to piston position on the compression stroke.

2   Before beginning this procedure, place the transmission in Neutral and apply the parking brake or block the rear wheels. Remove the spark plugs (see Chapter 1).

3   Disconnect the cable from the negative terminal of the battery (see Chapter 5). Remove the spark plugs (see Chapter 1).

4   In order to bring any piston to TDC, the crankshaft must be turned. When looking at the front of the engine, normal crankshaft rotation is clockwise.

5   Turn the crankshaft with a socket and ratchet attached to the bolt threaded into the front of the crankshaft. Turn the bolt in a clockwise direction.

6   Install a compression gauge in the number one spark plug hole and rotate the crankshaft until pressure registers on the gauge, which indicates the cylinder has started the compression stroke. Once the compression stroke has begun, TDC for the number one cylinder is obtained when the piston reaches the top of the cylinder on the compression stroke.

7   To bring the piston to the top of the cylinder, insert a long wooden dowel into the number one spark plug hole until it touches the top of the piston. Use the dowel (as a feeler gauge) to tell where the top of the piston is located in the cylinder while slowly rotating the crankshaft. As the piston rises, the dowel will be pushed out. The point at which the dowel stops moving outward is TDC.

8   If you go past TDC, rotate the crankshaft counterclockwise approximately 1/8-turn, then slowly rotate the crankshaft clockwise again until TDC is reached.

9   After the number one piston has been positioned at TDC on the compression stroke, TDC for any of the remaining pistons can be achieved by rotating the engine clockwise, 120-degrees at a time, and following the firing order.

4   Valve covers - removal and installation

Removal

1   Disconnect the cable from the negative battery terminal (see Chapter 5).

2   Remove the ignition coils (see Chapter 1).

3   Disconnect the engine wiring harness from the valve cover and position it aside.

4   On 2009 and earlier 3.6L models, remove the upper intake manifold when removing the right side valve cover; remove the intake manifold on 3.0L models when removing the left side valve cover or either valve cover on 2013 and later 3.6L models (see Section 7). On 2009 and earlier 3.6L models, detach the power steering fluid reservoir and move it out of the way.

5   Remove the valve cover bolts, then detach the cover from the cylinder head.

Note: *If the cover is stuck to the engine, bump one end with a block of wood and a hammer. If that doesn't work, try to slip a putty knife under it to break the seal. Don't try to pry the cover off as damage to the sealing surface could occur, leading to oil leaks in the future.*

Installation

Refer to illustrations 4.9, 4.10, 4.12a, 4.12b and 4.12c

6   The mating surfaces of each cylinder head and valve cover must be perfectly clean when the covers are reinstalled. Use a gasket scraper to remove all traces of sealant and old gasket material, then clean the mating surfaces with brake system cleaner. If there's sealant or oil on the mating surfaces when the cover is reinstalled, oil leaks may develop.

7   Clean the mounting bolt threads with a die to remove any corrosion and restore damaged threads. Make sure the threaded holes in the cylinder head are clean - run a tap into them to remove corrosion and restore damaged threads.

8   The manufacturer recommends installing tool EN-46101 to the spark plug tube holes to allow the spark plug tube seals to slide over the tubes without tearing. However, applying a small amount of engine oil to the spark plug tube seal inner faces, and using care when installing the seals, the valve cover with the

4.10 Place some RTV sealant at the joint where the timing chain cover and cylinder head meet

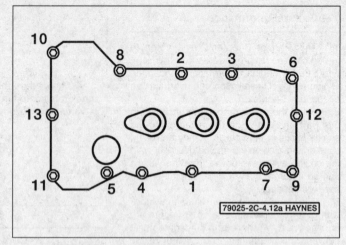

4.12a Valve cover bolt tightening sequence - left side (front, 2009 and earlier 3.6L and all 3.0L)

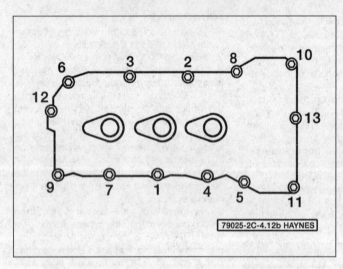

4.12b Valve cover bolt tightening sequence - left side (front, 2013 and later 3.6L engines)

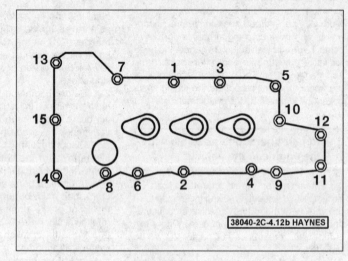

4.12c Valve cover bolt tightening sequence - right side (rear, all models)

seals attached can slide over the spark plug tubes without the special tools.

9   Position the gasket inside the cover groove **(see illustration)**. If the gasket will not stay in place, apply a thin coat of RTV sealant to the cover flange so the gasket adheres to the cover.

Note: *If the gasket wasn't leaking, is not damaged and is still pliable, it can be re-used.*

10   Apply an 8 mm wide by 4 mm high bead of RTV sealant to the joints between the timing chain cover and cylinder head mating faces where the valve cover will sit **(see illustration)**.

11   Inspect the valve cover bolt grommets for damage. If the grommets aren't damaged, they can be reused. Carefully position the valve cover(s) on the cylinder head and install the grommets and bolts.

12   Tighten the bolts in the proper sequence **(see illustrations)**, to the torque listed in this Chapter's Specifications.

13   The remainder of installation is the reverse of removal.

14   Start the engine and check carefully for oil leaks as the engine warms up.

5   Rocker arms and hydraulic lash adjusters - removal, inspection and installation

Removal

Refer to illustrations 5.2 and 5.4

1   Remove the valve covers (see Section 4).

2   Remove the camshafts **(see illustration)** (see Section 13).

3   Place markings on each rocker arm to ensure they are returned to the same position, then lift each one from the engine. Once the rocker arms are removed, the hydraulic lash adjusters can be lifted from the cylinder head.

Note: *It is important that any rocker arms or lash adjusters being used again be returned to their original position on the engine.*

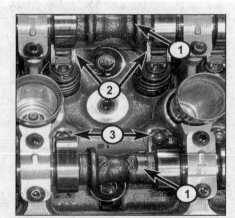

5.2 Remove the camshafts (1) to remove the rockers (2). (3) is the lash adjuster socket on the rocker

Inspection

4   Check each rocker arm for wear, cracks and other damage, especially where the valve stems contact the rocker arm **(see illustration)**.

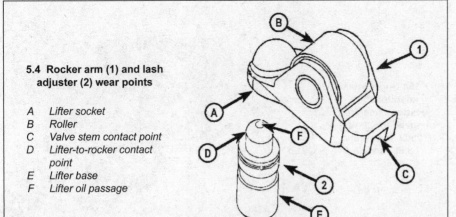

5.4 Rocker arm (1) and lash adjuster (2) wear points

A   *Lifter socket*
B   *Roller*
C   *Valve stem contact point*
D   *Lifter-to-rocker contact point*
E   *Lifter base*
F   *Lifter oil passage*

6.4 Thread the air hose adapter into the spark plug hole - adapters are commonly available at auto parts stores

5   Check the rollers for binding and roughness. If the bearings are worn or damaged, replacement of the entire rocker arm will be necessary.

Note: *Keep in mind that there is no valve adjustment on these engines, so excessive wear or damage in the valve train can easily result in excessive valve clearance, which in turn will cause valve noise when the engine is running.*

6   Make sure the oil passage hole in each hydraulic lash adjuster is not blocked.

Installation

7   Apply some camshaft/lifter pre-lube to the pivot pockets, roller and the slot that contacts the valve on each rocker arm. Apply clean engine oil into the bores for the hydraulic lash adjusters and install them into the cylinder head. Position the rocker arms onto the lash adjusters.

8   The remainder of installation is the reverse of removal.

9   Before starting and running the engine, change the oil and install a new oil filter (see Chapter 1).

6 Valve springs, retainers and seals - replacement

Refer to illustrations 6.4, 6.5, 6.7a, 6.7b and 6.16

Note: *Broken valve springs and defective valve stem seals can be replaced without removing the cylinder head. Two special tools and a compressed air source are normally required to perform this operation, so read through this Section carefully and rent or buy the tools before beginning the job.*

1   Remove the spark plugs (see Chapter 1).

2   Remove the valve covers (see Section 4), camshafts (see Section 13) and rocker arms (see Section 5).

3   Thread an adapter into the spark plug hole and connect an air hose from a compressed air source to it. Most auto parts stores can supply the air hose adapter.

Note: *Many cylinder compression gauges utilize a screw-in fitting that may work with your air hose quick-disconnect fitting. If a cylinder compression gauge fitting is used, it will be necessary to remove the Schrader valve from the end of the fitting before using it in this procedure.*

4   Apply compressed air to the cylinder.

The valves should be held in place by the air pressure **(see illustration)**.

5   Using a socket and a hammer, gently tap on the top of each valve spring retainer several times (this will break the seal between the valve keeper and the spring retainer and allow the keeper to separate from the valve spring retainer as the valve spring is compressed), then use a valve-spring compressor to compress the spring. Remove the keepers with small needle-nose pliers or a magnet **(see illustration)**. **Note:** Several different types of tools are available for compressing the valve springs with the head in place. One type grips the lower spring coils and presses on the retainer as the knob is turned; the lever-type utilizes the rocker arm bolt for leverage. Both types work very well, although the lever type is usually less expensive.

6   Remove the valve spring and retainer. **Note:** If air pressure fails to retain the valve in the closed position during this operation, the valve face or seat may be damaged. If so, the cylinder head will have to be removed for repair.

7   Remove the old valve stem seals, noting differences between the intake and exhaust seals **(see illustrations)**.

8   Wrap a rubber band or tape around the

6.5 After compressing the valve spring, remove the keepers with a magnet or needle-nose pliers

6.7a A screwdriver can be used to pry off the valve seals

6.7b Ensure the intake (1) and exhaust (2) seals are installed on the correct valve stems

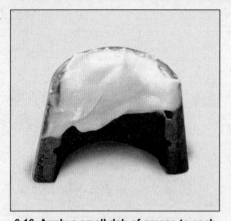

6.16 Apply a small dab of grease to each keeper as shown here before installation - it'll hold them in place on the valve stem as the spring is released

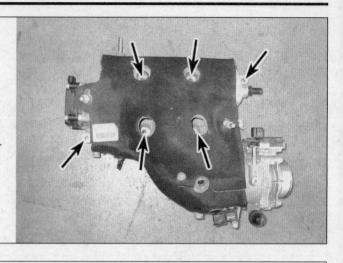

7.16a Upper intake manifold bolt locations (manifold removed for clarity) – 2009 and earlier 3.6L models

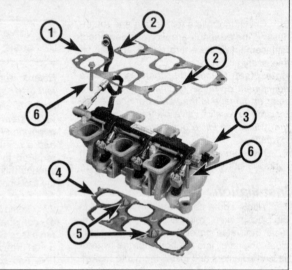

7.16b Typical intake manifold components – 2009 and earlier 3.6L models

1 *Upper intake manifold gasket*
2 *Locating tabs*
3 *Lower intake manifold*
4 *Gasket*
5 *Locating tabs*
6 *Lower intake manifold-to-cylinder head bolts (2 of 6 shown)*

top of the valve stem so the valve won't fall into the combustion chamber, then release the air pressure.

9   Inspect the valve stem for damage. Rotate the valve in the guide and check the end for eccentric movement, which would indicate that the valve is bent.

10   Move the valve up-and-down in the guide and make sure it does not bind. If the valve stem binds, either the valve is bent or the guide is damaged. In either case, the head will have to be removed for repair.

11   Reapply air pressure to the cylinder to retain the valve in the closed position, then remove the tape or rubber band from the valve stem.

12   If you're working on an exhaust valve, fit the new exhaust valve seal on the valve stem and press it down over the valve guide. Don't force the seal against the top of the guide.

13   If you're working on an intake valve, install a new intake valve stem seal over the valve stem and press it down over the valve guide. Don't force the intake valve seal against the top of the guide.

Note: *Do not install an exhaust valve seal on an intake valve, as high oil consumption will result.*

14   Install the spring and retainer in position over the valve stem.

15   Compress the valve spring assembly only enough to install the keepers in the valve stem.

16   Position the keepers in the valve stem groove. Apply a small dab of grease to the inside of each keeper to hold it in place if necessary **(see illustration)**. Remove the pressure from the spring tool and make sure the keepers are seated.

17   Disconnect the air hose and remove the adapter from the spark plug hole.

18   Repeat the above procedure on the remaining cylinders.

19   Install the rocker arm assemblies, camshafts, timing chains and the valve covers (see appropriate Sections).

20   Start the engine, then check for oil leaks and unusual sounds coming from the valve

cover area. Allow the engine to idle for at least five minutes before revving the engine.

7   Intake manifold - removal and installation

Warning: *Wait until the engine is completely cool before starting this procedure.*

Removal

1   Relieve the fuel pressure (see Chapter 4).

2   Disconnect the cable from the negative battery terminal (see Chapter 5).

2009 and earlier 3.6L models

Upper intake manifold

Refer to illustrations 7.16a and 7.16b

3   Remove the engine cover.

4   Remove the air outlet tube from the air filter housing (see Chapter 4).

5   Move the PCV tube from the air filter inlet.

6   Disconnect the electrical connectors to the intake manifold tuning valve, throttle body

and Manifold Absolute Pressure (MAP) sensor.

7   Disconnect the brake booster vacuum hose from the intake manifold.

8   Disconnect the wiring from the throttle body.

9   Disconnect the PCV tube from the intake manifold and move it out of the way.

10   Disconnect the quick-connect fitting at the tube from the EVAP canister purge solenoid.

11   Move the EVAP tube aside.

12   Remove the ECM bracket without disconnecting the connectors and place the ECM and bracket out of the way (see Chapter 6).

13   Disconnect then fuel injector electrical connectors and move them out of the way (see Chapter 4).

14   Remove the fasteners from the engine wiring harness and position it out of the way.

15   Verify that all interfering wiring and hoses have been disconnected and moved clear of the intake manifold. Use tape and a marker to identify components to avoid confusion later.

16   Completely loosen but do not remove the six mounting bolts from the top of the upper intake manifold, then remove the manifold and upper intake gasket as a unit **(see illustrations)**. Once the assembly has been removed, discard the gasket.

7.18 Lower intake manifold bolts – 2009 and earlier 3.6L models

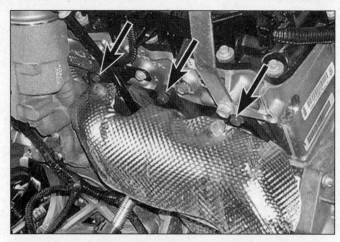

8.9 Exhaust manifold upper fastener locations (left side shown, right side similar) – 2009 and earlier 3.6L models only

Lower intake manifold

Refer to illustration 7.18

17   Disconnect the fuel feed line from the fuel rail, then remove the fuel rail and injectors (see Chapter 4).

18   Remove the lower intake manifold mounting bolts **(see illustration)**, then lift off the manifold.

3.0L models and 2013 and later 3.6L models

Note: *3.0L use a single one-piece intake manifold.*

19   Remove the oil filler cap, then lift the engine cover up and off of the ballstuds. Install the oil filler cap.

20   Disconnect the air filter housing outlet duct and remove the air filter housing (see Chapter 4).

21   Disconnect the brake booster hose vacuum and the PCV hose from the manifold.

22   Disconnect the electrical connectors to the EVAP purge solenoid valve and the throttle body then remove the hose from the EVAP solenoid (see Chapter 6).

23   Remove the fuel line cover fasteners and cover (see Chapter 4).

24   Remove the engine electrical harness clips and retainers from the intake manifold.

25   Remove the intake manifold sound insulator (if equipped). Remove the mounting bolts and remove the intake manifold.

26   Remove and discard the intake manifold gasket.

Installation

Note: *The sealing surfaces of the intake manifold and cylinder heads must be perfectly clean when the manifold is installed.*

27   Carefully remove all traces of old gasket material. After the gasket surfaces are cleaned and free of any gasket material, wipe the mating surfaces with a cloth saturated with solvent. If there is old sealant or oil on the mating surfaces when the manifold is reinstalled, oil or vacuum leaks may develop. Use a vacuum cleaner to remove any gasket material that

falls into the intake ports in the heads.

28   Use a tap of the correct size to chase the threads in the bolt holes, then use compressed air (if available) to remove the debris from the holes.

Warning: *Wear safety glasses or a face shield to protect your eyes when using compressed air.*

29   Install new gaskets on the cylinder heads, making sure that they're correctly aligned.

2009 and earlier 3.6L models

30   Install the lower intake manifold. Evenly tighten all the mounting bolts to the torque listed in this Chapter's Specifications.

31   Install new gaskets to the top of the lower intake manifold, then replace the upper intake manifold.

32   Install the bolts, then evenly tighten them to the torque listed in this Chapter's Specifications.

33   The remainder of installation is the reverse of removal. Check the coolant level, adding as necessary (see Chapter 1).

3.0L models and 2013 and later 3.6L models

34   Install the intake manifold and tighten the bolts in a criss-cross pattern, to the torque listed in this Chapter's Specifications.

35   The remainder of installation is the reverse of removal. Check the coolant level, adding as necessary (see Chapter 1).

---

8   Exhaust manifolds - removal and installation

Removal

Refer to illustration 8.9

Warning: *Use caution when working around the exhaust manifolds - the sheetmetal heat shields can be sharp on the edges. Also, the engine should be cold when this procedure is followed.*

Note: *On 3.0L models and 2013 and later 3.6L models, the exhaust manifold is incorporated*

into the cylinder head; the catalytic converter flange bolts directly to the cylinder head.

1   Disconnect the cable from the negative battery terminal (see Chapter 5). Raise the vehicle and support it securely on jackstands.

2   Apply penetrating oil to the studs and nuts that attach the catalytic converters to the exhaust manifolds (they're usually rusty).

3   Remove the catalytic converter bolts and disconnect the converters from the exhaust manifolds.

4   If you're working on a left (front) manifold, remove the oil dipstick tube.

5   If you're working on the right (rear) manifold, remove the air filter inlet (see Chapter 4), then disconnect the brake fluid level sensor electrical connector (see Chapter 9).

6   Remove both torque strut mounts and brackets (see Section 18).

7   Disconnect the oxygen sensor electrical connectors, then unclip the wiring harness.

8   Remove the exhaust manifold heat shield fasteners and remove the shield.

9   Remove the exhaust manifold mounting bolts **(see illustration)**. The lower bolts for the right (rear) manifold can be removed from under the vehicle.

10   Lift off the exhaust manifolds and discard the gaskets.

Installation

11   Check the manifold for cracks and make sure the bolt threads are clean and undamaged. The manifold and cylinder head mating surfaces must be clean before the manifolds are reinstalled - use a gasket scraper to remove all carbon deposits and gasket material.

Note: *The cylinder heads are made of aluminum, therefore aggressive scraping is not suggested and will damage the sealing surfaces. Also, if any of the manifold bolts are broken, it indicates a warped manifold. Have the manifold machined at a cylinder head reconditioning shop prior to installing the manifold to the vehicle. If this is not done, the manifold may not seal properly and the new manifold bolts will probably break.*

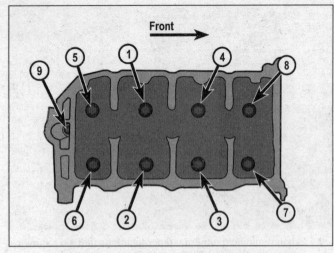

9.17a RH (rear) cylinder head bolt tightening sequence. Bolts 1 to 8 are M11 bolts. Bolt 9 is an M8 bolt

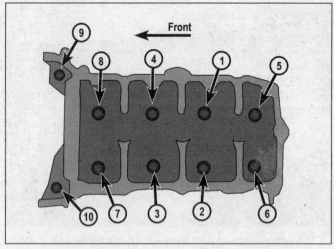

9.17b LH (front) cylinder head bolt tightening sequence. Bolts 1 to 8 are M11 bolts. Bolts 9 and 10 are M8 bolts

12   Place the manifold on the cylinder head and install the mounting bolts finger tight.

13   When tightening the manifold bolts, work from the center to the ends using a torque wrench. Tighten the bolts to the torque listed in this Chapter's Specifications. If required, bend the exposed end of the exhaust manifold gasket back against the cylinder head.

14   Apply an anti-seize compound to the threads of the outer heat shield bolts, then install the bolts and tighten them to the torque listed in this Chapter's Specifications.

15   The remainder of installation is the reverse of removal.

16   Start the engine and check for exhaust leaks.

9   Cylinder heads - removal and installation

Warning: *The engine must be completely cool before beginning this procedure.*
Note: *It will be necessary to purchase a new set of 11 mm cylinder head bolts.*

Removal

1   Reliever the fuel system pressure (see Chapter 4), then disconnect the cable from the negative battery terminal (see Chapter 5).

2   Drain the cooling system (see Chapter 1).

3   On 3.0L models and 2013 and later 3.6L models, remove the power brake booster pump and secure it out of the way (see Chapter 9).

4   Remove the intake manifold(s) (see Section 7), the valve covers (see Section 4) and the upper (secondary) timing chains (see Section 12).

5   Disconnect the ground wires from the heads. If you're removing the left (front) cylinder head, remove the dipstick tube and disconnect the electrical connector from the Engine Coolant Temperature (ECT) sensor. If you're removing the right (rear) head, detach

the power steering hose clip and disconnect the oxygen sensor electrical connector.

6   On 2009 and earlier 3.6L engines, disconnect the catalytic converters from the exhaust manifolds (see Section 8). The heads can be removed with the exhaust manifolds attached to them.

7   Remove the camshaft position actuator solenoid valves (see Section 12).

8   On 2013 and later 3.6L models, when removing the left side (front) cylinder head, remove the high-pressure fuel pump from the cylinder head (see Chapter 4) then remove the oil dipstick tube fastener and tube.

9   Loosen the head bolts in 1/4-turn increments in the reverse order of the tightening sequence **(see illustrations 9.17a and 9.17b)** until they can be removed by hand.
Note: *There will be different lengths and sizes of bolts in different locations. Make notes of where each bolt belongs to aid in assembly.*

10   Carefully lift the heads off the engine block. If you feel resistance, don't pry the head off. To dislodge the head, put a prybar into an intake port and pry with caution. Store the heads on wooden blocks to prevent damage.

Installation

Refer to illustrations 9.17a and 9.17b

11   The mating surfaces of the cylinder heads and block must be perfectly clean when the heads are reinstalled. Gasket removal solvents are available at auto parts stores and may prove helpful.

12   Use a gasket scraper to remove all traces of carbon and old gasket material, then wipe the mating surfaces with a cloth saturated with brake system cleaner. If there is oil on the mating surfaces when the heads are reinstalled, the gaskets may not seal correctly and leaks may develop. When working on the block, use a vacuum cleaner to remove any debris that falls into the cylinders. **Note:** *The cylinder heads are made of aluminum, therefore aggressive scraping is not suggested and*

will damage the sealing surfaces.

13   Check the block and head mating surfaces for nicks, deep scratches and other damage. If damage is slight, it can be removed with emery cloth. If it is excessive, machining may be the only alternative.

14   Use a tap of the correct size to chase the threads in the head bolt holes in the block. If a tap is not available, spray a liberal amount of brake cleaner into each hole. Use compressed air (if available) to remove the debris from the holes. Corrosion, sealant and damaged threads will affect torque readings, so be sure the threads are clean.
Warning: *Wear safety glasses or a face shield to protect your eyes when using compressed air.*

15   Position the new gaskets over the dowels in the block.

16   Carefully position the heads on the block without disturbing the gaskets.
Warning: *DO NOT reuse 8 mm or 11 mm head bolts - always replace them.*

17   Before installing the new 8 mm head bolt(s), coat the threads with a medium-strength thread locking compound. Then install the 8 mm head bolt(s) finger tight (bolt 9 on the RH cylinder head, bolts 9 and 10 on the LH head).

18   Install new 11 mm head bolts (bolts 1 through 8) and tighten them finger tight. Following the recommended sequence **(see illustrations)** , tighten the bolts to the Step 1 torque listed in this Chapter's Specifications. Then tighten them in sequence to the Step 2 angle of rotation listed in this Chapter's Specifications.

19   Tighten the 8 mm bolt(s) to the torque and angle of rotation listed in this Chapter's Specifications.

20   The remainder of installation is the reverse of removal.

21   Change the engine oil and filter, and refill the cooling system (see Chapter 1). Start the engine and check for proper operation and coolant or oil leaks.

10.9 The use of a three jaw puller will be necessary to remove the crankshaft pulley - always place the puller jaws around the hub, not the outer inertia ring

10.13a Use a torque wrench to tighten the new crankshaft pulley bolt to the Step 1 torque . . .

10   Crankshaft pulley - removal and installation

Removal

Refer to illustration 10.9

Note: *This procedure requires a special pulley installation tool as well as a new crankshaft pulley bolt. Read through the entire procedure and obtain the tools and materials before proceeding.*

1   Disconnect the cable from the negative battery terminal (see Chapter 5).
2   Raise the vehicle and support it securely on jackstands.
3   Remove the drivebelt (see Chapter 1).
4   Install an engine support fixture or an engine hoist to the engine lifting bracket on the right side. **Note:** *As an alternative, the engine may be supported from below the oil pan with a floor jack and a block of wood. If this is done, position the wood and the jack to avoid damaging the oil pan.*
5   Tighten the engine support chain to remove the weight from the engine mounts.
6   Remove the right side engine mount (see Section 18).
7   Carefully lower the engine about two inches for clearance.
8   Remove the crankshaft pulley bolt. Prevent the crankshaft from turning by removing the starter motor (see Chapter 5) and wedging a large screwdriver or prybar in the driveplate ring gear teeth.
9   Install a large three-jaw puller on the pulley **(see illustration)**, and remove the pulley. **Caution:** *The hooks of the puller must contact only the center hub - not the outer (inertia) ring. Use the proper adapter so as not to damage the end of the crankshaft or the threads in the end of the crankshaft.*
Note: *Make note of how far the pulley is pressed onto the crankshaft. It must be installed in the same position. Use the puller*

10.13b . . . then use a torque angle gauge to finish the tightening sequence

to draw the pulley from the crankshaft.

Installation

Refer to illustrations 10.13a and 10.13b

10   Lubricate the inside of the pulley bore with clean engine oil. **Note:** *Don't let oil contact the outside of the pulley's snout or the oil seal in the timing chain cover. The pulley must only be installed into a dry seal.*
11   Obtain a crankshaft pulley installation tool. They are available at automotive tool dealers and most auto parts stores. Use only this tool to press the pulley into place. **Caution:** *Don't try to hammer the pulley on or try to press it into place using a bolt screwed into the crankshaft.*
12   Install the pulley in its original position. The installation tool should be bottomed on the end of the crankshaft. Remove the installation tool.
13   Install the NEW crankshaft pulley bolt. Have an assistant wedge the large screwdriver or prybar in the driveplate ring gear teeth, then tighten the bolt to the torque and angle of rotation listed in this Chapter's Specifications **(see illustrations)**.
14   The remainder of installation is the reverse of removal.

11.2 If you don't have a seal removal tool, thread a self tapping screw partially into the seal, then use pliers as a lever to pull it from the engine

11   Crankshaft front oil seal - removal and installation

Removal

Refer to illustration 11.2

1   Remove the crankshaft pulley (see Section 10).
2   Note how the seal is installed - the new one must be installed to the same depth and facing the same way. Carefully pry the oil seal out of the cover **(see illustration)**.
3   If the seal is being replaced with the timing chain cover removed, support the cover on top of two blocks of wood and drive the seal out from the rear with a hammer and punch. **Caution:** *Be careful not to scratch, gouge or distort the area that the seal fits into or a leak will develop.*

Installation

4   Apply clean engine oil or multi-purpose grease to the outer edge of the new seal, then install it in the cover with the lip (spring side) facing IN. Drive the seal into place with a seal driver or a large socket and a hammer. Make sure the seal enters the bore squarely and stop when the front face is at the proper depth.

12.16 A 10 x 1.5 mm bolt can be threaded into the hole near the water pump to press the center portion of the cover away from the engine

12.17 Use a screwdriver to pry the cover from the engine at the indicated points

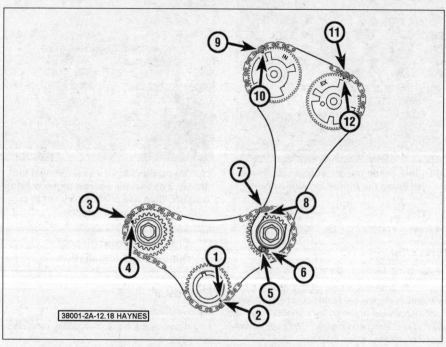

12.18 Stage one timing chain positions

1   Crankshaft sprocket timing mark
2   Primary camshaft drive chain timing plated link for the crankshaft
3   Primary camshaft drive chain timing link for the right (rear bank) primary camshaft intermediate drive sprocket
4   Right (rear bank) primary camshaft intermediate drive chain sprocket timing mark (window) for the right secondary camshaft drive chain
5   Left (front bank) primary camshaft intermediate drive chain sprocket timing window
6   Left (front bank) secondary timing chain timing (plated) link for the left primary camshaft intermediate drive chain sprocket (behind the hole in the sprocket)

7   Primary camshaft drive chain sprocket timing (plated) link for the primary camshaft intermediate drive chain sprocket
8   Left (front bank) primary camshaft intermediate drive chain sprocket timing mark for the primary camshaft drive chain
9   Left (front bank) intake secondary camshaft timing drive chain timing (plated) link
10 Left (front bank) intake camshaft sprocket actuator timing mark (circle)
11 Left (front bank) exhaust camshaft drive chain timing (plated) link
12 Left (front bank) exhaust camshaft sprocket actuator timing mark (circle)

Note: *Don't let oil get on the rubber part of the seal or on the part of the crankshaft pulley that touches the seal. The seal must be installed dry.*
5   Check the surface on the pulley hub that the oil seal rides on. If the surface has been grooved from long-time contact with the seal, the pulley should be replaced with a new one.
6   Lubricate only the inside of the pulley hub with clean engine oil and install the crankshaft pulley (see Section 10).
7   The remainder of installation is the reverse of removal.

12 Timing chains and camshaft sprockets - removal, inspection and installation

Warning: *Wait until the engine is completely cool before beginning this procedure.*
Caution: *The timing system is complex, and severe engine damage will occur if you make any mistakes. Do not attempt this procedure unless you are highly experienced with this type of repair. If you are at all unsure of your abilities, be sure to consult an expert. Double-check all your work and be sure everything is correct before you attempt to start the engine.*
Caution: *Before removing the timing chain(s) from their sprockets(s), check to make sure that all of the sprockets are equipped with timing marks like the ones shown in the illustrations. There are two stages or positions that the engine must be rotated to for alignment of all the timing marks.*
Note: *The engine shown in the illustrations has only one variable camshaft sprocket on each cylinder head. However, procedures for models with dual variable sprockets are identical.*

Removal

Refer to illustrations 12.16, 12.17, 12.18, 12.19 and 12.21

Note: *This procedure requires three special tools to lock the camshafts on each cylinder*

12.19 Typical LH (front) cylinder head with camshaft locking tool EN-48383-1 installed

12.21 RH (rear) cylinder head timing chain tensioner (1), outer guide pivot bolt (2) and inner guide bolts (3)

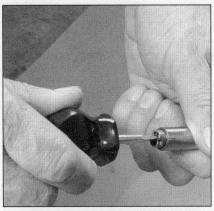

12.29a Remove the plunger from the tensioner and reset it by screwing the bottom of the plunger shaft into the top half of the plunger shaft with a screwdriver

head in position: tool numbers EN 46383-1, 46383-2 and 46383-3 are available through SPX Tools. Read through the entire procedure and obtain the tools before proceeding.

1   Disconnect the cable from the negative battery terminal (see Chapter 5).

2   Drain the oil then remove the oil filter (see Chapter 1).

3   Remove the spark plugs (see Chapter 1) to make it easier to rotate the engine.

4   Remove the drivebelt and tensioner (see Chapter 1).

5   Remove the intake manifold (see Section 7).

6   Remove the valve covers (see Section 4).

7   Remove the water pump (see Chapter 3).

8   Remove the power steering reservoir bracket. Unbolt the power steering pump and move it out of the way.

Note: *It is not necessary to disconnect the power steering lines.*

9   Remove the alternator (see Chapter 5).

10   Remove the crankshaft pulley (see Section 10).

11   Remove the Camshaft Position (CMP) sensor fasteners, and remove the sensors from the front cover (see Chapter 6).

12   Disconnect the electrical connectors from the camshaft position actuator valves. Remove the bolts, then remove the four actuator solenoid valves.

13   Remove the engine mount bracket bolts and bracket (see Section 18).

14   Remove the drivebelt idler pulley bolt and pulley from the front of the cover.

15   Remove the thermostat housing (see Chapter 3).

16   Remove all of the bolts from the timing chain cover (**see illustration 12.47** for locations). Screw a 10 mm x 1.5 mm bolt in the threaded hole near the top of the front cover. Tighten this screw while prying at the correct locations to break the cover loose from the engine block (**see illustration**).

17   Use a screwdriver in the indicated areas around the perimeter of the cover to pry it from the engine block (**see illustration**).

Note: *Special tool EN-48589, available from SPX Tools, can also be used to rotate*

the crankshaft without using the crankshaft pulley bolt.

18   Reinstall the crankshaft pulley bolt and use it to rotate the crankshaft clockwise until the timing mark on the primary timing chain sprocket aligns with the mark on the oil pump housing in the 5 o'clock position; this is stage one position (**see illustration**).

Note: *If the RH (rear bank) chain is the only chain being removed, rotate the engine until the timing mark on the primary chain sprocket aligns with the mark on the oil pump housing in the 9 o'clock position.*

19   In this position, the flat segment on the rear of each camshaft on the LH (front) cylinder head should be facing up and parallel with the valve cover mating surface, and the flat segment on the rear of each camshaft. If this is not the case, rotate the crankshaft clockwise one full turn and re-align the crankshaft timing marks. Install the special tool to the camshafts of the LH cylinder head (**see illustration**).

20   Remove the timing chain tensioner bolts for the RH (rear) cylinder head. Carefully remove the tensioner and discard the gasket.

Note: *The tensioner is under spring pressure. Use care, as the tensioner can fly apart once the bolts are removed.*

21   Remove the RH (rear) cylinder head timing chain guides, then remove the chain (**see illustration**). Remove the special holding tools from the rear of the camshafts.

22   Repeat the process to remove the lower timing chain or the primary timing chain for the LH (front) cylinder head.

Note: *Use care, as the tensioners are under spring pressure. Do not attempt to remove the timing chain guide from the oil pump housing. At the time of publication, this was only supplied as a complete assembly with the oil pump.*

23   If necessary, after noting their installed position, remove the bolts retaining the idler sprockets and remove the sprockets.

24   If necessary, mark the relationship between the camshaft sprockets and their respective camshaft. Hold the camshaft steady with an open ended wrench placed on the hex in the camshaft and remove the camshaft sprocket bolt. Remove the sprocket

12.29b Slowly push the plunger back into the tensioner body and secure it in place with a paper clip

from the camshaft.

Inspection

Refer to illustrations 12.29a and 12..29b

25   Inspect the timing chain cover for damage. Remove all old sealant from the cover and the engine block.

26   Remove and discard the seal for the coolant passage from the rear of the housing.

27   Clean all components with suitable solvent and inspect the guides and timing chain sprockets for wear or damage. Any gouges or deformities on the guides or tensioners will require replacement parts. Ensure the rear face of the tensioner is clean and there are no marks that would affect the sealing quality once reassembled. Remove any old gasket material from the tensioner mating faces on the engine.

28   Inspect the timing chains for wear, stiff or loose links and binding.

29   Inspect the tensioners for damage or wear. Reset the tensioner by removing the plunger from the tensioner body and screw the rear of the plunger into the plunger shaft using a flat bladed screwdriver. Install the plunger to the body of the tensioner, and while maintaining pressure on the plunger, insert a paper clip into the hole in the tensioner body to maintain

12.32 Alignment mark between the LH timing chain and idler sprocket (visible through hole)

the tensioner in the retracted position (**see illustrations**). Repeat this on the other two timing chain tensioners.

Note: *The wire must be inserted into the tensioner to maintain it in the compressed position. If this is not done, the spring pressure will not tension the chain once it is installed to the engine.*

Installation

Refer to illustrations 12.32, 12.37, 12.39a, 12.39b, 12.40a, 12.40b, 12.41, 12.43, 12.45, 12.46, 12.47a and 12.47b

Caution: *Before starting the engine, carefully rotate the crankshaft by hand through at least two full revolutions (use a socket and breaker bar on the crankshaft pulley center bolt). If you feel any resistance, STOP! There is something wrong - most likely valves are contacting the pistons. You must find the problem before proceeding. Check your work and see if any updated repair information is available.*

30   Align the crankshaft sprocket with the Woodruff key and slide the sprocket onto the crankshaft. Ensure the timing mark on the sprocket is visible and the timing marks are aligned. The crankshaft should be in the stage one position with the crankshaft sprocket

12.39a Crankshaft sprocket in the correct 9 o'clock position

12.37 Lower timing chain sprocket alignment marks (A). Tensioner (1), upper timing chain guide (2) and lower timing chain guide (3) locations

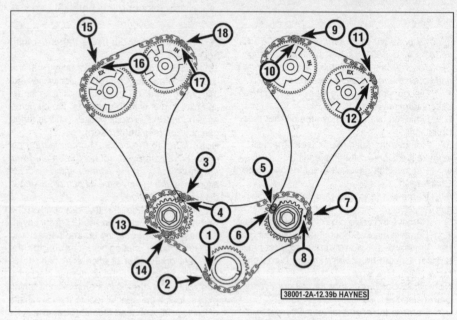

12.39b Stage two timing chain positions

1   Camshaft sprocket timing mark
2   Primary camshaft drive chain timing plated link for the crankshaft
3   Primary camshaft drive chain timing link for the right primary camshaft intermediate drive sprocket
4   Right primary camshaft intermediate drive chain sprocket timing mark (window) for the right secondary camshaft drive chain
5   Left primary camshaft intermediate drive chain sprocket timing window
6   Left secondary timing chain timing (plated) link for the left primary camshaft intermediate drive chain sprocket (behind the hole in the sprocket)
7   Primary camshaft drive chain sprocket timing (plated) link for the primary camshaft intermediate drive chain sprocket
8   Left primary camshaft intermediate drive chain sprocket timing mark for the primary camshaft drive chain
9   Left intake secondary camshaft timing drive chain timing (plated) link

10   Left intake camshaft sprocket actuator timing mark (circle)
11   Left exhaust camshaft drive chain timing (plated) link
12   Left exhaust camshaft sprocket actuator timing mark (circle)
13   Right secondary camshaft timing drive chain timing (plated) link for the right primary camshaft intermediate drive chain sprocket
14   Right primary camshaft intermediate drive chain sprocket timing (window) mark for the right secondary camshaft timing drive chain
15   Right exhaust secondary camshaft timing drive chain timing (plated) link
16   Right exhaust camshaft sprocket actuator timing mark (triangle)
17   Right intake camshaft timing drive chain timing (plated) link
18   Right intake camshaft sprocket actuator timing mark (triangle)

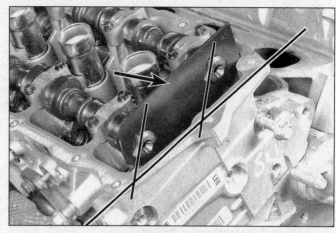

12.40a RH (rear) cylinder head with camshaft retaining tool installed

12.40b LH (front) cylinder head with camshaft retaining tool installed

mark aligned with the mark on the oil pump housing that is in the 5 o'clock position **(see illustration 12.18)**.

Caution: *If resistance is encountered when installing the sprocket, do not hammer it onto the crankshaft. It may eventually move onto the shaft, but it may be cracked in the process and fail later, causing extensive engine damage.*

31 Install special retaining tool #EN-48383-1 onto the LH camshafts **(see illustration 12.19)** (if not already done), making sure to fully seat the tool on the ends of the camshaft.

Caution: *Do not rotate the camshaft(s) more than 10-degrees to install the special tool.*

32 Install the left (inner) secondary chain around the sprocket on the left camshaft intermediate drive idler with the timing (plated) link aligned to the alignment access hole in the idler outer sprocket **(see illustration)**.

33 Continue installing the chain over both camshaft actuator sprockets aligning the plated links with the timing marks on the actuators **(see illustration 12.18)**, making sure there are 10 links between the two plated links on the secondary chain. Check that all of the timing marks are still aligned correctly.

34 Install the outer LH (front) timing chain guide that runs between the idler sprocket and the exhaust camshaft, then install the inner LH (front) timing chain guide to the engine.

Tighten the bolts to the torque listed in this Chapter's Specifications.

Note: *Ensure the inner timing chain guide is not fouling on the mounting pad for the tensioner before tightening the bolts.*

35 Using a new gasket, install the tensioner for the LH (front) timing chain. Ensure the tensioner is aligned with the inner guide and tighten the tensioner bolts to the torque listed in this Chapter's Specifications. Release the pin from the tensioner. Grasp the inner guide and force it into the tensioner. This will allow the tensioner to tension the chain and check that the timing chain alignment marks are all correct.

36 Install the RH (rear) timing chain idler sprocket to the engine ensuring that the small sprocket is facing outward along with the markings "RB Front." Tighten the idler sprocket bolt to the torque listed in this Chapter's Specifications.

37 Ensure the crankshaft sprocket is still aligned with the timing mark on the oil pump housing that is in the 5 o'clock position and install the lower, or primary timing chain. Ensure the shiny chain links align with the arrows on both large idler sprockets and also with the timing mark on the crankshaft sprocket. Install timing chain guide to the lower chain running between the two idler sprockets, tightening the bolts to the torque listed in this Chapter's

Specifications. Using a new gasket, install the tensioner and tighten the tensioner bolts. Remove the pin from the tensioner, force the tensioner guide and pushrod into the tensioner body to unlock it. Allow the tensioner to tension the timing chain. Check that the timing chain alignment marks are all correct. Remove the camshaft special locking tools from the camshafts on both cylinder heads **(see illustration)**.

38 Remove the special retaining tool from the LH camshafts.

39 Rotate the crankshaft 115-degrees, until the timing mark on the crankshaft sprocket is aligned with the mark on the oil pump housing that is in the 9 o'clock position or stage two position **(see illustrations)**.

40 Install the special locking tool EN-48383-2 onto the LH cylinder head camshafts. Rotate the camshafts for the RH cylinder head and install the other special locking tool to the RH (rear) cylinder head (EN-48383-2). Ensure the tools are correctly seated against the camshafts **(see illustrations)**.

Note: *The camshaft flats for the RH cylinder head must be parallel to the valve cover gasket face on the rear of the cylinder head.*

41 Install the RH cylinder head timing chain, aligning the bright (or marked) chain links with the letter R stamped on the camshaft sprockets. The lower bright link on the chain must align with the hole in the face of the inner idler sprocket. Check that all of the timing marks are still aligned correctly **(see illustration)**.

42 Install the inner RH timing chain guide that runs between the idler sprocket and the intake camshaft, then install the outer RH timing chain guide to the engine. Tighten the bolts to the torque listed in this Chapter's Specifications.

43 Using a new gasket with the tab of the gasket facing towards the center of the engine, install the tensioner for the RH timing chain. Ensure the tensioner is aligned with the outer guide and tighten the tensioner bolts to the torque listed in this Chapter's Specifications. Remove the pin from the tensioner, force the tensioner guide and pushrod into the tensioner

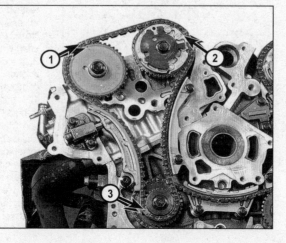

12.41 RH (rear) cylinder head timing chain alignment marks: (1) exhaust camshaft, (2) intake camshaft, (3) idler sprocket

12.43 Force the tensioner guide and chain into the tensioner body to unlock the tensioner

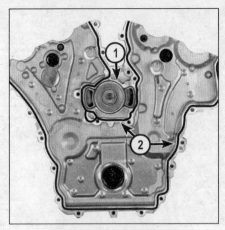

12.45 Seal location (1) and sealant location (2)

12.46 Fabricate two timing chain cover aligning studs from M8 x 1.25 mm bolts with the heads removed and use the plastic sleeves that were between the original timing cover bolts and the timing cover to ensure the cover is in the correct position.

body to unlock it. Allow the tensioner to tension the timing chain **(see illustration)**. Check that the timing chain alignment marks are all correct, then remove the camshaft retaining tools and rotate the engine slowly by hand until oil is forced into the three timing chain tensioners. **Caution:** *If you encounter any resistance while rotating the engine, STOP! There is something wrong - most likely valves are contacting the pistons. You must find the problem before proceeding. Check your work and see if any updated repair information is available.* **Note:** While doing this, ensure that the timing chains don't jump teeth on any of the timing gears.

44   Carefully clean all old sealant from the block and the timing chain cover, being careful to avoid scratching the sealing surfaces. Remove all residue with brake system cleaner. Remove the special locking tools from the camshafts.

45   Install a new seal to the inside of the timing cover and apply a bead of RTV sealant to the faces of the cover that will mate with the engine **(see illustration)**. Slide the timing chain cover over the crankshaft and the two studs, then install the bolts hand tight into the cover.

46   Install two 8 mm studs into the engine block to guide the timing chain cover into the correct position **(see illustration)**.

47   Remove the two 8 mm studs before tightening the timing chain cover bolts in the correct sequence **(see illustrations)**.

48   The remainder of installation is the reverse of removal.

13   Camshafts - removal, inspection and installation

Note: *The camshaft bearing caps are marked with a number and letter and a raised arrow. The number signifies the bearing journal (cylinder number) position from the front of the engine. The letter indicates whether it originates from the intake camshaft (I), or the exhaust (E) camshaft and the arrow always points toward the front of the engine.*

Note: *It is necessary to obtain special timing chain holding tools EN 48383-2 and EN 48383-3, available from SPX Tools, to perform this procedure. If you don't use this tool set or a similar fixture, you'll still be able to do the job, however you'll have to remove the timing chain cover to ensure that the chains stay properly attached to their lower sprockets. If you decide to fabricate timing chain holding tools yourself, make sure that they hold the timing chains and*

upper sprockets securely and tightly in place.

Removal

1   Disconnect the cable from the negative terminal of the battery (see Chapter 5).

2   Remove the valve cover(s) (see Section 4).

3   On 3.0L models, if you're removing the left side (front cylinder bank) camshafts, remove the high-pressure fuel pump (see Chapter 4).

4   On 3.6L models, remove the lower intake manifold (see Section 7).

5   Remove the Camshaft Position (CMP) sensors (see Chapter 6).

6   Remove the camshaft position actuator solenoid valves (see Chapter 6).

7   Use a ratchet and socket on the crankshaft pulley bolt to rotate the engine until the flats at the rear of the camshafts you're working on are facing up and are parallel to the surface of the top of the cylinder head. This puts the camshafts in a low-tension position.

8   Check the camshaft endplay by mounting a dial indicator to the front of the engine with the plunger against the end of the camshaft. Pry the camshaft toward the rear of the engine and zero the indicator. Pry it toward the front of the engine and note the reading. Compare this with the endplay listed in this this Chapter's Specifications. Repeat the test on the other camshafts. A camshaft with excessive endplay should be replaced.

Note: *Use care to avoid damaging the camshaft when prying it. It can easily be chipped.*

9   Hold the camshafts with an open-end wrench, then loosen (but don't remove) the camshaft position actuator (sprocket) bolts.

10   Install special tool EN 48313 (or equivalent) to hold the timing chain and sprockets in position. Make sure the tools are engaged securely. Make absolutely sure there's no way an actuator (sprocket) can come loose before proceeding. Make matchmarks on the actuators (sprockets) and the chain in case they get moved.

11   Remove the camshaft actuator (sprocket) bolts, then detach the actuators from the camshafts.

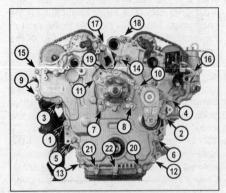

12.47a Timing chain cover bolt tightening sequence – 2009 and earlier 3.6L models and 3.0L models

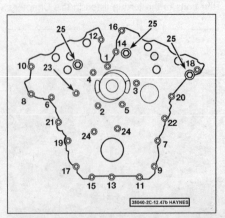

12.47b Timing chain cover bolt tightening sequence – 2013 and later 3.6L models (bolt 25 may not apply to all models)

12 Before removing the bearing caps, arrange to store them in a clearly labeled box to ensure that they're installed in their original locations.

13 Remove the camshaft bearing cap bolts in gradual steps in the reverse of the tightening sequence **(see illustrations 13.24 and 13.25)**.

14 Inspect the camshafts for wear and damage. Using a micrometer, measure the journal diameters and lobe heights, comparing your measurements to the values listed in this Chapter's Specifications.

Inspection

Refer to illustrations 13.16a and 13.16b

15 After the camshaft has been removed, clean it with solvent and dry it, then inspect the bearing journals for uneven wear, pitting and evidence of seizure. If the journals are damaged, the camshaft bearings are probably damaged as well. Both the shaft and bearings will have to be replaced.

16 Measure the bearing journals with a micrometer **(see illustration)** to determine whether they are excessively worn or out-of-round. Measure the camshaft lobes also to check for wear. Measure the camshaft lobes at their highest point, then subtract the measurement of the lobe at its smallest diameter - the difference is the lobe lift **(see illustration)**. Refer to this Chapter's Specifications.

17 Inspect the camshaft lobes for heat discoloration, score marks, chipped areas, pitting and uneven wear. If the lobes are in good condition and if the lobe lift measurements are as specified, you can reuse the camshaft.

18 Check the camshaft bearings in the block for wear and damage. Look for galling, pitting and discolored areas. Inspect the housing journals for damage and replace them if necessary.

19 The inside diameter of each bearing can be determined with an inside micrometer or a bore gauge and outside micrometer. Subtract the camshaft bearing journal diameter(s) from the corresponding bearing inside diameter(s) to obtain the bearing oil clearance. If it's excessive, new bearings will be required

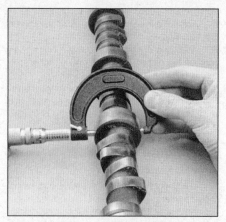

13.16a Measure the camshaft bearing journals with a micrometer

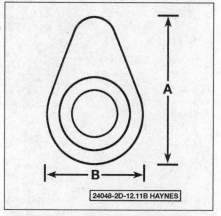

24048-2D-12.11B HAYNES

13.16b Measure the camshaft lobe maximum diameter (A) and the minimum diameter (B). Subtract B from A; the difference is the lobe lift

regardless of the condition of the originals. Refer to this Chapter's Specifications.

20 Camshaft bearing replacement requires special tools and expertise that place it outside the scope of the home mechanic. Take the block to an automotive machine shop to ensure the job is done correctly.

Installation

Refer to illustrations 13.24 and 13.25

21 Ensure the sealing rings on the front of the camshafts are in place and lubricate the camshaft bearing journals and cam lobes with camshaft installation lube.

22 Working on the rear-bank cylinder head, lubricate the bearing journals and place the camshafts onto the bearing journals. Position the camshafts so the flat portions on the rear of the camshaft are parallel to the valve cover gasket face on the top of the cylinder head. This puts the camshafts in a low-tension position

23 Apply engine oil to the inside of the bearing caps, then install the front bearing cap first, ensuring the thrust face is correctly positioned. Install the remaining bearing caps and hand tighten the bolts.

Note: *Ensure the bearing caps are returned to their original positions and the arrows face the center of the cylinder head.*

24 Tighten the bearing cap bolts for both camshafts in the correct sequence **(see illustration)** to the torque listed in this Chapter's Specifications. Once the torque procedure is complete, loosen bolts 1, 2 and 3, 4 and re-torque them.

25 Repeat Steps 21 through 24 to install the front-bank cylinder head camshafts, then tighten the bearing cap bolts in sequence **(see illustration)** to the torque listed in this Chapter's Specifications.

Caution: *If there is any question about the timing chain holding tool retaining the sprockets and chain during the previous Steps, remove the timing chain cover to verify that all chains and sprockets are correctly positioned before starting the engine.*

26 The remainder of installation is the reverse of removal.

27 Before starting and running the engine, change the oil and install a new oil filter (see Chapter 1).

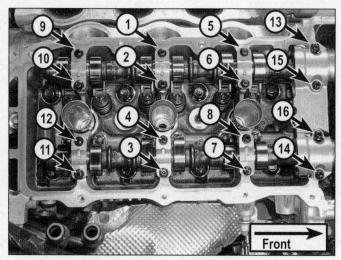

13.24 Rear-bank camshaft bearing cap TIGHTENING sequence

13.25 Front bank camshaft bearing cap TIGHTENING sequence

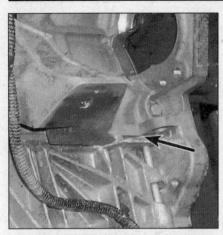

14.9 Pry point between the oil pan and cylinder block. There is another one at the front, and one on the opposite side of the engine

14 Oil pan - removal and installation

Removal

Refer to illustration 14.9

1   Raise the vehicle and support it securely on jackstands.

2   Remove the engine torque mounts (see Section 18).

3   Remove the splash shield fasteners and shield from the under the vehicle.

4   Drain the engine oil, then remove the oil filter (see Chapter 1).

5   On 3.6L models, install an engine support fixture of the type that straddles the top of the engine onto the fender edges and attaches to the top of the engine to take its weight. These can be rented at most rental yards. Bolt it securely to the engine and tighten it to lift the weight of the engine off of its mounts.

6   Disconnect the catalytic converter from exhaust manifold and remove the pipes.

7   Remove the air conditioning compressor without disconnecting the lines and secure the

compressor out of the way (see Chapter 3).

8   Remove the timing chain front cover (see Section 12).

9   Remove the oil pan mounting bolts, then remove the oil pan by prying only at the proper points **(see illustration)**.

10   The oil pickup can be removed after the bolts in the bottom of the oil pan are removed. The pickup seal must be replaced if removed.

Installation

Refer to illustration 14.16

11   Thoroughly clean the mounting surfaces of the oil pan and engine block of old gasket material and sealer. Wipe the gasket surfaces clean with a rag soaked in brake system cleaner.

12   Check the oil pump pickup for cracks or signs of leakage. Ensure the screen in the bottom of the pickup is not blocked or damaged.

13   Install a new oil seal to the oil pump pickup. Install the oil pickup and tighten the bolts in the bottom of the oil pan. Install the baffle to the inside of the oil pan, and tighten the retaining bolts.

14   Fabricate two aligning studs from M8 x1.25 mm bolts with the heads removed and screw them into the oil pan bolt holes labeled 1 and 2 to allow for correct oil pan alignment.

15   Apply a 3 mm wide bead of oxygen sensor-safe RTV sealant to the oil pan mating face with the engine block. Also apply some sealant to the corners of the block where the front cover and the rear cover meet the engine block. Install the oil pan and tighten the bolts finger-tight.

16   Remove the aligning studs, then tighten the oil pan bolts, in sequence **(see illustration)**, to the torque listed in this Chapter's Specifications. **Note:** *The two bolts holding the oil pan to the rear main oil seal housing are a different torque than the other bolts (see this Chapter's Specifications).*

17   The remainder of installation is the reverse of removal.

18   Install a new oil filter. Add the proper type and quantity of oil (see Chapter 1). Start the engine and check for leaks before placing the vehicle back in service.

15 Oil pump - removal, inspection and installation

Removal

Refer to illustration 15.2

1   Remove the timing chains (see Section 12). After noting the installed position, remove the timing chain sprocket from the crankshaft.

2   Remove the oil pump retaining bolts and slide the pump off the end of the crankshaft **(see illustration)**.

Inspection

3   Remove the timing chain guide from the oil pump housing, then remove the oil pump cover and withdraw the rotors from the pump body. Remove the snap-ring from the side of the pump housing and withdraw the cap, spring and plunger from the oil pump housing. These components are the oil pressure relief valve assembly. Clean the components with solvent, dry them thoroughly and inspect for any obvious damage. Also check the bolt holes for damaged threads and the splined surfaces on the crankshaft sprocket for any apparent damage. If any of the components are scored, scratched or worn, replace the entire oil pump assembly. There are no serviceable parts currently available.

Installation

4   If re-using the oil pump, assemble the pressure relief valve components and the rotors to the oil pump housing. Prior to installing the cover, prime the pump by pouring clean motor oil between the rotors. Install the cover and tighten the retaining bolts.

5   Position the oil pump over the end of the crankshaft and align the flats on the crankshaft with the flats on the oil pump drive gear. Make sure the pump is fully seated against the block.

6   Install the oil pump mounting bolts and tighten them to the torque listed in this Chapter's Specifications.

7   The remainder of installation is the

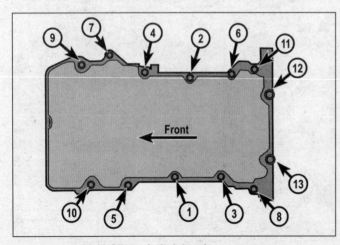

14.16 Oil pan bolt tightening sequence

15.2 Oil pump retaining bolts

16.2a Most driveplates have locating dowels - if the one you're working on doesn't have one, make some marks to ensure proper alignment on reassembly

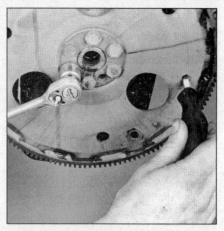

16.2b A large screwdriver wedged in one of the holes in the driveplate can be used to keep the driveplate from turning as the mounting bolts are removed

17.4 Rear main oil seal housing bolts

17.6 Alignment dowels installed in the rear main oil seal housing bolt holes

reverse of removal.
8   Add oil and coolant as necessary. Run the engine and check for oil and coolant leaks. Check the oil pressure (see Chapter 2D).

16   Driveplate - removal and installation

Removal

Refer to illustrations 16.2a and 16.2b
1   Raise the vehicle and support it securely on jackstands. Remove the transaxle (see Chapter 7A).
2   Remove the bolts that secure the driveplate to the crankshaft **(see illustration)**. If the crankshaft turns, insert a screwdriver through one of the holes in the driveplate **(see illustration)**.
Note: *If there is a retaining ring between the bolts and the driveplate, note which side faces the driveplate when removing it.*
3   Remove the driveplate from the crankshaft.
Caution: *When removing a driveplate, wear gloves to protect your fingers - the edges of the ring gear teeth may be sharp.*
4   Clean the driveplate to remove grease and oil. Inspect the surface for cracks, and check for cracked and broken ring gear teeth. Lay the driveplate on a flat surface to check for warpage.
5   Clean and inspect the mating surfaces of the driveplate and the crankshaft. If the crankshaft rear seal is leaking, replace it before reinstalling the driveplate (see Section 17).

Installation

6   Position the driveplate against the crankshaft. Align the marks made during removal. Note that some engines have an alignment dowel or staggered bolt holes to ensure correct installation. Before installing the bolts, apply thread locking compound to the threads and place the retaining ring in position on the driveplate.

7   Wedge a screwdriver through the ring gear teeth to keep the driveplate from turning as you tighten the bolts to the torque listed in this Chapter's Specifications. If the front pump seal/O-ring is leaking, now would be a very good time to replace it.
8   The remainder of installation is the reverse of removal.

17   Rear main oil seal - replacement

Refer to illustrations 17.4 and 17.6
Note: *At the time of publication, the rear main oil seal is supplied as an assembly consisting of the oil seal and housing. The manufacturer recommends using special studs EN 46109, seal alignment tool EN 47839 and handle J 42183 to ensure correct alignment of the seal. These tools are available through SPX Tools.*
1   Disconnect the cable from the negative battery terminal (see Chapter 5).
2   Remove the driveplate (see Section 16).
3   Remove the oil pan (see Section 14).
4   Remove the retaining bolts, then remove and discard the seal housing from the engine **(see illustration)**. It may be necessary to use a knife to cut the sealant away from the joint between the rear main oil seal housing and the engine block. **Note:** *There are specific points provided at the corners for prying the seal housing.* **Caution:** *Use care not to scratch the machined faces of the cylinder block, rear main seal housing or oil pan when scraping old gasket material from components.*
5   Thoroughly clean the mounting surfaces of the oil pan, rear main oil seal housing and engine block of old gasket material and sealer. Wipe the gasket surfaces clean with a rag soaked in brake system cleaner.
6   Install the two dowels into the oil seal housing bolt holes **(see illustration)**.
7   Apply a 1/8-inch wide bead of RTV sealant around the sealing area of the oil seal housing, passing on the inside of the bolt holes.

8   Install the special alignment tool (EN 47839) on the rear of the crankshaft, attaching it with the two supplied screws.
9   Install the new oil seal housing over the alignment tool and studs onto the rear of the engine block. Remove the guide studs but leave the alignment tool in place.
10   Install the housing bolts and tighten them to the torque listed in this Chapter's Specifications.
11   Remove the alignment tool.
12   The remainder of installation is the reverse of removal.
13   Add oil as necessary (see Chapter 1). Run the engine and check for leaks.

18   Engine mounts - check and replacement

1   Engine mounts seldom require attention, but broken or deteriorated mounts should be replaced immediately or the added strain placed on the driveline components may cause damage.

Check

2   During the check, the engine must be raised slightly to remove the weight from the mount.

3   Raise the vehicle and support it securely on jackstands, then position the jack under the engine oil pan. Place a large block of wood between the jack head and the oil pan, then carefully raise the engine just enough to take the weight off the mount. Do not use the jack to support the entire weight of the engine.

4   Check the mount to see if the rubber is cracked, hardened or separated from the metal plates. Sometimes the rubber will split right down the center.

5   Check for relative movement between the mount plates and the engine or frame (use a large screwdriver or pry bar to attempt to move the mount). If movement is noted, check the tightness of the mount fasteners first before condemning the mount. Usually when engine mounts are broken, they are very obvious as the engine will easily move away from the mount when pried or under load.

Replacement

Note: *The engine/transaxle assembly must be balanced in its weight distribution among the several powertrain mounts before the mounting bolts are tightened. Loosen all the mount bolts, then shake the engine from side to side and front-to-rear as much as possible to settle it. Tighten all the mounts in this order: transaxle mount-to-transaxle bolts (rear first, then middle, then front); engine mount-to-* bracket bolts (center first); shake the engine again front-to-rear; rear mount through-bolt; then front mount through-bolt.

Right-side mount

6   Remove the engine cover.

7   Remove the air filter housing (see Chapter 4).

8   Raise the vehicle and support it securely on jackstands.

2009 and earlier 3.6L models

9   Remove the engine mount struts (see Step 21).

10   Disconnect the catalytic converter pipe from the rear exhaust manifold and move it out of the way.

11   Remove the right front wheel and the right inner splash shield (see Chapter 11).

12   Remove the engine mount-to-subframe nuts.

13   Using a floor jack, place a block of wood between the floor jack and the oil pan and support the engine.

14   Remove the engine mount bracket bolts and bracket.

15   Remove the mount upper nuts, then using the floor jack, raise the engine slightly and remove the mount.

3.0L models and 2013 and later 3.6L models

16   Before removing the mount, mark the mount location using a permanent marker so the mount can be installed in the same location.

17   Support the engine. This can be done by using an engine hoist, an engine support fixture that attaches between the fenders and holds the engine from above (the preferred method), or a floor jack with a block of wood placed under the oil pan.

18   Remove the mount-to-body bolts and nut, then remove the mount-to-engine bracket bolts from the right engine mount.

19   Using the support fixture, engine hoist or jack, raise the engine up far enough to remove the mount.

20   Installation is the reverse of removal. Use new fasteners and apply thread-locking compound to the engine mount studs, then tighten the nuts to the torque listed in this Chapter's Specifications.

Engine torque strut mounts – 3.6L models

Warning: *The engine must be completely cool before beginning this procedure.*

21   The engine torque struts mount between the engine and the upper radiator support on the body. They can be replaced without jacking up the engine, by removing the through-bolts and the bolts attaching the struts to the mounts on the radiator support.

Note: *When reinstalling the mounts, you may have to use a prybar to rotate the engine toward or away from the radiator to align the through-bolts.*

Note: *To replace the transaxle rear mount, the front mount or the left side mounts, see Chapter 7A.*

Notes

Notes

Chapter 2 Part D
General engine overhaul procedures

Contents

Specifications

General

Displacement
2.4L four-cylinder engines (LAF)	146 cubic inches
3.0L V6 engines (LF1 and LFW)	183 cubic inches
3.4L V6 engines (LNJ)	204 cubic inches

3.6L V6 engines
(LY7) 2009 and earlier	217 cubic inches
(LFX) 2013 and later	220 cubic inches

Bore and stroke
2.4L engines	3.46 x 3.86 inches
3.0L engines	3.50 x 3.16 inches
3.4L engines	3.62 x 3.31 inches
3.6L engines	3.70 x 3.37 inches
Cylinder compression	Lowest cylinder must be within 70 percent of highest cylinder

Minimum compression pressure
2.4L engines	100 psi
3.0L engines	140 psi
3.4L engines	100 psi

3.6L V6 engines
(LY7) 2009 and earlier	140
(LFX) 2013 and later	100

Compression ratio

2.4L engines
2013 and earlier models	10:1
2014 and later models	11.16:1
3.0L engines	11.7:1
3.4L engines	9.6:1

3.6L V6 engines
(LY7) 2009 and earlier	10.2:1
(LFX) 2013 and later	11.5:1

Minimum oil pressure (engine at operating temperature)

2.4L engines	30 to 70 psi @ 1,000 rpm
3.0L engines	10 psi @ idle
	30 psi @ 2,000 rpm
3.4L engines	30 to 35 psi @ 1,850 rpm
3.6L engines	10 psi @ idle
2009 and earlier (LY7)	30 psi @ 2,000 rpm
2013 and later (LFX)	20 psi @ 2,000 rpm

Camshaft (3.4L V6 models)

Camshaft bearing journal diameter	1.868 to 1.869 inches
Bearing oil clearance	0.002 to 0.004 inch
Lobe lift	0.273 inch

Torque specifications   Ft-lbs (unless otherwise indicated)

Note: *One foot-pound (ft-lb) of torque is equivalent to 12 inch-pounds (in-lbs) of torque. Torque values below approximately 15 foot-pounds are expressed in inch-pounds, because most foot-pound torque wrenches are not accurate at these smaller values.*

Subframe mounting bolts*	114
Driveplate-to-torque converter bolts	
2.4L engines	45
3.4L engines	52
3.0L engines*	
Step 1	22
Step 2	Tighten an additional 45 degrees
3.6L engines*	
2009 and earlier engines (LY7)	
Step 1	22
Step 2	Tighten an additional 45 degrees
2013 and later engines (LFX)	
Step 1	44
Step 2	Tighten an additional 45 degrees
Camshaft thrust plate screws (3.4L models)	89 in-lbs
Connecting rod bearing cap bolts*	
2.4L engines	
Step 1	18
Step 2	Tighten an additional 100 degrees
3.4L engines	
Step 1	15
Step 2	Tighten an additional 75 degrees
3.0L and 3.6L engines*	
Step 1	
3.0L and 2009 and earlier 3.6L engines (LY7)	15
2013 and later 3.6L engines (LFX)	18
Step 2	Tighten an additional 110 degrees
Main bearing cap bolts*	
2.4L engines	
Lower crankcase-to-block	
Step 1	15
Step 2	Tighten an additional 70 degrees
Lower crankcase perimeter bolt	18
3.4L engines	
Step 1	37
Step 2	Tighten an additional 77 degrees
3.0L and 3.6L engines	
Inner	
Step 1	15
Step 2	Tighten an additional 80 degrees
Outer*	
Step 1	132 in-lbs
Step 2	Tighten an additional 110 degrees
Side*	
Step 1	22
Step 2	Tighten an additional 60 degrees
Subframe mounting bolts	See Chapter 10

** Bolts must be replaced with **NEW** bolts*

1.1 An engine block being bored. An engine rebuilder will use special machinery to recondition the cylinder bores

1.2 If the cylinders are bored, the machine shop will normally hone the engine on a machine like this

1   General information - engine overhaul

Refer to illustrations 1.1, 1.2, 1.3, 1.4, 1.5 and 1.6

Included in this portion of Chapter 2 are general information and diagnostic testing procedures for determining the overall mechanical condition of your engine.

The information ranges from advice concerning preparation for an overhaul and the purchase of replacement parts and/or components to detailed, step-by-step procedures covering removal and installation.

The following Sections have been written to help you determine whether your engine needs to be overhauled and how to remove and install it once you've determined it needs to be rebuilt. For information concerning in-vehicle engine repair, see Chapter 2A, 2B or 2C.

It's not always easy to determine when, or if, an engine should be completely over-hauled, because a number of factors must be considered.

High mileage is not necessarily an indication that an overhaul is needed, while low mileage doesn't preclude the need for an overhaul. Frequency of servicing is probably the most important consideration. An engine that's had regular and frequent oil and filter changes, as well as other required maintenance, will most likely give many thousands of miles of reliable service. Conversely, a neglected engine may require an overhaul very early in its service life.

Excessive oil consumption is an indication that piston rings, valve seals and/or valve guides are in need of attention. Make sure that oil leaks aren't responsible before deciding that the rings and/or guides are bad. Perform a cylinder compression check to determine the extent of the work required (see Section 3). Also check the vacuum readings under various conditions (see Section 4).

Check the oil pressure with a gauge installed in place of the oil pressure sending unit and compare it to this Chapter's Specifications (see Section 2). If it's extremely low, the bearings and/or oil pump are probably worn out.

Loss of power, rough running, knocking or metallic engine noises, excessive valve train noise and high fuel consumption rates may also point to the need for an overhaul, especially if they're all present at the same time. If a complete tune-up doesn't remedy the situation, major mechanical work is the only solution.

An engine overhaul involves restoring the internal parts to the specifications of a new engine. During an overhaul, the piston rings are replaced and the cylinder walls are reconditioned (rebored and/or honed) **(see illustrations 1.1 and 1.2)**. If a rebore is done by an automotive machine shop, new oversize pistons will also be installed. The main bearings, connecting rod bearings and camshaft bearings are generally replaced with new ones and, if necessary, the crankshaft may be reground to restore the journals

1.3 A crankshaft having a main bearing journal ground

1.4 A machinist checks for a bent connecting rod, using specialized equipment

(see illustration 1.3). Generally, the valves are serviced as well, since they're usually in less-than-perfect condition at this point. While the engine is being overhauled, other components, such as the distributor, starter and alternator, can be rebuilt as well. The end result should be similar to a new engine that will give many trouble free miles. **Note:** *Critical cooling system components such as the hoses, drivebelts, thermostat and water pump should be replaced with new parts when an engine is overhauled. The radiator should be checked carefully to ensure that it isn't clogged or leaking (see Chapter 3). If you purchase a rebuilt engine or short block, some rebuilders will not warranty their engines unless the radiator has been professionally flushed. Also, we don't recommend overhauling the oil pump - always install a new one when an engine is rebuilt.*

Overhauling the internal components on today's engines is a difficult and time-consuming task which requires a significant amount of specialty tools and is best left to a professional engine rebuilder **(see illustrations 1.4, 1.5 and 1.6)**. A competent engine rebuilder will handle the inspection of your old parts and offer advice concerning the reconditioning or replacement of the original engine, never purchase parts or have machine work done on other components until the block has been thoroughly inspected by a professional machine shop. As a general rule, time is the primary cost of an overhaul, especially since the vehicle may be tied up for a minimum of two weeks or more. Be aware that some engine builders only have the capability to rebuild the engine you bring them while other rebuilders have a large inventory of rebuilt exchange engines in stock. Also be aware that many machine shops could take as much as two weeks time to completely rebuild your engine depending on shop workload. Sometimes it makes more sense to simply exchange your engine for another engine that's already rebuilt to save time.

2   Oil pressure check

Refer to illustrations 2.2 and 2.3

1   Low engine oil pressure can be a sign of an engine in need of rebuilding. A "low oil pressure" indicator (often called an "idiot light") is not a test of the oiling system. Such indicators only come on when the oil pressure is dangerously low. Even a factory oil pressure gauge in the instrument panel is only a relative indication, although much better for driver information than a warning light. A better test is with a mechanical (not electrical) oil pressure gauge.

2   Locate the engine oil pressure sending unit on the engine block **(see illustration)**: on 2.4L models, it's located above the starter motor just in front of the oil filter housing; on 3.0L and 3.6L engines, it's located on the front of the engine, to the lower rear of the alternator, near the oil filter; on 3.4L engines, it's

1.5 A bore gauge being used to check the main bearing bore

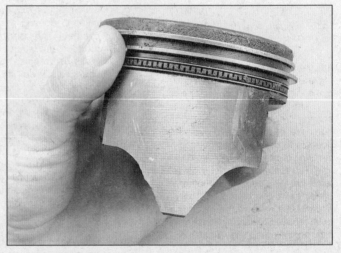

1.6 Uneven piston wear like this indicates a bent connecting rod

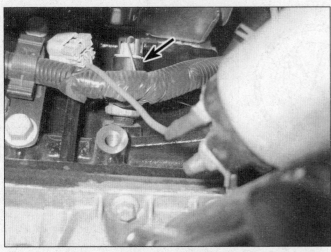

2.2 Location of the oil pressure sending unit on 3.4L V6 engines, other engines similar

2.3 Remove the oil pressure sending unit and install an oil pressure gauge

located above and to the rear of the oil filter on the front side of the engine.

Note: *On some models, a heat shield will have to be removed for access to the sending unit.*

3　Disconnect the wiring, then unscrew and remove the oil pressure sending unit. Screw in the hose for your oil pressure gauge **(see illustration)**. If necessary, install an adapter fitting. Use Teflon tape or thread sealant on the threads of the adapter and/or the fitting on the end of your gauge's hose.

4　Connect an accurate tachometer to the engine, according to the tachometer manufacturer's instructions.

5　Check the oil pressure with the engine running (normal operating temperature) at the specified engine speed, and compare it to this Chapter's Specifications. If it's extremely low, the bearings and/or oil pump are probably worn out.

3　Cylinder compression check

Refer to illustration 3.7

1　A compression check will tell you what mechanical condition the upper end of your engine (pistons, rings, valves, head gaskets) is in. Specifically, it can tell you if the compression is down due to leakage caused by worn piston rings, defective valves and seats or a blown head gasket. **Note:** *The engine must be at normal operating temperature and the battery must be fully charged for this check.*

2　Begin by cleaning the area around the spark plugs before you remove them (compressed air should be used, if available). The idea is to prevent dirt from getting into the cylinders as the compression check is being done.

3　On 3.0L V6 models, remove the upper intake manifold (see Chapter 2C).

4　On all models except 3.4L V6 engines, remove all of the ignition coils and spark plugs from the engine (see Chapter 1).

5　On 3.4L models, disable the ignition system by disconnecting the primary (low voltage) wires from the coil pack (see Chapter 5), then remove the spark plugs from the engine (see Chapter 1).

6　Disable the fuel system by removing the fuel pump fuse (see Chapter 4), then remove the air intake duct (see Chapter 4) and block the throttle plate wide open.

7　Install a compression gauge in the number one cylinder spark plug hole **(see illustration)**.

8　Crank the engine over at least seven compression strokes and watch the gauge. The compression should build up quickly in a healthy engine. Low compression on the first stroke, followed by gradually increasing pressure on successive strokes, indicates worn piston rings. A low compression reading on the first stroke, which doesn't build up during successive strokes, indicates leaking valves or a blown head gasket (a cracked head could also be the cause). Deposits on the undersides of the valve heads can also cause low compression. Record the highest gauge reading obtained.

9　Repeat the procedure for the remaining cylinders and compare the results to this Chapter's Specifications.

10　Add some engine oil (about three squirts from a plunger-type oil can) to each cylinder, through the spark plug hole, and repeat the test.

11　If the compression increases after the oil is added, the piston rings are definitely worn. If the compression doesn't increase significantly, the leakage is occurring at the valves or head gasket. Leakage past the valves may be caused by burned valve seats and/or faces or warped, cracked or bent valves.

12　If two adjacent cylinders have equally low compression, there's a strong possibility that the head gasket between them is blown. The appearance of coolant in the combustion chambers or the crankcase would verify this condition.

13　If one cylinder is slightly lower than the others, and the engine has a slightly rough idle, a worn lobe on the camshaft could be the cause.

14　If the compression is unusually high, the combustion chambers are probably coated with carbon deposits. If that's the case, the cylinder head(s) should be removed and decarbonized.

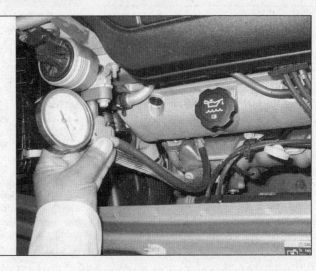

3.7 Use a compression gauge with a threaded fitting for the spark plug hole, not the type that requires hand pressure to maintain the seal

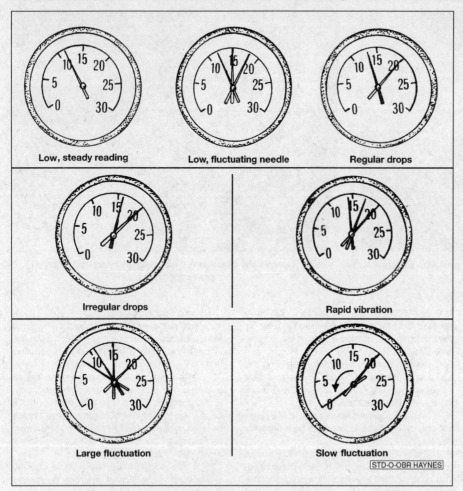

Low, steady reading	Low, fluctuating needle	Regular drops
Irregular drops		Rapid vibration
Large fluctuation		Slow fluctuation

STD-O-OBR HAYNES

4.6 Typical vacuum gauge readings

timing light and eliminate all other possible causes, utilizing the tests provided in this Chapter before you remove the timing chain cover to check the timing marks.

8   If the reading is three to eight inches below normal and it fluctuates at that low reading, suspect an intake manifold gasket leak at an intake port or a faulty fuel injector.

9   If the needle has regular drops of about two-to-four inches at a steady rate, the valves are probably leaking. Perform a compression check or leak-down test to confirm this.

10   An irregular drop or down-flick of the needle can be caused by a sticking valve or an ignition misfire. Perform a compression check or leak-down test and read the spark plugs.

11   A rapid vibration of about four in-Hg vibration at idle combined with exhaust smoke indicates worn valve guides. Perform a leak-down test to confirm this. If the rapid vibration occurs with an increase in engine speed, check for a leaking intake manifold gasket or head gasket, weak valve springs, burned valves or ignition misfire.

12   A slight fluctuation, say one inch up and down, may mean ignition problems. Check all the usual tune-up items and, if necessary, run the engine on an ignition analyzer.

13   If there is a large fluctuation, perform a compression or leak-down test to look for a weak or dead cylinder or a blown head gasket.

14   If the needle moves slowly through a wide range, check for a clogged PCV system, incorrect idle fuel mixture, throttle body or intake manifold gasket leaks.

15   Check for a slow return after revving the engine by quickly snapping the throttle open until the engine reaches about 2,500 rpm and let it shut. Normally the reading should drop to near zero, rise above normal idle reading (about 5 in-Hg over) and return to the previous idle reading. If the vacuum returns slowly and doesn't peak when the throttle is snapped shut, the rings may be worn. If there is a long delay, look for a restricted exhaust system (often the muffler or catalytic converter). An easy way to check this is to temporarily disconnect the exhaust ahead of the suspected part and redo the test.

15   If compression is way down or varies greatly between cylinders, it would be a good idea to have a leak-down test performed by an automotive repair shop. This test will pinpoint exactly where the leakage is occurring and how severe it is.

4   Vacuum gauge diagnostic checks

Refer to illustration 4.6

1   A vacuum gauge provides inexpensive but valuable information about what is going on in the engine. You can check for worn rings or cylinder walls, leaking head or intake manifold gaskets, incorrect carburetor adjustments, restricted exhaust, stuck or burned valves, weak valve springs, improper ignition or valve timing and ignition problems.

2   Unfortunately, vacuum gauge readings are easy to misinterpret, so they should be used in conjunction with other tests to confirm the diagnosis.

3   Both the absolute readings and the rate of needle movement are important for accurate interpretation. Most gauges measure vacuum in inches of mercury (in-Hg). The following references to vacuum assume the

diagnosis is being performed at sea level. As elevation increases (or atmospheric pressure decreases), the reading will decrease. For every 1,000 foot increase in elevation above approximately 2,000 feet, the gauge readings will decrease about one inch of mercury.

4   Connect the vacuum gauge directly to the intake manifold vacuum, not to ported (throttle body) vacuum. Be sure no hoses are left disconnected during the test or false readings will result.

5   Before you begin the test, allow the engine to warm up completely. Block the wheels and set the parking brake. With the transaxle in Park, start the engine and allow it to run at normal idle speed. **Warning:** *Keep your hands and the vacuum gauge clear of the fans.*

6   Read the vacuum gauge; an average, healthy engine should normally produce about 17 to 22 in-Hg with a fairly steady needle. Refer to the following vacuum gauge readings and what they indicate about the engine's condition **(see illustration)**:

7   A low steady reading usually indicates a leaking gasket between the intake manifold and cylinder head(s) or throttle body, a leaky vacuum hose, late ignition timing or incorrect camshaft timing. Check ignition timing with a

5   Engine rebuilding alternatives

The do-it-yourselfer is faced with a number of options when purchasing a rebuilt engine. The major considerations are cost, warranty, parts availability and the time required for the rebuilder to complete the project. The decision to replace the engine block, piston/connecting rod assemblies and crankshaft depends on the final inspection results of your engine. Only then can you make a cost effective decision whether to have your engine overhauled or simply purchase an exchange engine for your vehicle.

Some of the rebuilding alternatives include:

6.1 After tightly wrapping water-vulnerable components, use a spray cleaner on everything, with particular concentration on the greasiest areas, usually around the valve cover and lower edges of the block. If one section dries out, apply more cleaner

6.2 Depending on how dirty the engine is, let the cleaner soak in according to the directions and hose off the grime and cleaner. Get the rinse water down into every area you can get at; then dry important components with a hair dryer or paper towels

6.3 Get an engine stand sturdy enough to firmly support the engine while you're working on it. Stay away from three-wheeled models; they have a tendency to tip over more easily, so get a four-wheeled unit

Individual parts - If the inspection procedures reveal that the engine block and most engine components are in reusable condition, purchasing individual parts and having a rebuilder rebuild your engine may be the most economical alternative. The block, crankshaft and piston/connecting rod assemblies should all be inspected carefully by a machine shop first.

Short block - A short block consists of an engine block with a crankshaft and piston/connecting rod assemblies already installed. All new bearings are incorporated and all clearances will be correct. The existing camshafts, valve train components, cylinder head and external parts can be bolted to the short block with little or no machine shop work necessary.

Long block - A long block consists of a short block plus an oil pump, oil pan, cylinder head, valve cover, camshaft and valve train components, timing sprockets and chain or gears and timing cover. All components are installed with new bearings, seals and gaskets incorporated throughout. The installation of manifolds and external parts is all that's necessary.

Low mileage used engines - Some companies now offer low mileage used engines which is a very cost effective way to get your vehicle up and running again. These engines often come from vehicles which have been in totaled in accidents or come from other countries which have a higher vehicle turn over rate. A low mileage used engine also usually has a similar warranty like the newly remanufactured engines.

Give careful thought to which alternative is best for you and discuss the situation with local automotive machine shops, auto parts dealers and experienced rebuilders before ordering or purchasing replacement parts.

6   Engine removal - methods and precautions

Refer to illustrations 6.1, 6.2, 6.3 and 6.4

If you've decided that an engine must be removed for overhaul or major repair work, several preliminary steps should be taken. Read all removal and installation procedures carefully prior to committing to this job.

Locating a suitable place to work is extremely important. Adequate work space, along with storage space for the vehicle, will be needed. If a shop or garage isn't available, at the very least a flat, level, clean work surface made of concrete or asphalt is required. A vehicle hoist is necessary for engine removal since the engine and transaxle are removed

as an assembly out the bottom of the vehicle.

Cleaning the engine compartment and engine before beginning the removal procedure will help keep tools clean and organized **(see illustrations 6.1 and 6.2)**.

An engine hoist will also be necessary. Make sure the hoist is rated in excess of the combined weight of the engine and transaxle. Safety is of primary importance, considering the potential hazards involved in removing the engine from the vehicle.

If you're a novice at engine removal, get at least one helper. One person cannot easily do all the things you need to do to remove a big heavy engine and transaxle assembly from the engine compartment. Also helpful is to seek advice and assistance from someone who's experienced in engine removal.

Plan the operation ahead of time. Arrange for or obtain all of the tools and equipment you'll need prior to beginning the job **(see illustrations 6.3 and 6.4)**. Some of

6.4 Since many of the fasteners on these engines are tightened using the angle torque method, a torque angle gauge is essential for proper assembly

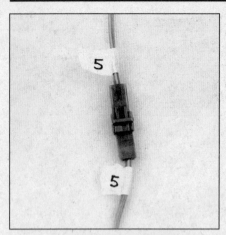

7.14 Label both ends of each wire and hose before disconnecting it

the equipment necessary to perform engine removal and installation safely and with relative ease are (in addition to a vehicle hoist and an engine hoist) a heavy duty floor jack (preferably fitted with a transaxle jack head adapter), complete sets of wrenches and sockets as described in the front of this manual, wooden blocks, plenty of rags and cleaning solvent for mopping up spilled oil, coolant and gasoline.

Plan for the vehicle to be out of use for quite a while. A machine shop can do the work that is beyond the scope of the home mechanic. Machine shops often have a busy schedule, so before removing the engine, consult the shop for an estimate of how long it will take to rebuild or repair the components that may need work.

7   Engine - removal and installation

Warning 1: *Gasoline is extremely flammable, so take extra precautions when you work on any part of the fuel system. Don't smoke or allow open flames or bare light bulbs near the work area, and don't work in a garage where a gas-type appliance (such as a water heater or clothes dryer) is present. Since gasoline is carcinogenic, wear fuel-resistant gloves when there's a possibility of being exposed to fuel, and, if you spill any fuel on your skin, rinse it off immediately with soap and water. Mop up any spills immediately and do not store fuel-soaked rags where they could ignite. The fuel system is under constant pressure, so, if any fuel lines are to be disconnected, the fuel pressure in the system must be relieved first (see Chapter 4 for more information). When you perform any kind of work on the fuel system, wear safety glasses and have a Class B type fire extinguisher on hand.*
Warning 2: *The engine must be completely cool before beginning this procedure.*
Note 1: *Engine removal on these models is a difficult job, especially for the do-it-yourself mechanic working at home. Because of the vehicle's design, the manufacturer states that*

the engine and transaxle have to be removed as a unit from the bottom of the vehicle, not the top. With a floor jack and jackstands, the vehicle can't be raised high enough and supported safely enough for the engine/transaxle assembly to slide out from underneath. The manufacturer recommends that removal of the engine transaxle assembly only be performed on a frame-contact type vehicle hoist.
Note 2: *Read through the entire Section before beginning this procedure. The engine and transaxle are removed as a unit from below, then separated outside the vehicle.*

Removal

Refer to illustrations 7.14, 7.26a, 7.26b and 7.33

Note: *Have the air conditioning system discharged and recovered by an authorized service facility before beginning this procedure.*

1   Park the vehicle on a frame-contact type vehicle hoist, then engage the arms of the hoist with the jacking points of the vehicle. Raise the hoist arms until they contact the vehicle, but not so much that the wheels come off the ground.
2   Position the steering wheel so the front wheels point straight ahead, then disconnect the cable from the negative battery terminal and remove the battery tray (see Chapter 5).
3   Relieve the fuel system pressure (see Chapter 4).
4   Remove the engine cover.
5   On 2.4L models, remove the intake manifold (see Chapter 2A).
6   Remove the air filter housing and the air intake duct (see Chapter 4).
7   Disconnect the fuel line from the fuel rail (see Chapter 4).
8   Drain the cooling system (see Chapter 1). Support the radiator/condenser assembly to the body with wire or large plastic tie-wraps, and remove the expansion tank and its hoses.
9   On 3.4L models, remove the battery and the battery box. Place the ECM (engine control module) on top of the engine and secure it so it won't be damaged. Also set the TCM (transmission control module) on the engine.
10   Disconnect the ECM wiring from the underhood fuse center.
11   Disconnect the electrical connector from the air conditioning compressor, then unbolt the compressor and secure the compressor out of the way (see Chapter 3).
12   Disconnect the shift cable from the transaxle (see Chapter 7).
13   Follow the heater hoses from the firewall and detach them from the pipes on the engine.
14   Clearly label and disconnect all vacuum lines, hoses, wiring harness connectors and fuel lines. Masking tape and/or a touch up paint applicator work well for marking items **(see illustration)**. Take photos or sketch the locations of retainers, clips and brackets. Move the wiring harnesses out of the way.
15   Remove the radiator hoses.
16   Disconnect the transaxle oil cooler lines from the transaxle. Seal the open ends to prevent contamination.

17   Loosen the wheel lug nuts and the drive-axle/hub nuts, then raise the vehicle on the hoist. Remove the wheels.
18   On 3.4L models, refer to Chapter 11 and remove the air deflector under the front bumper cover.
19   Unbolt the catalytic converters from the exhaust manifolds and the rear exhaust pipes. Remove the front sections of the exhaust system from the vehicle, then tie up the rear sections using wire.
20   On 3.4L AWD models, disconnect the transfer case vent hose. Disconnect the shift cable(s) from the transaxle (see Chapter 7). Also disconnect any wiring harness connectors from the transaxle and cable brackets from the engine.
21   Remove the inner fender splash shields (see Chapter 11).
22   Refer to Chapter 10 and disconnect the stabilizer bar links, tie-rod ends, and the steering intermediate shaft. **Caution:** *Don't allow the steering shaft to rotate after the intermediate shaft has been disconnected, as damage to the airbag clockspring could occur. To prevent this, run the seat belt through the steering wheel and click it into its latch.*
23   Disconnect the lower control arms from the steering knuckles (see Chapter 10) and remove the driveaxles (see Chapter 8).
24   On AWD models, remove the rear driveshaft.
25   Scribe or make paint marks where the subframe meets the chassis for installation alignment purposes. Lower the vehicle.
26   Support the engine/transaxle assembly from above with an engine hoist securely attached by heavy-duty chains to the engine lifting brackets **(see illustrations)**. Loosen all the engine and transaxle mount fasteners. With the hoist taking the weight off the mounts, remove the engine/transaxle mounts. **Warning:** *DO NOT place any part of your body under the engine when it's supported only by a hoist or other lifting device.*
27   Support the subframe with two floor jacks - one positioned under each side of the subframe. Remove the subframe bolts and discard them (see Chapter 10).
28   With an assistant to help, carefully and slowly lower each jack until the subframe is down far enough to be slid out from under the vehicle.
29   Recheck to be sure nothing is still connecting the engine to the vehicle; there are many ground wires, for example, that must be disconnected. Disconnect anything still remaining.
30   Inspect the engine/transaxle assembly thoroughly once more to make sure that nothing is still attached, then slowly lower the powertrain down out of the engine compartment and onto the floor. Check carefully to make sure nothing is hanging up as this is done.
31   Once the powertrain is on the floor, disconnect the lifting chains and roll the engine hoist out of the way, then raise the vehicle hoist until the vehicle clears the powertrain.
32   Connect an engine hoist to the engine, raise the engine/transaxle up a little and support the engine with blocks of wood. Support

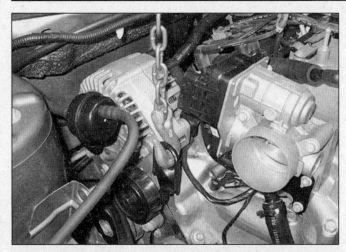

7.26a Attach one end of the chain to the lifting bracket at the right end of the engine . . .

7.26b . . . and the other end of the chain to the lifting bracket at the left end

the transaxle with a floor jack; preferably one with a transmission adapter. Secure the transaxle to the jack with safety chains.

33   Remove the starter (see Chapter 5), then mark the torque converter to the driveplate and remove the driveplate-to-torque converter bolts **(see illustration)**.

34   Remove the transaxle-to-engine mounting bolts and separate the engine from the transaxle.

35   Remove the driveplate (see Chapter 2A, 2B or 2C) and mount the engine on an engine stand **(see illustration 6.3)**.

Installation

36   Installation is the reverse of the removal procedure, noting the following points:

a) *Check the engine/transaxle mounts. If they're worn or damaged, replace them.*
b) *Replace the subframe mounting bolts with NEW ones.*
c) *Attach the transaxle to the engine following the procedure described in Chapter 7.*
d) *When installing the subframe, align the marks made during removal, then tighten the subframe mounting bolts to the torque listed in the Chapter 10 Specifications.*
e) *Refill the cooling system with the proper mixture of antifreeze. Refill the crankcase with the recommended engine oil (see Chapter 1).*
f) *Reconnect the battery (see Chapter 5).*
g) *Run the engine and check for proper operation and leaks. Shut off the engine and recheck fluid levels. Check the transaxle fluid level, adding as necessary (see Chapter 1).*

8   Engine overhaul - disassembly sequence

1   It's much easier to remove the external components if it's mounted on an engine stand. A stand can often be rented quite

cheaply from an equipment rental yard. Before the engine is mounted on a stand, the driveplate should be removed from the engine.

2   If a stand isn't available, it's possible to remove the external engine components with it blocked up on the floor. Be extra careful not to tip or drop the engine when working without a stand.

3   If you're going to obtain a rebuilt engine, all external components must come off first, to be transferred to the replacement engine. These components include:

Driveplate
Ignition system components
Emissions-related components
Engine mounts and mount brackets
Engine rear cover (spacer plate between driveplate and engine block), if equipped
Intake/exhaust manifolds
Fuel injection components
Oil filter
Spark plugs and ignition coil pack (and spark plug wires) or coil-over plug assemblies
Thermostat and housing assembly
Water pump

Note: *When removing the external components from the engine, pay close attention to details that may be helpful or important during installation. Note the installed position of gaskets, seals, spacers, pins, brackets, washers, bolts and other small items.*

4   If you're going to obtain a short block (assembled engine block, crankshaft, pistons and connecting rods), then remove the timing chain, cylinder heads, oil pan, oil pump pick-up tube, oil pump and water pump from your engine so that you can turn in your old short block to the rebuilder as a core. See *Engine rebuilding alternatives* for additional information regarding the different possibilities to be considered.

9   Camshaft (3.4L models) - removal, inspection and installation

Removal

Refer to illustrations 9.2 and 9.4

1   Remove the engine (see Section 7).
2   Remove the bolt and clamp holding the

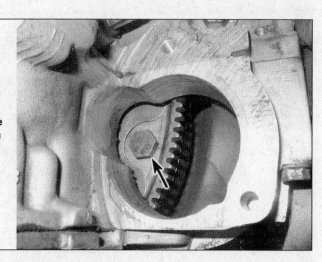

7.33 Remove the torque converter bolts through the starter opening

9.2 Remove the bolt and pull out the oil pump drive

9.4 Remove the retaining bolts and the camshaft thrust plate

oil pump drive and pull the oil pump drive straight up and out of the block **(see illustration)**.

3   Remove the timing chain and sprockets and the valve lifters (see Chapter 2A).

4   Remove the bolts holding the camshaft thrust plate to the block **(see illustration)** and remove the thrust plate.

5   Slide the camshaft straight out of the engine, using a long bolt (with the same thread as the camshaft sprocket bolt) screwed into the front of the camshaft as a handle. Support the shaft near the block and be very slow and careful not to scrape or nick the bearings.

Inspection

Refer to illustrations 9.7a and 9.7b

6   After the camshaft has been removed, clean it with solvent and dry it, then inspect the bearing journals for uneven wear, pitting and evidence of seizure. If the journals are damaged, the camshaft bearings are probably damaged as well. Both the shaft and bearings will have to be replaced.

7   Measure the bearing journals with a micrometer **(see illustration)** to determine

whether they are excessively worn or out-of-round. Measure the camshaft lobes also to check for wear. Measure the camshaft lobes at their highest point, then subtract the measurement of the lobe at its smallest diameter - the difference is the lobe lift **(see illustration)**. Refer to the Specifications listed in this Chapter.

8   Inspect the camshaft lobes for heat discoloration, score marks, chipped areas, pitting and uneven wear. If the lobes are in good condition and if the lobe lift measurements are as specified, you can reuse the camshaft.

9   Check the camshaft bearings in the block for wear and damage. Look for galling, pitting and discolored areas. Inspect the housing journals for damage and replace them if necessary.

10   The inside diameter of each bearing can be determined with an inside micrometer or a bore gauge and outside micrometer. Subtract the camshaft bearing journal diameter(s) from the corresponding bearing inside diameter(s) to obtain the bearing oil clearance. If it's excessive, new bearings will be required regardless of the condition of the originals. Refer to the Specifications listed in this Chapter.

11   Camshaft bearing replacement requires special tools and expertise that place it outside the scope of the home mechanic. Take the block to an automotive machine shop to ensure the job is done correctly.

Installation

12   Lubricate the camshaft bearing journals and cam lobes with a special camshaft installation lubricant.

13   Slide the camshaft into the engine, using a long bolt (the same thread as the camshaft sprocket bolt) screwed into the front of the camshaft as a handle. Support the cam near the block and be careful not to scrape or nick the bearings. Install the camshaft retainer plate and tighten the bolts to the torque listed in this Chapter's Specifications.

14   Dip the gear portion of the oil pump drive in engine oil and insert it into the block. It should be flush with its mounting boss before inserting the retaining bolt. **Note:** *Position a new O-ring on the oil pump driveshaft before installation.*

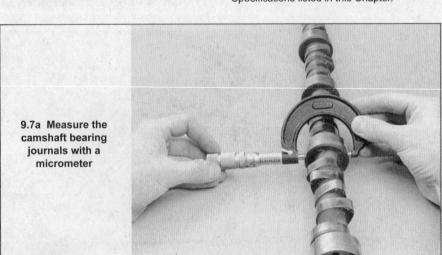

9.7a Measure the camshaft bearing journals with a micrometer

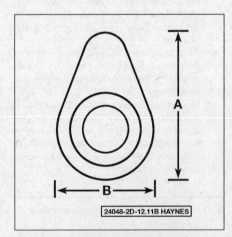

`24048-2D-12.11B HAYNES`

9.7b Measure the camshaft lobe maximum diameter (A) and the minimum diameter (B) - subtract B from A and the difference is the lobe lift

10.1 Before you try to remove the pistons, use a ridge reamer to remove the raised material (ridge) from the top of the cylinders

10.3 Checking the connecting rod endplay (side clearance)

15   Complete the installation of the timing chain and sprockets by referring to Chapter 2A.

---

10   Pistons and connecting rods - removal and installation

Removal

Refer to illustrations 10.1, 10.3 and 10.4

Note: *Prior to removing the piston/connecting rod assemblies, remove the cylinder head and oil pan (see Chapter 2A, 2B or 2C).*

1   Use your fingernail to feel if a ridge has formed at the upper limit of ring travel (about 1/4-inch down from the top of each cylinder). If carbon deposits or cylinder wear have produced ridges, they must be completely removed with a special tool **(see illustration)**. Follow the manufacturer's instructions provided with the tool. Failure to remove the ridges before attempting to remove the piston/connecting rod assemblies may result in piston breakage.

2   After the cylinder ridges have been removed, turn the engine so the crankshaft is facing up.

3   Before the main bearing caps and connecting rods are removed, check the connecting rod endplay with feeler gauges. Slide them between the first connecting rod and the crankshaft throw until the play is removed **(see illustration)**. Repeat this procedure for each connecting rod. The endplay is equal to the thickness of the feeler gauge(s). Check with an automotive machine shop for the endplay service limit (a typical endplay limit should measure between 0.005 to 0.015 inch [0.127 to 0.381 mm]). If the play exceeds the service limit, new connecting rods will be required. If new rods (or a new crankshaft) are installed, the endplay may fall under the minimum allowable. If it does, the rods will have to be machined to restore it.

If necessary, consult an automotive machine shop for advice.

4   Check the connecting rods and caps for identification marks. If they aren't plainly marked, use paint or marker to clearly identify each rod and cap (1, 2, 3, etc., depending on the cylinder they're associated with) **(see illustration)**.

5   Remove the connecting rod cap bolts. **Note:** *New connecting rod cap bolts must be used when reassembling the engine. Save the old bolts for the oil clearance check.*

6   Remove the number one connecting rod cap and bearing insert. Don't drop the bearing insert out of the cap.

7   Remove the bearing insert and push the connecting rod/piston assembly out through the top of the engine. Use a wooden or plastic hammer handle to push on the upper bearing surface in the connecting rod. If resistance is felt, double-check to make sure that all of the ridge was removed from the cylinder.

8   Repeat the procedure for the remaining cylinders.

9   After removal, reassemble the connecting rod caps and bearing inserts in their respective connecting rods and install the cap bolts finger tight. Leaving the old bearing inserts in place until reassembly will help prevent the connecting rod bearing surfaces from being accidentally nicked or gouged.

10   The pistons and connecting rods are now ready for inspection and overhaul at an automotive machine shop.

Piston ring installation

Refer to illustrations 10.13, 10.14, 10.15, 10.19a, 10.19b and 10.22

11   Before installing the new piston rings, the ring end gaps must be checked. It's assumed that the piston ring side clearance has been checked and verified correct.

12   Lay out the piston/connecting rod assemblies and the new ring sets so the ring sets will be matched with the same piston and cylinder during the end gap measurement and engine assembly.

13   Insert the top (number one) ring into the first cylinder and square it up with the cylinder walls by pushing it in with the top of the piston

10.4 If the connecting rods and caps are not marked, use paint to mark the caps to the rods by cylinder number (for example, this would be the No. 4 connecting rod)

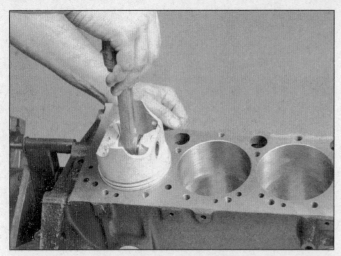

10.13 Install the piston ring into the cylinder then push it down into position using a piston so the ring will be square in the cylinder

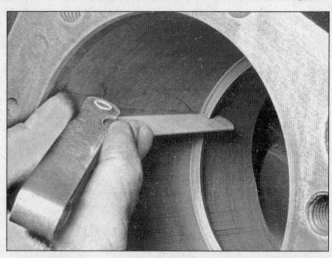

10.14 With the ring square in the cylinder, measure the ring end gap with a feeler gauge

(see illustration). The ring should be near the bottom of the cylinder, at the lower limit of ring travel.

14   To measure the end gap, slip feeler gauges between the ends of the ring until a gauge equal to the gap width is found **(see illustration)**. The feeler gauge should slide between the ring ends with a slight amount of drag. A typical ring gap should fall between 0.010 and 0.020 inch (0.25 to 0.50 mm) for compression rings and up to 0.030 inch (0.76 mm) for the oil ring steel rails. If the gap is larger or smaller than specified, double-check to make sure you have the correct rings before proceeding.

15   If the gap is too small, it must be enlarged or the ring ends may come in contact with each other during engine operation, which can cause serious damage to the engine. If necessary, increase the end gaps by filing the ring ends very carefully with a fine file. Mount the file in a vise equipped with soft jaws, slip the ring over the file with the ends contacting the file face and slowly move the ring to remove material from the ends. When per-

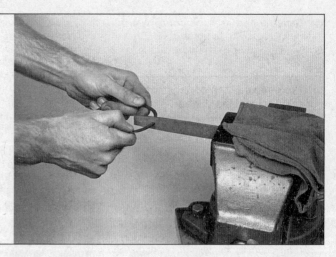

10.15 If the ring end gap is too small, clamp a file in a vise as shown and file the piston ring ends - be sure to remove all raised material

forming this operation, file only by pushing the ring from the outside end of the file towards the vise **(see illustration)**.

16   Excess end gap isn't critical unless it's greater than 0.040 inch (1.01 mm). Again,

double-check to make sure you have the correct ring type.

17   Repeat the procedure for each ring that will be installed in the first cylinder and for each ring in the remaining cylinders. Remember to keep rings, pistons and cylinders matched up.

18   Once the ring end gaps have been checked/corrected, the rings can be installed on the pistons.

19   The oil control ring (lowest one on the piston) is usually installed first. It's composed of three separate components. Slip the spacer/expander into the groove **(see illustration)**. If an anti-rotation tang is used, make sure it's inserted into the drilled hole in the ring groove. Next, install the upper side rail in the same manner **(see illustration)**. Don't use a piston ring installation tool on the oil ring side rails, as they may be damaged. Instead, place one end of the side rail into the groove between the spacer/expander and the ring land, hold it firmly in place and slide a finger around the piston while pushing the rail into the groove. Finally, install the lower side rail.

20   After the three oil ring components have been installed, check to make sure that both

10.19a Installing the spacer/expander in the oil ring groove

10.19b DO NOT use a piston ring installation tool when installing the oil control side rails

10.22 Use a piston ring installation tool to install the number 2 and the number 1 (top) rings - be sure the directional mark on the piston ring(s) is facing toward the top of the piston

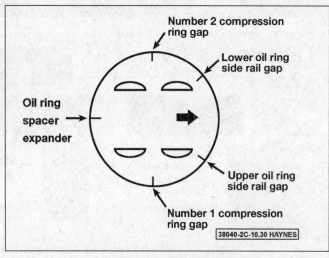

10.30 Piston ring end gap positions

the upper and lower side rails can be rotated smoothly inside the ring grooves.

21   The number two (middle) ring is installed next. It's usually stamped with a mark which must face up, toward the top of the piston. Do not mix up the top and middle rings, as they have different cross-sections. **Note:** *Always follow the instructions printed on the ring package or box - different manufacturers may require different approaches.*

22   Use a piston ring installation tool and make sure the identification mark is facing the top of the piston, then slip the ring into the middle groove on the piston **(see illustration)**. Don't expand the ring any more than necessary to slide it over the piston.

23   Install the number one (top) ring in the same manner. Make sure the mark is facing up. Be careful not to confuse the number one and number two rings.

24   Repeat the procedure for the remaining pistons and rings.

Installation

Note: *On 2.4L engines, install special tool EN-43966 or use long threaded dowels to install the piston and rod assembly with the lower crankcase installed.*

25   Before installing the piston/connecting rod assemblies, the cylinder walls must be perfectly clean, the top edge of each cylinder bore must be chamfered, and the crankshaft must be in place.

26   Remove the cap from the end of the number one connecting rod (refer to the marks made during removal). Remove the original bearing inserts and wipe the bearing surfaces of the connecting rod and cap with a clean, lint-free cloth. They must be kept spotlessly clean.

Connecting rod bearing oil clearance check

Refer to illustrations 10.30, 10.35, 10.37 and 10.41

27   Clean the back side of the new upper bearing insert, then lay it in place in the con-

necting rod.

28   Make sure the tab on the bearing fits into the recess in the rod. Don't hammer the bearing insert into place and be very careful not to nick or gouge the bearing face. Don't lubricate the bearing at this time.

29   Clean the back side of the other bearing insert and install it in the rod cap. Again, make sure the tab on the bearing fits into the recess in the cap, and don't apply any lubricant. It's critically important that the mating surfaces of the bearing and connecting rod are perfectly clean and oil free when they're assembled.

30   Position the piston ring gaps at the specified intervals around the piston as shown **(see illustration)**.

31   Lubricate the piston and rings with clean engine oil and attach a piston ring compressor to the piston. Leave the skirt protruding about 1/4-inch to guide the piston into the cylinder. The rings must be compressed until they're flush with the piston.

32   Rotate the crankshaft until the number one connecting rod journal is at BDC (bottom dead center) and apply a liberal coat of engine oil to the cylinder walls.

33   With the arrow on top of the piston facing the front (timing chain) of the engine, gently insert the piston/connecting rod assembly into the number one cylinder bore and rest the bottom edge of the ring compressor on the engine block.

34   Tap the top edge of the ring compressor to make sure it's contacting the block around its entire circumference.

35   Gently tap on the top of the piston with the end of a wooden or plastic hammer handle **(see illustration)** while guiding the end of the connecting rod into place on the crankshaft journal. The piston rings may try to pop out of the ring compressor just before entering the cylinder bore, so keep some downward force on the ring compressor. Work slowly, and if any resistance is felt as the piston enters the cylinder, stop immediately. Find out what's hanging up and fix it before proceeding. Do not, for any reason, force the piston into the cylinder - you might break a ring and/or the piston.

36   Once the piston/connecting rod assembly is installed, the connecting rod bearing oil clearance must be checked before the rod cap is permanently installed.

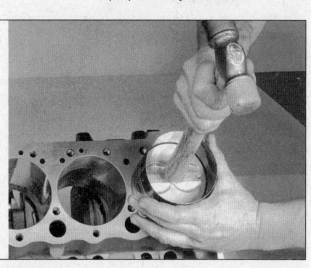

10.35 Use a plastic or wooden hammer handle to push the piston into the cylinder

ENGINE BEARING ANALYSIS

Debris

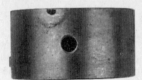

Babbitt bearing embedded with debris from machinings

Microscopic detail of debris

Microscopic detail of gouges

Overplated copper alloy bearing gouged by cast iron debris

Aluminum bearing embedded with glass beads

Microscopic detail of glass beads

Damaged lining caused by dirt left on the bearing back

Misassembly

Result of a lower half assembled as an upper - blocking the oil flow

Excessive oil clearance is indicated by a short contact arc

Polished and oil-stained backs are a result of a poor fit in the housing bore

Result of a wrong, reversed, or shifted cap

Overloading

Damage from excessive idling which resulted in an oil film unable to support the load imposed

Damaged upper connecting rod bearings caused by engine lugging; the lower main bearings (not shown) were similarly affected

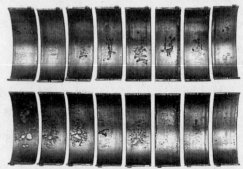

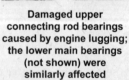

The damage shown in these upper and lower connecting rod bearings was caused by engine operation at a higher-than-rated speed under load

Misalignment

A poorly finished crankshaft caused the equally spaced scoring shown

A tapered housing bore caused the damage along one edge of this pair

A bent connecting rod led to the damage in the "V" pattern

A warped crankshaft caused this pattern of severe wear in the center, diminishing toward the ends

Lubrication

Result of dry start: The bearings on the left, farthest from the oil pump, show more damage

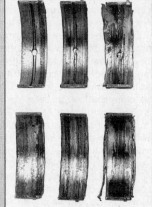

Result of a low oil supply or oil starvation

Severe wear as a result of inadequate oil clearance

Corrosion

Microscopic detail of corrosion

Corrosion is an acid attack on the bearing lining generally caused by inadequate maintenance, extremely hot or cold operation, or inferior oils or fuels

Microscopic detail of cavitation

Example of cavitation - a surface erosion caused by pressure changes in the oil film

Damage from excessive thrust or insufficient axial clearance

Bearing affected by oil dilution caused by excessive blow-by or a rich mixture

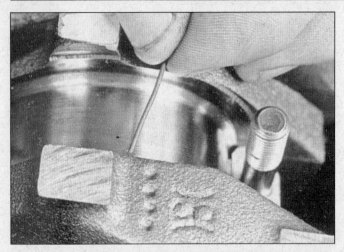

10.37 Place Plastigage on each connecting rod bearing journal parallel to the crankshaft centerline

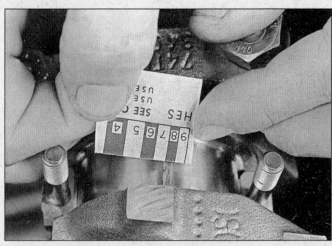

10.41 Use the scale on the Plastigage package to determine the bearing oil clearance - be sure to measure the widest part of the Plastigage and use the correct scale; it comes with both standard and metric scales

37   Cut a piece of the appropriate size Plastigage slightly shorter than the width of the connecting rod bearing and lay it in place on the number one connecting rod journal, parallel with the journal axis **(see illustration)**.

38   Clean the connecting rod cap bearing face and install the rod cap. Make sure the mating mark on the cap is on the same side as the mark on the connecting rod **(see illustration 10.4)**.

39   Install the old rod bolts at this time, and tighten them to the torque listed in this Chapter's Specifications. **Note:** *Use a thin-wall socket to avoid erroneous torque readings that can result if the socket is wedged between the rod cap and the bolt. If the socket tends to wedge itself between the fastener and the cap, lift up on it slightly until it no longer contacts the cap. DO NOT rotate the crankshaft at any time during this operation.*

40   Remove the fasteners and detach the rod cap, being very careful not to disturb the Plastigage. If you're working on a 3.6L engine, discard the cap bolts at this time as they cannot be reused. **Note:** *You MUST use new connecting rod bolts on 3.6L engines.*

41   Compare the width of the crushed Plastigage to the scale printed on the Plastigage envelope to obtain the oil clearance **(see illustration)**. The connecting rod oil clearance is usually about 0.001 to 0.002 inch. Consult an automotive machine shop for the clearance specified for the rod bearings on your engine.

42   If the clearance is not as specified, the bearing inserts may be the wrong size (which means different ones will be required). Before deciding that different inserts are needed, make sure that no dirt or oil was between the bearing inserts and the connecting rod or cap when the clearance was measured. Also, recheck the journal diameter. If the Plastigage was wider at one end than the other, the journal may be tapered. If the clearance still exceeds the limit specified, the bearing will have to be replaced with an undersize bearing. **Caution:** *When installing a new crankshaft always use a standard size bearing.*

Final installation

43   Carefully scrape all traces of the Plastigage material off the rod journal and/or bearing face. Be very careful not to scratch the bearing - use your fingernail or the edge of a plastic card.

44   Make sure the bearing faces are perfectly clean, then apply a uniform layer of clean moly-base grease or engine assembly lube to both of them. You'll have to push the piston into the cylinder to expose the face of the bearing insert in the connecting rod.

45   Slide the connecting rod back into place on the journal, install the rod cap, install the *new* bolts and tighten them to the torque listed in this Chapter's Specifications. **Caution:** *Install new connecting rod cap bolts on 3.6L engines. Do NOT reuse old bolts on these engines - they have stretched and cannot be reused. Also, don't apply any lubricant to the bolts.*

46   Repeat the entire procedure for the remaining pistons/connecting rods.

47   The important points to remember are:

a) *Keep the back sides of the bearing inserts and the insides of the connecting rods and caps perfectly clean when assembling them.*

b) *Make sure you have the correct piston/rod assembly for each cylinder.*

c) *The arrow on the piston must face the front (timing chain) of the engine.*

d) *Lubricate the cylinder walls liberally with clean oil.*

e) *Lubricate the bearing faces when installing the rod caps after the oil clearance has been checked.*

48   After all the piston/connecting rod assemblies have been correctly installed, rotate the crankshaft a number of times by hand to check for any obvious binding.

49   As a final step, check the connecting rod endplay as described in Step 3. If it was correct before disassembly and the original crankshaft and rods were reinstalled, it should still be

correct. If new rods or a new crankshaft were installed, the endplay may be inadequate. If so, the rods will have to be removed and taken to an automotive machine shop for resizing.

11   Crankshaft - removal and installation

Removal

Refer to illustrations 11.1 and 11.3

Note: *The crankshaft can be removed only after the engine has been removed from the vehicle. It's assumed that the driveplate, crankshaft pulley, timing chain, oil pan, oil pump body, oil filter and piston/connecting rod assemblies have already been removed. The rear main oil seal retainer must be unbolted and separated from the block before proceeding with crankshaft removal.*

Note: *On 3.4L engines, the crankshaft position sensor must be removed from the side of the block (see Chapter 6).*

1   Before the crankshaft is removed, measure the endplay. Mount a dial indicator with the indicator in line with the crankshaft and just touching the end of the crankshaft as shown **(see illustration)**.

2   Pry the crankshaft all the way to the rear and zero the dial indicator. Next, pry the crankshaft to the front as far as possible and check the reading on the dial indicator. The distance traveled is the endplay. A typical crankshaft endplay will fall between 0.003 to 0.010 inch (0.076 to 0.254 mm). If it is greater than that, check the crankshaft thrust surfaces for wear after it's removed. If no wear is evident, new main bearings should correct the endplay.

3   If a dial indicator isn't available, feeler gauges can be used. Gently pry the crankshaft all the way to the front of the engine. Slip feeler gauges between the crankshaft and the front face of the thrust bearing or washer to

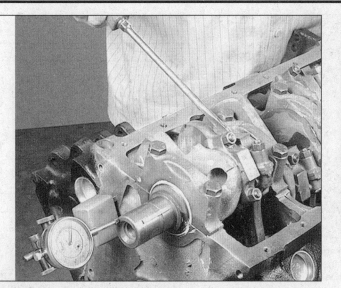

11.1 Checking crankshaft endplay with a dial indicator

11.3 Checking the crankshaft endplay with feeler gauges at the thrust bearing journal

determine the clearance **(see illustration)**.

4   On four-cylinder models, loosen the lower crankcase and perimeter bolts and the lower crankcase bolts 1/4-turn at a time each, until they can be removed by hand. Follow the reverse of the tightening sequence **(see illustrations 11.32 and 11.33)**.

Caution: *The lower crankcase bolts must be replaced with new ones upon installation. Save the old bolts, however, as they will be used for the main bearing oil clearance check.*

5   Remove the lower crankcase. Try not to drop the bearing inserts if they come out with the lower crankcase.

6   If you're working on a 3.4L engine, loosen the main bearing cap bolts 1/4-turn at a time each, until they can be removed by hand.

7   On 3.0L and 3.6L V6 models, remove the side bearing cap bolts followed by the outermost bearing cap bolts and finally the innermost cap bolts.

8   Remove the main bearing caps. Pull each main bearing cap straight up and off the cylinder block. Gently tap the main bearing cap with a soft-face hammer, if necessary. Try not to drop the bearing inserts if they come out with the caps.

9   Carefully lift the crankshaft out of the engine. It may be a good idea to have an assistant available, since the crankshaft is quite heavy and awkward to handle. With the bearing inserts in place inside the engine block and main bearing caps, reinstall the main bearing caps onto the engine block and tighten the bolts finger tight. Make sure the caps are in the exact order they were removed with the arrow pointing toward the front (timing chain and front cover) of the engine.

Installation

10   Crankshaft installation is the first step in engine reassembly. It's assumed at this point that the engine block and crankshaft have been cleaned, inspected and repaired or reconditioned.

11   Position the engine block with the bottom facing up.

12   Remove the bolts and lift off the main bearing caps.

13   If they're still in place, remove the original bearing inserts from the block and from the main bearing caps. Wipe the bearing surfaces of the block and main bearing caps with a clean, lint-free cloth. They must be kept spotlessly clean. This is critical for determining the correct bearing oil clearance.

Main bearing oil clearance check

Refer to illustrations 11.20 and 11.24

14   Without mixing them up, clean the back sides of the new upper main bearing inserts (with grooves and oil holes) and lay one in each main bearing saddle in the engine block. Each upper bearing (engine block) has an oil groove and oil hole in it. **Caution:** *The oil holes in the block must line up with the oil holes in the engine block inserts.* The thrust washer or thrust bearing insert must be installed in the correct location. **Caution:** *Do not hammer the bearing insert into place and don't nick or gouge the bearing faces. DO NOT apply any lubrication at this time.*

Note: *On four-cylinder models, the thrust*

bearing is located on the engine block number 2 journal.

Note: *Clean the back sides of the lower main bearing inserts and lay them in the corresponding location in the main bearing caps (V6) or the lower crankcase saddles (four-cylinder). Make sure the tab on the bearing insert fits into the recess in the block or main bearing caps.*

15   Clean the faces of the bearing inserts in the block and the crankshaft main bearing journals with a clean, lint-free cloth.

16   Check or clean the oil holes in the crankshaft, as any dirt here can go only one way - straight through the new bearings.

17   Once you're certain the crankshaft is clean, carefully lay it in position in the cylinder block.

18   Before the crankshaft can be permanently installed, the main bearing oil clearance must be checked.

19   Cut several strips of the appropriate size of Plastigage. They must be slightly shorter than the width of the main bearing journal.

20   Place one piece on each crankshaft main bearing journal, parallel with the journal axis as shown **(see illustration)**.

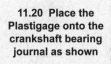

11.20 Place the Plastigage onto the crankshaft bearing journal as shown

11.24 Use the scale on the Plastigage package to determine the bearing oil clearnace - be sure to measure the widest part of the Plastigage and use the correct scale; it comes with both standard and metric scales

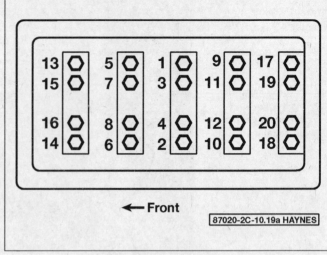

11.32 Lower crankcase bolt tightening sequence - 2.4L engines

21   Clean the faces of the bearing inserts in the main bearing caps or lower crankcase. Install the caps (V6 models) or the lower crankcase (four-cylinder models) without disturbing the Plastigage.

22   Apply clean engine oil to all bolt threads prior to installation, install all bolts finger-tight, then tighten them to the torque listed in this Chapter's Specifications. DO NOT rotate the crankshaft at any time during this operation.

23   Remove the bolts and carefully lift the main bearing caps straight up and off the block. Do not disturb the Plastigage or rotate the crankshaft.

24   Compare the width of the crushed Plastigage on each journal to the scale printed on the Plastigage envelope to determine the main bearing oil clearance **(see illustration)**. Check with an automotive machine shop for the crankshaft bearing oil clearance for your engine.

25   If the clearance is not as specified, the bearing inserts may be the wrong size (which means different ones will be required). Before

deciding if different inserts are needed, make sure that no dirt or oil was between the bearing inserts and the caps or block when the clearance was measured. If the Plastigage was wider at one end than the other, the crankshaft journal may be tapered. If the clearance still exceeds the limit specified, the bearing insert(s) will have to be replaced with an undersize bearing insert(s). **Caution:** *When installing a new crankshaft, always install a standard bearing insert set.*

26   Carefully scrape all traces of the Plastigage material off the main bearing journals and/or the bearing insert faces. Be sure to remove all residue from the oil holes. Use your fingernail or the edge of a plastic card - don't nick or scratch the bearing faces.

Final installation

Refer to illustrations 11.32, 11.33, 11.34 and 11.35

27   Carefully lift the crankshaft out of the cylinder block.

28   Clean the bearing insert faces in the cylinder block, then apply a thin, uniform layer of moly-base grease or engine assembly lube to each of the bearing surfaces. Be sure to coat the thrust faces as well as the journal face of the thrust bearing.

29   Make sure the crankshaft journals are clean, then lay the crankshaft back in place in the cylinder block.

30   Clean the bearing insert faces and apply the same lubricant to their faces only. Clean the engine block. Thoroughly clean the bearing cap mating surfaces (V6 engines) or the mating surface of the lower crankcase (four-cylinder engines). The surfaces must be free of oil residue.

31   Prior to installation, apply clean engine oil to the NEW bolt threads, wiping off any excess, then install all bolts finger-tight. **Caution:** *Install NEW main bearing cap bolts. Do NOT reuse old bolts on these engines - they have stretched and cannot be reused.*

32   On four-cylinder models, install the lower

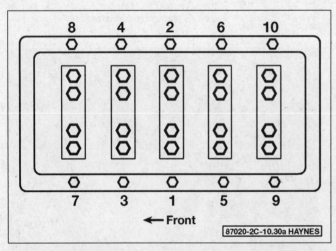

11.33 Lower crankcase perimeter bolt tightening sequence - 2.4L engines; all bolts must be replaced with new ones whenever they have been loosened

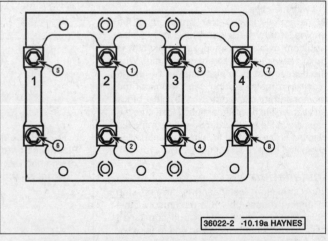

11.34 3.4L V6 crankshaft main bearing cap TIGHTENING sequence; all bolts must be replaced with new ones whenever they have been loosened

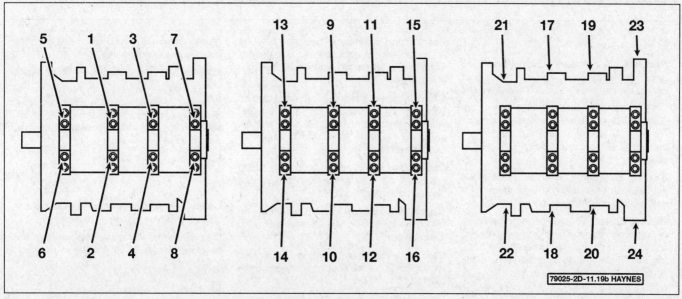

11.35 3.0L and 3.6L V6 crankshaft main bearing cap and side bolt TIGHTENING sequence; all bolts must be replaced with new ones whenever they have been loosened

crankcase and tighten the bolts in sequence **(see illustration)** to the torque listed in this Chapter's Specifications.

33   On four-cylinder models, install the lower crankcase perimeter bolts and tighten them in sequence **(see illustration)**, to the torque listed in this Chapter's Specifications.

34   On 3.4L models, tighten the new main bearing cap bolts in sequence **(see illustrations)** to the torque listed in this Chapter's Specifications.

35   On 3.0L and 3.6L models, install new side bolts and tighten them, in sequence **(see illustration)**, to the torque listed in this Chapter's Specifications.

Note: *The side bolts have a special seal. If they are not replaced with new ones, oil leaks can develop. The four corner bolts are longer than the four inner ones.*

36   Recheck the crankshaft endplay with a feeler gauge or a dial indicator. The endplay should be correct if the crankshaft thrust faces aren't worn or damaged and if new bearings have been installed.

37   Rotate the crankshaft a number of times by hand to check for any obvious binding. It should rotate with a running torque of 50 in-lbs or less. If the running torque is too high, correct the problem at this time.

38   Install the new rear main oil seal (see Chapter 2A, 2B or 2C).

---

12   Engine overhaul - reassembly sequence

1   Before beginning engine reassembly, make sure you have all the necessary new

parts, gaskets and seals as well as the following items on hand:

> *Common hand tools*
> *A 1/2-inch drive torque wrench*
> *New engine oil*
> *Gasket sealant*
> *Thread locking compound*

2   If you obtained a short block it will be necessary to install the cylinder head, the oil pump and pick-up tube, the oil pan, the water pump, the timing belt and timing cover, and the valve cover (see Chapter 2A or 2B). In order to save time and avoid problems, the external components must be installed in the following general order:

> *Thermostat and housing cover*
> *Water pump*
> *Intake and exhaust manifolds*
> *Fuel injection components*
> *Emission control components*
> *Spark plugs*
> *Ignition coils or coil pack and spark plug*
>   *wires*
> *Oil filter*
> *Engine mounts and mount brackets*
> *Driveplate*

---

13   Initial start-up and break-in after overhaul

Warning: *Have a fire extinguisher handy when starting the engine for the first time.*

1   Once the engine has been installed in the vehicle, double-check the engine oil and coolant levels.

2   With the spark plugs out of the engine

and the ignition system and fuel pump disabled, crank the engine until oil pressure registers on the gauge or the light goes out.

3   Install the spark plugs, hook up the plug wires and restore the ignition system and fuel pump functions.

4   Start the engine. It may take a few moments for the fuel system to build up pressure, but the engine should start without a great deal of effort.

5   After the engine starts, it should be allowed to warm up to normal operating temperature. While the engine is warming up, make a thorough check for fuel, oil and coolant leaks.

6   Shut the engine off and recheck the engine oil and coolant levels.

7   Drive the vehicle to an area with minimum traffic, accelerate from 30 to 50 mph, then allow the vehicle to slow to 30 mph with the throttle closed. Repeat the procedure 10 or 12 times. This will load the piston rings and cause them to seat properly against the cylinder walls. Check again for oil and coolant leaks.

8   Drive the vehicle gently for the first 500 miles (no sustained high speeds) and keep a constant check on the oil level. It is not unusual for an engine to use oil during the break-in period.

9   At approximately 500 to 600 miles, change the oil and filter.

10   For the next few hundred miles, drive the vehicle normally. Do not pamper it or abuse it.

11   After 2,000 miles, change the oil and filter again and consider the engine broken in.

COMMON ENGINE OVERHAUL TERMS

B

Backlash - The amount of play between two parts. Usually refers to how much one gear can be moved back and forth without moving the gear with which it's meshed.

Bearing Caps - The caps held in place by nuts or bolts which, in turn, hold the bearing surface. This space is for lubricating oil to enter.

Bearing clearance - The amount of space left between shaft and bearing surface. This space is for lubricating oil to enter.

Bearing crush - The additional height which is purposely manufactured into each bearing half to ensure complete contact of the bearing back with the housing bore when the engine is assembled.

Bearing knock - The noise created by movement of a part in a loose or worn bearing.

Blueprinting - Dismantling an engine and reassembling it to EXACT specifications.

Bore - An engine cylinder, or any cylindrical hole; also used to describe the process of enlarging or accurately refinishing a hole with a cutting tool, as to bore an engine cylinder. The bore size is the diameter of the hole.

Boring - Renewing the cylinders by cutting them out to a specified size. A boring bar is used to make the cut.

Bottom end - A term which refers collectively to the engine block, crankshaft, main bearings and the big ends of the connecting rods.

Break-in - The period of operation between installation of new or rebuilt parts and time in which parts are worn to the correct fit. Driving at reduced and varying speed for a specified mileage to permit parts to wear to the correct fit.

Bushing - A one-piece sleeve placed in a bore to serve as a bearing surface for shaft, piston pin, etc. Usually replaceable.

C

Camshaft - The shaft in the engine, on which a series of lobes are located for operating the valve mechanisms. The camshaft is driven by gears or sprockets and a timing chain. Usually referred to simply as the cam.

Carbon - Hard, or soft, black deposits found in combustion chamber, on plugs, under rings, on and under valve heads.

Cast iron - An alloy of iron and more than two percent carbon, used for engine blocks and heads because it's relatively inexpensive and easy to mold into complex shapes.

Chamfer - To bevel across (or a bevel on) the sharp edge of an object.

Chase - To repair damaged threads with a tap or die.

Combustion chamber - The space between the piston and the cylinder head, with the piston at top dead center, in which air-fuel mixture is burned.

Compression ratio - The relationship between cylinder volume (clearance volume) when the piston is at top dead center and cylinder volume when the piston is at bottom dead center.

Connecting rod - The rod that connects the crank on the crankshaft with the piston. Sometimes called a con rod.

Connecting rod cap - The part of the connecting rod assembly that attaches the rod to the crankpin.

Core plug - Soft metal plug used to plug the casting holes for the coolant passages in the block.

Crankcase - The lower part of the engine in which the crankshaft rotates; includes the lower section of the cylinder block and the oil pan.

Crank kit - A reground or reconditioned crankshaft and new main and connecting rod bearings.

Crankpin - The part of a crankshaft to which a connecting rod is attached.

Crankshaft - The main rotating member, or shaft, running the length of the crankcase, with offset throws to which the connecting rods are attached; changes the reciprocating motion of the pistons into rotating motion.

Cylinder sleeve - A replaceable sleeve, or liner, pressed into the cylinder block to form the cylinder bore.

D

Deburring - Removing the burrs (rough edges or areas) from a bearing.

Deglazer - A tool, rotated by an electric motor, used to remove glaze from cylinder walls so a new set of rings will seat.

E

Endplay - The amount of lengthwise movement between two parts. As applied to a crankshaft, the distance that the crankshaft can move forward and back in the cylinder block.

F

Face - A machinist's term that refers to removing metal from the end of a shaft or the face of a larger part, such as a flywheel.

Fatigue - A breakdown of material through a large number of loading and unloading cycles. The first signs are cracks followed shortly by breaks.

Feeler gauge - A thin strip of hardened steel, ground to an exact thickness, used to check clearances between parts.

Free height - The unloaded length or height of a spring.

Freeplay - The looseness in a linkage, or an assembly of parts, between the initial application of force and actual movement. Usually perceived as slop or slight delay.

Freeze plug - See Core plug.

G

Gallery - A large passage in the block that forms a reservoir for engine oil pressure.

Glaze - The very smooth, glassy finish that develops on cylinder walls while an engine is in service.

H

Heli-Coil - A rethreading device used when threads are worn or damaged. The device is installed in a retapped hole to reduce the thread size to the original size.

I

Installed height - The spring's measured length or height, as installed on the cylinder head. Installed height is measured from the spring seat to the underside of the spring retainer.

J

Journal - The surface of a rotating shaft which turns in a bearing.

K

Keeper - The split lock that holds the valve spring retainer in position on the valve stem.

Key - A small piece of metal inserted into matching grooves machined into two parts fitted together - such as a gear pressed onto a shaft - which prevents slippage between the two parts.

Knock - The heavy metallic engine sound, produced in the combustion chamber as a result of abnormal combustion - usually detonation. Knock is usually caused by a loose or worn bearing. Also referred to as detonation, pinging and spark knock. Connecting rod or main bearing knocks are created by too much oil clearance or insufficient lubrication.

L

Lands - The portions of metal between the piston ring grooves.

Lapping the valves - Grinding a valve face and its seat together with lapping compound.

Lash - The amount of free motion in a gear train, between gears, or in a mechanical assembly, that occurs before movement can begin. Usually refers to the lash in a valve train.

Lifter - The part that rides against the cam to transfer motion to the rest of the valve train.

M

Machining - The process of using a machine to remove metal from a metal part.

Main bearings - The plain, or babbit, bearings that support the crankshaft.

Main bearing caps - The cast iron caps, bolted to the bottom of the block, that support the main bearings.

O

O.D. - Outside diameter.

Oil gallery - A pipe or drilled passageway in the engine used to carry engine oil from one area to another.

Oil ring - The lower ring, or rings, of a piston; designed to prevent excessive amounts of oil from working up the cylinder walls and into the combustion chamber. Also called an oil-control ring.

Oil seal - A seal which keeps oil from leaking out of a compartment. Usually refers to a dynamic seal around a rotating shaft or other moving part.

O-ring - A type of sealing ring made of a special rubberlike material; in use, the O-ring is compressed into a groove to provide the sealing action.

Overhaul - To completely disassemble a unit, clean and inspect all parts, reassemble it with the original or new parts and make all adjustments necessary for proper operation.

P

Pilot bearing - A small bearing installed in the center of the flywheel (or the rear end of the crankshaft) to support the front end of the input shaft of the transmission.

Pip mark - A little dot or indentation which indicates the top side of a compression ring.

Piston - The cylindrical part, attached to the connecting rod, that moves up and down in the cylinder as the crankshaft rotates. When the fuel charge is fired, the piston transfers the force of the explosion to the connecting rod, then to the crankshaft.

Piston pin (or wrist pin) - The cylindrical and usually hollow steel pin that passes through the piston. The piston pin fastens the piston to the upper end of the connecting rod.

Piston ring - The split ring fitted to the groove in a piston. The ring contacts the sides of the ring groove and also rubs against the cylinder wall, thus sealing space between piston and wall. There are two types of rings: Compression rings seal the compression pressure in the combustion chamber; oil rings scrape excessive oil off the cylinder wall.

Piston ring groove - The slots or grooves cut in piston heads to hold piston rings in position.

Piston skirt - The portion of the piston below the rings and the piston pin hole.

Plastigage - A thin strip of plastic thread, available in different sizes, used for measuring clearances. For example, a strip of plastigage is laid across a bearing journal and mashed as parts are assembled. Then parts are disassembled and the width of the strip is measured to determine clearance between journal and bearing. Commonly used to measure crankshaft main-bearing and connecting rod bearing clearances.

Press-fit - A tight fit between two parts that requires pressure to force the parts together. Also referred to as drive, or force, fit.

Prussian blue - A blue pigment; in solution, useful in determining the area of contact between two surfaces. Prussian blue is commonly used to determine the width and location of the contact area between the valve face and the valve seat.

R

Race (bearing) - The inner or outer ring that provides a contact surface for balls or rollers in bearing.

Ream - To size, enlarge or smooth a hole by using a round cutting tool with fluted edges.

Ring job - The process of reconditioning the cylinders and installing new rings.

Runout - Wobble. The amount a shaft rotates out-of-true.

S

Saddle - The upper main bearing seat.

Scored - Scratched or grooved, as a cylinder wall may be scored by abrasive particles moved up and down by the piston rings.

Scuffing - A type of wear in which there's a transfer of material between parts moving against each other; shows up as pits or grooves in the mating surfaces.

Seat - The surface upon which another part rests or seats. For example, the valve seat is the matched surface upon which the valve face rests. Also used to refer to wearing into a good fit; for example, piston rings seat after a few miles of driving.

Short block - An engine block complete with crankshaft and piston and, usually, camshaft assemblies.

Static balance - The balance of an object while it's stationary.

Step - The wear on the lower portion of a ring land caused by excessive side and back-clearance. The height of the step indicates the ring's extra side clearance and the length of the step projecting from the back wall of the groove represents the ring's back clearance.

Stroke - The distance the piston moves when traveling from top dead center to bottom dead center, or from bottom dead center to top dead center.

Stud - A metal rod with threads on both ends.

T

Tang - A lip on the end of a plain bearing used to align the bearing during assembly.

Tap - To cut threads in a hole. Also refers to the fluted tool used to cut threads.

Taper - A gradual reduction in the width of a shaft or hole; in an engine cylinder, taper usually takes the form of uneven wear, more pronounced at the top than at the bottom.

Throws - The offset portions of the crankshaft to which the connecting rods are affixed.

Thrust bearing - The main bearing that has thrust faces to prevent excessive endplay, or forward and backward movement of the crankshaft.

Thrust washer - A bronze or hardened steel washer placed between two moving parts. The washer prevents longitudinal movement and provides a bearing surface for thrust surfaces of parts.

Tolerance - The amount of variation permitted from an exact size of measurement. Actual amount from smallest acceptable dimension to largest acceptable dimension.

U

Umbrella - An oil deflector placed near the valve tip to throw oil from the valve stem area.

Undercut - A machined groove below the normal surface.

Undersize bearings - Smaller diameter bearings used with re-ground crankshaft journals.

V

Valve grinding - Refacing a valve in a valve-refacing machine.

Valve train - The valve-operating mechanism of an engine; includes all components from the camshaft to the valve.

Vibration damper - A cylindrical weight attached to the front of the crankshaft to minimize torsional vibration (the twist-untwist actions of the crankshaft caused by the cylinder firing impulses). Also called a harmonic balancer.

W

Water jacket - The spaces around the cylinders, between the inner and outer shells of the cylinder block or head, through which coolant circulates.

Web - A supporting structure across a cavity.

Woodruff key - A key with a radiused backside (viewed from the side).

Notes

Chapter 3
Cooling, heating and air conditioning systems

Contents

Specifications

General

Radiator cap pressure rating	Refer to pressure specification on cap
Thermostat rating (opening to fully open temperature range).................	188 to 206-degrees F (87 to 97-degrees C)
Cooling system capacity..	See Chapter 1
Refrigerant type..	R-134a
Refrigerant capacity..	Refer to HVAC specification tag

Torque specifications

Note: *One foot-pound (ft-lb) of torque is equivalent to 12 inch-pounds (in-lbs) of torque. Torque values below approximately 15 foot-pounds are expressed in inch-pounds, because most foot-pound torque wrenches are not accurate at these smaller values.*

	Ft-lbs (unless otherwise indicated)	Nm
Thermostat water outlet housing bolts		
2.4L engines..	88 in-lbs	10
3.0L and 3.6L engines..	89 in-lbs	10
3.4L engines..	18	
Air conditioning condenser refrigerant line nut(s)		
2009 and earlier models..	15	
2010 and later models...	16	
Water pump mounting bolts		
2.4L engines..	18	
3.0L and 3.6L engines*		
Step 1	89 in-lbs	10
Step 2	89 in-lbs	10
Step 3	Tighten an additional 45-degrees	
3.4L engines..	18	
Water pump pulley bolts*		
3.0L and 3.6L engines..	89 in-lbs	
3.4L engines..	18	
Water pump cover bolts (2.4L engines).............................	18	
Water pump sprocket bolts (2.4L engines)	89 in-lbs	
Air conditioning compressor refrigerant line nut		
(2009 and earlier models)	15	
Air conditioning compressor bolt/nut		
2009 and earlier models..	37	
2010 and later models		
2.4L engines	16	
V6 engines	43	
Refrigerant pressure sensor	35 in-lbs	
Thermal Expansion Valve (TXV) mounting fastener...............	41 in-lbs	

* *Use new bolts*

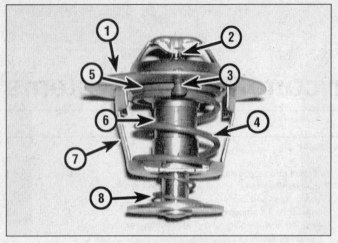

1.4 Typical thermostat

1	Flange	5	Valve seat
2	Piston	6	Valve
3	Jiggle pin	7	Frame
4	Main coil spring	8	Secondary coil spring

2.5 Use a hydrometer (available at most auto parts stores) to test the condition of your coolant

1   General information

Engine cooling system

Refer to illustration 1.4

All vehicles covered by this manual employ a pressurized engine cooling system with thermostatically controlled coolant circulation. The cooling system consists of a radiator, an expansion tank, a pressure cap (located on the expansion tank), a thermostat, one or two electric cooling fans and an impeller-type water pump.

The expansion tank functions somewhat differently than a conventional recovery tank. Designed to separate any trapped air in the coolant, it is an integral part of the cooling system and has a pressure cap on top. The radiator on these models does not have a pressure cap. **Warning:** *Unlike a conventional coolant recovery tank, the pressure cap on the expansion tank should never be opened after the engine has warmed up, because of the danger of severe burns caused by steam or scalding coolant.*

The water pump mounted on the engine block pumps coolant through the engine. The coolant flows around each cylinder and toward the rear of the engine. Cast-in coolant passages direct coolant around the intake and exhaust ports, near the spark plug areas and in close proximity to the exhaust valve guides.

A wax-pellet type thermostat controls engine coolant temperature **(see illustration)**. During warm up, the closed thermostat prevents coolant from circulating through the radiator. As the engine nears normal operating temperature, the thermostat opens and allows hot coolant to travel through the radiator, where it's cooled before returning to the engine.

The cooling system is sealed by a pressure-type cap on the coolant expansion tank, which raises the boiling point of the coolant and increases the cooling efficiency of the radiator. If the system pressure exceeds the cap pressure relief value, the excess pressure in the system forces the spring-loaded valve inside the cap off its seat and allows the pressure to escape through the overflow tube.

Engine cooling fans

These models are equipped with electric cooling fans. The fans are controlled by relays and the main computer for the engine. The relays are located in the underhood fuse box. The computer uses information from various sensors and the air conditioning system to control the fans.

Heating system

A typical heating system consists of a blower fan and heater core located in a housing under the dash, the hoses connecting the heater core to the engine cooling system and the heater/air conditioning control head on the dashboard. Hot engine coolant is circulated through the heater core. When the heater mode is activated, a flap door in the housing opens to expose the heater core to the passenger compartment through air ducts. A fan switch on the control head activates the blower motor, which forces air through the core, heating the air.

Air conditioning system

The air conditioning system consists of a condenser mounted in front of the radiator, an evaporator mounted adjacent to the heater core, a compressor mounted on the engine, a receiver-drier located near the condenser and the plumbing connecting all of the above components.

A blower fan forces the warmer air of the passenger compartment through the evaporator core (sort of a radiator-in-reverse), transferring the heat from the air to the refrigerant. The liquid refrigerant boils off into low pressure vapor, taking the heat with it when it leaves the evaporator.

2   Antifreeze - general information

Refer to illustration 2.5

Warning: *Do not allow antifreeze to come in contact with your skin or painted surfaces of the vehicle. Rinse off spills immediately with plenty of water. Antifreeze is highly toxic if ingested. Never leave antifreeze lying around in an open container or in puddles on the floor; children and pets are attracted by its sweet smell and may drink it. Check with local authorities about disposing of used antifreeze. Many communities have collection centers which will see that antifreeze is disposed of safely. Never dump used antifreeze on the ground or pour it into drains.*
Caution: *Do not mix coolants of different colors. Doing so might damage the cooling system and/or the engine. Read the warning label in the engine compartment for additional information.*
Note: *Non-toxic antifreeze is now manufactured and available at local auto parts stores, but even this type must be disposed of properly.*

The cooling system should be filled with a water/ethylene glycol based antifreeze solution, which will prevent freezing down to at least -20-degrees F (even lower in cold climates). It also provides protection against corrosion and increases the coolant boiling point. The engines in these vehicles have aluminum heads, and the 3.6L V6 engine also has an

3.7 Pull the engine cover up to remove it

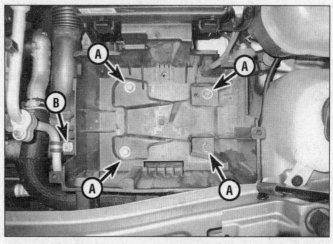

3.8 Battery tray mounting fasteners (A) and the air conditioning line bracket (B)

aluminum cylinder block. The manufacturer recommends that the correct type of coolant be used and strongly urges that coolant types not be mixed (see the Chapter 1 Specifications).

Drain, flush and refill the cooling system at the intervals listed in the maintenance schedule (see Chapter 1).

Before adding antifreeze to the system, inspect all hose connections. Antifreeze can leak through very minute openings.

The exact mixture of antifreeze to water which you should use depends on the relative weather conditions. The mixture should contain at least 50-percent antifreeze, but should never contain more than 70-percent antifreeze. Consult the mixture ratio chart on the container before adding coolant. **Note:** *Premixed antifreeze, with water already added, is now commonly available at most auto parts stores. The percentage is usually a 50/50 mix of antifreeze to water. Be sure to read the label on the antifreeze container closely to be certain that you are filling the cooling system with the proper mix of antifreeze and water.*

Hydrometers are available at most auto parts stores to test the coolant **(see illustration)**. Use antifreeze that meets factory specifications for engines with aluminum components (see Chapter 1).

3 Thermostat - check and replacement

Check

1   Before assuming the thermostat is to blame for a cooling system problem, check the coolant level, drivebelt tension (see Chapter 1) and temperature gauge operation.
2   If the engine seems to be taking a long time to warm up, based on heater output or temperature gauge operation, the thermostat is probably stuck open. Replace the thermostat with a new one.
3   If the engine runs hot, use your hand to check the temperature of the lower radiator hose. If the hose isn't hot, but the engine is, the thermostat is probably stuck closed, preventing the coolant inside the engine from escaping to

the radiator. Replace the thermostat. **Caution:** *Don't drive the vehicle without a thermostat. The computer will stay in open loop and emissions and fuel economy will suffer.*
4   If the lower radiator hose is hot, it means that the coolant is flowing and the thermostat is open. Consult the *Troubleshooting* Section at the front of this manual for cooling system diagnosis.

Replacement

Warning: *The engine must be completely cool before beginning this procedure.*
5   Disconnect the cable from the negative battery terminal (see Chapter 5).
6   Drain the cooling system (see Chapter 1). If the coolant is relatively new or in good condition (see Chapter 1), save it and reuse it. Read the **Warning** in Section 2.

3.4L V6 engine

Refer to illustrations 3.7, 3.8, 3.9 and 3.13

7   Remove the top engine cover **(see illustration)**. On 2007 and later models, remove the oil filler cap and tube, disconnect the EGR electrical connector, then pull the engine cover up to remove it.
8   Remove the battery (see Chapter 5) and the battery tray:
   a) *Detach the air conditioning line from the battery box* **(see illustration)**.
   b) *Remove the battery tray mounting bolts.*
   c) *Lift the tray up slightly and detach the cooling air duct.*
   d) *Remove the battery tray.*
9   Remove the exhaust crossover pipe from in front of the thermostat housing **(see illustration)**. **Note:** *Use penetrating oil on the exhaust pipe fasteners and allow it to soak a few minutes before removing them.*
10   Unscrew the thermostat housing bolts, then detach it.
11   Remove the thermostat and seal.
12   Clean any old sealant from the thermostat housing bolts and make sure that the housing mating surfaces are clean.

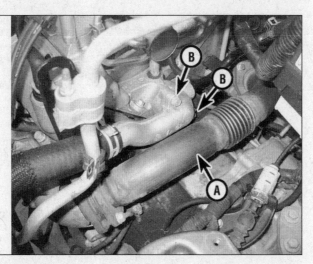

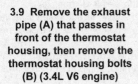

3.9 Remove the exhaust pipe (A) that passes in front of the thermostat housing, then remove the thermostat housing bolts (B) (3.4L V6 engine)

3.13 Install a new rubber gasket onto the thermostat

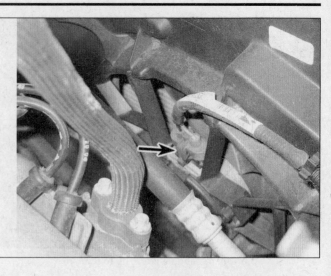

4.1 To test either fan motor, disconnect the electrical connector and use jumper wires to connect the fan directly to the battery and ground - if the fan still doesn't work, replace the motor

13   Install a new thermostat and seal **(see illustration)**. Place the housing into position and use a small amount of sealant on the thermostat housing bolts and install them finger tight, then tighten the bolts to the torque listed in this Chapter's Specifications.

14   The remainder of installation is the reverse of removal. Be sure that the exhaust crossover pipe is 1/4-inch away from the thermostat housing when installed. Tighten the exhaust pipe mounting nuts to the torque listed in this Chapter's Specifications. Proceed to Step 21.

3.0L and 3.6L V6 engine

15   The thermostat housing is mounted to the engine on the opposite end from the drivebelt.

16   Remove the heater inlet and outlet pipes by removing the pipe mounting flange from the thermostat housing.

17   Unscrew the thermostat housing bolts, then detach the housing. Note how the thermostat is positioned in the housing.

18   Clean the housing and engine mating surfaces.

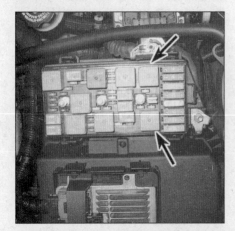

4.3 Release the tabs and remove the cover from the underhood fuse and relay box to access the cooling fan relays (check the underside of the cover for the relay locations on your model)

19   Install a new thermostat and seal **(see illustration 3.13)**. Install the thermostat housing with the bolts finger tight at first, then tighten the bolts to the torque listed in this Chapter's Specifications.

20   The remainder of installation is the reverse of removal.

2.4L engine

21   Disconnect the radiator outlet hose and the coolant expansion tank outlet hose, from the thermostat housing.

22   Unscrew the bolts and lift off the thermostat housing cover.

23   Remove the thermostat from the housing.

24   Install a new rubber seal onto the replacement thermostat. Place the thermostat into the housing, spring end first.

25   Install the thermostat housing cover and mounting bolts, and then tighten the bolts to the torque listed in this Chapter's Specifications.

All engines

26   Refill the cooling system (see Chapter 1).

27   Start the engine and allow it to reach normal operating temperature, then check for leaks and proper thermostat operation (as described in Steps 3 and 4).

---

4   Engine cooling fans - check and replacement

Warning: *To avoid possible injury or damage, DO NOT operate the engine with a damaged fan. Do not attempt to repair fan blades - replace a damaged fan with a new one.*

Check

Refer to illustrations 4.1 and 4.3

1   If the engine is getting hot (or overheating) and neither of the cooling fans are coming on, check their fuses first. If the fuses are okay, unplug the electrical connector for each fan motor and apply battery voltage to the terminals of each motor **(see illustration)**. Use a fused jumper wire on one terminal and another

jumper wire connected to a good ground on the other terminal. If either fan motor doesn't come on, replace it. **Caution:** *Do not apply battery power to the harness side of the connector.*

2   If the fan motors are okay, check the fan relay(s).

3   Locate the fan relays in the engine compartment fuse/relay box **(see illustration)**.

4   Test the relays (see Chapter 12).

5   If the fuses, motors and relays are functional, check all wiring and connections to the fan motors. If no obvious problems are found, have the cooling fan system diagnosed by a dealer service department or other repair shop with the proper diagnostic equipment.

Replacement

Refer to illustrations 4.9a, 4.9b, 4.10, 4.11, 4.14a and 4.14b

Warning: *Wait until the engine is completely cool before beginning this procedure.*

6   On all models except 2.4L models, refer to Chapter 11 and remove the front bumper cover. On Torrent models, it may be necessary to remove the hood latch also (see Chapter 11).

7   Drain the engine coolant (see Chapter 1).

8   Disconnect the fan electrical connectors and detach any wiring harnesses attached to the fan shroud.

9   On all models except 2.4L models, remove the front bumper impact absorber **(see illustration)**. If you're working on a 2008 or later model, detach the electrical connector from the bumper support and remove the support **(see illustration)**. On 2007 and earlier models, the bumper support can remain installed.

10   Remove the small air duct for the battery box and the additional shrouding from the front of the air conditioning condenser **(see illustration)**.

11   Remove the condenser/radiator/fan assembly upper mounting brackets **(see illustration)**.

12   Disconnect the radiator hoses from the radiator.

13   Disconnect the transaxle cooler lines from the radiator (see Chapter 7) and detach them from the fan assembly as necessary.

14   Unbolt the fan assembly from the radia-

4.9a Front bumper impact absorber fasteners

4.9b The front bumper support mounting fasteners (A) and electrical connector (B)

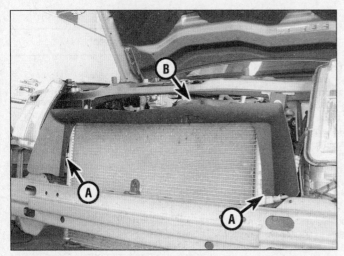

4.10 Plastic shroud mounting fasteners (A) and the air duct for the battery box (B)

4.11 Mounting brackets for the condenser/radiator/fan assembly

tor **(see illustration)**. On 2008 and later models, move the condenser/radiator/fan assembly slightly up then forward on the right side, pull the fan shroud assembly up slightly to separate it from the condenser and radiator, then remove it out the side. On 2007 and earlier models, separate the fan shroud assembly, then remove it out the top. Note the rubber mounts on the bottom of the radiator and inspect them for damage. Replace them if necessary **(see illustration)**. **Note:** *Support the radiator and air conditioning condenser to avoid any damage and stress on the refrigerant lines to the condenser.*

15   To detach the fan blade from the motor, remove the motor shaft clip or bolt, depending on design. **Note:** *On clip-type models only, mark the relationship of the fan to the fan motor shaft if the fan is going to be reinstalled. On nut retained fans, the fan cannot be reused. Heat the new fan's hub with HOT tap water (120-degrees for 1 minute) before installation or the fan could crack and cause damage. Also, a special nut may be used to retain the fan.*

16   To detach the fan motor from the shroud, remove the mounting screws. **Note:** *On some designs, the fan motor is riveted to the shroud and requires drilling the rivets out to remove the*

fan motor. When working with this type, tape the openings of the fan motors to protect them from debris, if necessary. Use locking nuts and bolts to attach the motor to the fan shroud.

4.14a Fan shroud assembly right-side mounting bolt (left side similar)

4.14b Rubber mount at the bottom of the radiator

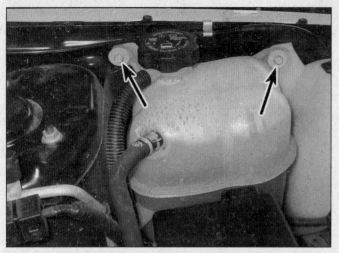

5.3 Coolant expansion tank mounting bolts

6.2 Location of the condenser-to-radiator mounting clip and tab on one side (2005 model shown - other models similar)

17　Installation is the reverse of removal. Fill the cooling system with the proper mixture of antifreeze and water (see Chapter 1) and check for leaks after the vehicle has reached normal operating temperature. Check the transaxle fluid and add more as needed (see Chapter 1).

5　Coolant expansion tank - removal and installation

Refer to illustration 5.3

Warning: *Wait until the engine is completely cool before beginning this procedure.*

1　Carefully remove the pressure cap after the engine has completely cooled. Remove as much coolant as possible from the tank using a suction device. Alternatively, you can drain about half of the coolant from the system (see Chapter 1) before disconnecting the hoses. The coolant may be saved and reused if it's in good condition.

2　Disconnect the upper hose from the tank and immediately plug it to prevent leakage. Check the hose for cracks or hardness and replace it if necessary.

3　Remove the tank mounting bolts **(see illustration)**.

4　Lift the tank up, disconnect the hose from the bottom of the tank and plug it immediately.

5　Clean the inside of the tank with soapy water and a brush to remove any deposits. Inspect the tank carefully. If you find any damage, replace it.

6　Installation is the reverse of removal. Refill the cooling system (see Chapter 1).

6　Radiator - removal and installation

Warning: *Wait until the engine is completely cool before beginning this procedure.*

Removal

Refer to illustration 6.2

1　Follow Steps 6 through 14 in Section 4.

2　On 2007 and earlier models, unclip the condenser **(see illustration)**. On 2009 and earlier models, remove the air conditioning condenser-to-radiator mounting bolts near the top of the radiator and on 2010 and later models, the condenser must be completely removed (see Section 16).

3　Carefully lift the condenser to release the lower retaining tabs and separate it from the radiator. Be careful not to damage the fins on either unit.

4　Inspect the radiator for leaks and damage. If it needs repair, have a radiator shop or dealer service department perform the work, as special techniques are required.

5　Bugs and dirt can be removed from the radiator by spraying it with a garden hose nozzle from the back side. The radiator should be flushed out with a garden hose before installation.

6　Check the rubber mounts on the bottom of the radiator for wear or deterioration and replace them if necessary.

Installation

7　Installation is the reverse of the removal procedure. Carefully guide the condenser/radiator/fan assembly into position and make certain that the rubber mounts on the bottom are properly seated into the support.

8　Tighten the radiator and air conditioning condenser bracket bolts securely.

9　After installation, fill the cooling system with the proper mixture of antifreeze and water (see Chapter 1).

10　Start the engine and check for leaks. Allow the engine to reach normal operating temperature, then recheck the coolant level and add more if required.

11　Check the transaxle fluid and add more as needed (see Chapter 1).

7　Water pump - check

Refer to illustration 7.3

1　A failure in the water pump can cause serious engine damage due to overheating.

2　The water pump on these models is driven by the drivebelt.

3　Water pumps are generally equipped with weep (or vent) holes **(see illustration)**. If a failure occurs in the pump seal, coolant will leak from the hole. Use a flashlight and small mirror to find the hole on the water pump to check for leaks.

4　If the water pump shaft bearings fail, there may be a howling sound at the pump while it's running. Shaft wear can be felt with the drivebelt removed if the water pump pulley is rocked up and down. Don't mistake drivebelt slippage, which causes a squealing sound, for water pump bearing failure.

5　Even a pump that exhibits no outward signs of a problem, such as noise or leakage, can still be due for replacement. Removal for close examination is the only sure way to tell.

6　A quick water pump performance check is to turn the heater on. If the pump is failing, it might not be able to efficiently circulate hot water all the way to the heater core as it should.

8　Water pump - replacement

Refer to illustration 8.3

Warning: *Wait until the engine is completely cool before beginning this procedure.*

1　Disconnect the cable from the negative battery terminal (see Chapter 5).

2　Drain the cooling system (see Chapter 1). If the coolant is relatively new or in good condition, save it and reuse it. Read the **Warning** in Section 2.

7.3 Typical weep hole on the underside of a water pump

8.3 Water pump pulley bolts (3.4L engine model shown)

3.0L, 3.4L and 3.6L engines

3   On 3.4L engine models, loosen, but do not remove, the water pump pulley bolts **(see illustration)**.

4   Remove the drivebelt (see Chapter 1).

5   On 3.0L and 3.6L engine models, hold the water pump pulley with a strap wrench, or equivalent, then remove the pulley mounting bolts and pulley.

6   On 3.4L engine models, remove the pulley mounting bolts and the pulley from the water pump flange.

7   Remove the water pump mounting bolts.

8   Carefully remove the pump from the engine. If the water pump is stuck, gently tap it with a soft-faced hammer to break the seal.

9   Thoroughly clean the mating surface on the engine. Do the same on the water pump if its going to be reinstalled.

10   Compare the replacement pump with the old one to make sure that they're identical.

11   Place the gasket/seal and water pump into position. Install the mounting bolts until they are all finger tight.

12   With everything correctly in place, tighten the water pump mounting bolts to the torque listed in this Chapter's Specifications.

13   The remainder of installation is the reverse of removal. Tighten the water pump pulley bolts to the torque listed in this Chapter's Specifications. Refill the cooling system with the proper concentration of antifreeze (see Chapter 1). Start the engine and allow it to reach normal operating temperature while inspecting the system for leaks.

2.4L engines

14   Remove the air filter housing (see Chapter 4).

15   Remove the intake manifold (see Chapter 2A).

16   Remove the exhaust manifold heat shield (see Chapter 2A).

17   Remove the coolant heater (if equipped).

18   Loosen the right front wheel lug nuts. Raise the vehicle and support it securely on jackstands. Remove the wheel and the inner fender splash shield (see Chapter 11).

19   Remove the catalytic converter (see Chapter 6).

20   Remove the thermostat housing and the water pipe by twisting it away from the water pump to separate them. Then, separate the water pipe from the thermostat housing and discard the O-ring seals for the pipe and housing (see Section 3).

21   Remove the cover over the water pump sprocket that's located left of the drivebelt tensioner **(see illustration)**.

22   Remove the drain plug at the bottom of the water pump and drain excess coolant from the pump into a container. Reinstall the plug after draining.

23   Install a special holding tool onto the water pump sprocket (tool J-43651), available from specialty tool manufacturers and some dealer service departments. A tool can be fabricated if necessary. Proper installation of either tool should result in absolutely no sprocket movement.

Note: *The special tool is designed to lock the sprocket into position, allowing the balance shaft chain to remain in its timed state while the water pump is being replaced. If you're using a homemade tool like the one shown in the illustration, remove one of the water pump-to-sprocket bolts before installing the tool.*

24   Remove all water pump-to-sprocket bolts.

25   Remove the two bolts attaching the water pump to the front and two bolts attaching it to the rear of the engine block and remove the pump from the engine. If the water pump is stuck, gently tap it with a soft-faced hammer to break the seal.

26   Clean the bolt threads and the threaded holes and remove any corrosion or sealant. Remove all traces of old gasket material from the sealing surfaces. Remove the sealing ring from the water pump (if the same pump is to be installed).

27   Install a new sealing ring into the groove in the pump. To install the water pump, place a threaded stud loosely into the pump flange to serve as a guide pin; this will help to align the water pump sprocket with the water pump flange as the pump is installed.

28   Install the water pump mounting bolts and tighten them loosely. Install two bolts into the water pump sprocket and tighten them loosely. Remove the threaded stud (guide pin) and install the third water pump sprocket bolt.

29   Tighten the water pump mounting bolts (two in the front and two in the rear of the engine block) to the torque listed in this Chapter's Specifications.

30   Tighten the water pump sprocket bolts to the torque listed in this Chapter's Specifications.

31   Install the water pump access cover and tighten the bolts to the torque listed in this Chapter's Specifications.

32   Install the thermostat housing and water pipe and tighten the bolts to the torque listed in this Chapter's Specifications. Lubricate the seal lightly with silicon gel before installing the water pipe into the water pump and the thermostat housing

33   The remainder of installation is the reverse of removal. Refill the cooling system with the proper concentration of antifreeze (see Chapter 1). Start the engine and allow it to reach normal operating temperature while inspecting the system for leaks.

9   Coolant temperature indicator - check

Warning: *Wait until the engine is completely cool before beginning this procedure.*

1   The coolant temperature indicator system consists of the temperature gauge (on the dash), a sensor mounted on the engine and the vehicle's main computer. The Engine Coolant Temperature (ECT) sensor provides a signal to the Powertrain Control Module (PCM - the vehicle's computer) (see Chapter 6). The PCM controls the temperature gauge.

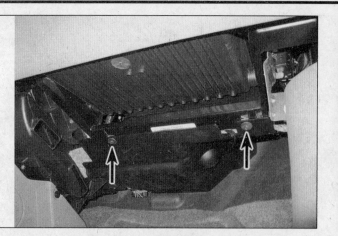

10.1 Remove the fasteners for the insulator panel and detach it from the stud on the right side of the panel, then remove it

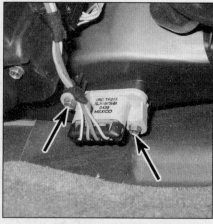

10.3 Mounting screws for the blower motor resistor/module

2   If the temperature gauge goes above normal and begins to read hot, check the coolant level in the system (see Chapter 1). Also, check that the coolant mixture is correct (see Section 2). Finally, refer to the *Troubleshooting* Section at the beginning of this book before assuming that the temperature indicator is faulty.

3   Start the engine and warm it up for 10 minutes. If the temperature gauge has not moved from the C position, check the wiring harness connections going to the instrument cluster.

4   If there is a problem with the ECT sensor, it is very likely that the CHECK ENGINE light will come on and the sensor or circuit will need repair (see Chapter 6). Due to the complexity of this system, further diagnosis and repair should be referred to a dealership service department or other qualified repair shop.

10   Blower motor resistor/module and blower motor - replacement

Refer to illustration 10.1

Warning: *The models covered by this manual are equipped with Supplemental Restraint*

Systems (SRS), more commonly known as airbags. Always disable the airbag system before working in the vicinity of any airbag system component to avoid the possibility of accidental deployment of the airbag, which could cause personal injury (see Chapter 12).

1   Working in the passenger compartment under the glove box, remove the insulator panel **(see illustration)**.

Blower motor resistor/module

Refer to illustration 10.3

2   Disconnect the electrical connector from the blower motor resistor/module. It's mounted near the blower assembly.

3   Remove the mounting screws and remove the resistor/module from the blower housing **(see illustration)**.

4   Installation is the reverse of removal.

Blower motor

Refer to illustration 10.6

5   Disconnect the blower motor electrical connector.

6   Remove the blower motor mounting screws, then remove the blower motor assembly **(see illustration)**.

7   Remove the blower motor fan retaining clip and remove the blower fan from the motor.

8   Installation is the reverse of removal.

11   Heater/air conditioning control assembly - removal and installation

2009 and earlier models

Refer to illustrations 11.4 and 11.5

1   Turn the ignition key to the ON position and turn the temperature control knob to the extreme cold position, then turn the key off and remove it.

2   Disconnect the cable from the negative battery terminal (see Chapter 5).

3   Refer to Chapter 11 and remove the instrument panel center bezel.

4   Disconnect the electrical connectors from the rear of the unit and unclip the cable from the temperature control knob as equipped **(see illustration)**.

5   Working at the back of the trim, remove

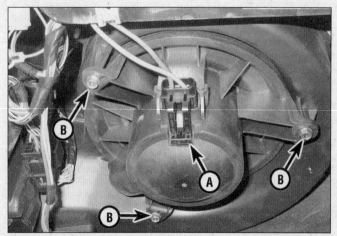

10.6 Disconnect the blower motor electrical connector (A) and remove the three mounting screws (B)

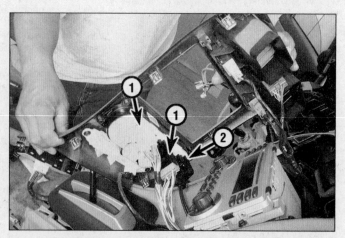

11.4 Heater/air conditioning control assembly mounting details:

1   *Electrical connectors*   2   *Cable connector*

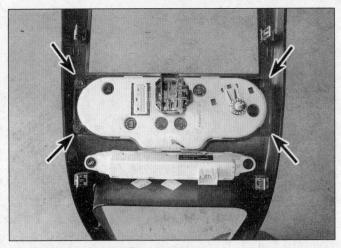

11.5 Heater/air conditioning control assembly mounting fasteners

12.3 Disconnect the heater hoses from inside the engine compartment

the four mounting screws for the heater control assembly and separate it from the trim (see illustration).

6   Installation is the reverse of removal. Be sure to install the cable connector when the control knob is in the extreme cold position.

2010 and later models

7   Disconnect the cable from the negative battery terminal (see Chapter 5).

8   Remove the center dash trim panel/radio/air control panel (see Chapter 11).

9   Disconnect the electrical connectors to the radio/air control panel assembly.

Note: The radio control module and heating/air conditioning control module are integrated and are not serviceable separately.

10   Remove the upper trim panel from the instrument panel (see Chapter 11).

11   Disengage the control assembly module retaining clips holding and pull out the control module out from the instrument panel.

12   Installation is the reverse of removal.

Caution: If the heating/air conditioning control module is being replaced, the module must be programmed using a factory scan tool or the HVAC system will not function properly.

13   Installation is the reverse of removal.

---

12   Heater core - replacement

Warning: The models covered by this manual are equipped with Supplemental Restraint Systems (SRS), more commonly known as airbags. Always disable the airbag system before working in the vicinity of any airbag system component to avoid the possibility of accidental deployment of the airbag, which could cause personal injury (see Chapter 12).

Note: This is a time-consuming and complex procedure. Make sure that you have sufficient time as well as the proper equipment and parts, and read through the entire procedure before you begin.

2007 and earlier models

Refer to illustrations 12.3, 12.9a, 12.9b

1   Disconnect the cable from the negative battery terminal (see Chapter 5).

2   Drain the cooling system (see Chapter 1).

3   Disconnect the heater hoses from the heater core inlet and outlet pipes at the firewall (see illustration). Plug the heater core pipes to prevent coolant spillage when the heater housing is removed.

4   Remove the entire instrument panel (see Chapter 11).

5   Remove the control cable from the shifter.

6   Press the cable clip tabs and release the cable from the shifter.

7   Disconnect all wiring from the shift assembly.

8   Unbolt the shifter and set it aside in order to gain access to the heater core.

9   Remove the large duct from the front of the heater unit (see illustrations).

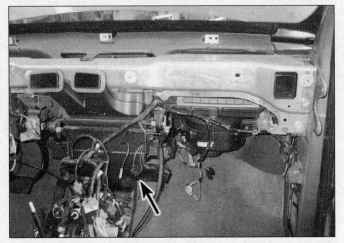

12.9a The heater core is behind a duct and a cover

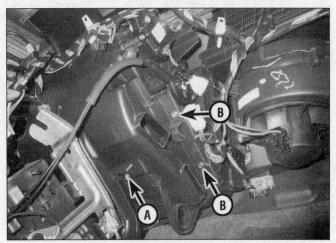

12.9b Remove the plastic duct from the front of the heater assembly by removing the screws (A) (on both sides) after first removing the shifter and its bracket - then remove the heater core cover screws (B)

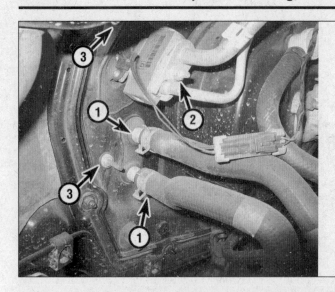

12.17 Heater/air conditioning unit mounting details:

1   *Heater core hose connections*
2   *Refrigerant line fitting retaining nut*
3   *Exterior mounting nuts (one nut not completely in view in this photo - vicinity given)*

12.27a Heater/air conditioning unit-to-instrument panel support fastener locations (left side)

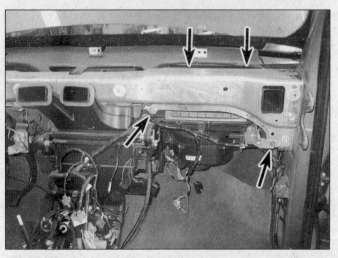

12.27b Heater/air conditioning unit-to-instrument panel support fastener locations (right side) (some hidden from view in this photo - vicinity given)

12.28 Instrument panel support bolts (some hidden from view in this photo - vicinity given)

10   Remove the heater core cover **(see illustration 12.9b)**

11   Remove the small plastic cover from the heater core pipes.

12   Remove the foam seal from the heater core pipes.

13   Remove the heater core by pulling on the end tank. **Note:** *Pull on the end tank only. It will help to spray the pipes and the seal with a lubricant or soapy water first.*

2008 and later models

Refer to illustrations 12.17, 12.27a, 12.27b and 12.28

14   Have the air conditioning system refrigerant discharged and recovered by an air conditioning technician.

15   Disconnect the cable from the negative battery terminal (see Chapter 5).

16   Drain the cooling system (see Chapter 1).

17   Remove the retaining nut for the refrigerant line fittings, then disconnect them **(see illustration)**. Discard all seals. Plug all open

lines and fittings to prevent contamination of the air conditioning system.

18   Disconnect the heater hoses from the heater core inlet and outlet pipes at the firewall. Plug the heater core pipes to prevent coolant spillage when the heater housing is removed.

19   Remove the heater/air conditioning unit mounting nuts at the firewall.

20   Remove the instrument panel (see Chapter 11).

21   Detach the control cable from the shifter (see Chapter 7).

22   Disconnect all wiring from the shifter assembly.

23   Remove the four nuts for the shifter assembly, then remove the assembly.

24   Remove the defroster, center and floor air ducts coming from the heater/air conditioning unit.

25   Remove the steering column (see Chapter 10).

26   Remove the brake pedal assembly. **Note:** *Refer to Steps 22 and 23 of Section*

11 *(power brake booster removal) in Chapter 9. Move the power brake booster forward slightly.*

27   Remove the mounting fasteners that attach the heater/air conditioning unit to the instrument panel support **(see illustrations)**.

28   Remove the instrument panel support **(see illustration)**. **Caution:** *Wear gloves when removing the instrument panel support to protect against sharp metal edges.*

29   Disconnect the electrical connectors for the blower motor and module (see Section 10).

30   Disconnect all electrical connectors and wire harnesses from the heater/air conditioning unit.

31   Remove the mounting fasteners for the heater/air conditioning unit, then carefully pull it directly away from the firewall.

32   Remove the heater core cover.

33   Pull the heater core from the heater/air conditioning unit.

13.1 Look for the evaporator drain hose on the firewall - make sure it isn't clogged

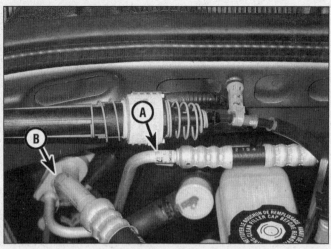

13.7 Evaporator inlet (A) and outlet (B) lines

All models

34   Installation is the reverse of removal.

35   Reconnect the cable to the negative terminal of the battery (see Chapter 5). Refill the cooling system (see Chapter 1).

13   Air conditioning and heating system - check and maintenance

Refer to illustration 13.1

Warning: *The air conditioning system is under high pressure. Do not loosen any hose fittings or remove any components until after the system has been discharged by a dealer service department or service station. Always wear eye protection when disconnecting air conditioning system fittings.*

1   The following maintenance checks should be performed on a regular basis to ensure the air conditioner continues to operate at peak efficiency.

a) *Check the drivebelt. If it's worn or deteriorated, replace it (see Chapter 1).*

b) *Check the system hoses. Look for cracks, bubbles, hard spots and deterioration. Inspect the hoses and all fittings for oil bubbles and seepage. If there's any evidence of wear, damage or leaks, replace the hose(s).*

c) *Inspect the condenser fins for leaves, bugs and other debris. Use a fin comb or compressed air to clean the condenser.*

d) *Make sure the system has the correct refrigerant charge.*

e) *Check the evaporator housing drain tube for blockage* **(see illustration).**

2   It's a good idea to operate the system for about 10 minutes at least once a month, particularly during the winter. Long term non-use can cause hardening, and subsequent failure, of the seals.

3   Because of the complexity of the air conditioning system and the special equipment necessary to service it, in-depth troubleshooting and repairs are not included in this manual (refer to the *Haynes Automotive Heating and Air Conditioning Repair Manual*). However, simple checks and component replacement procedures are provided in this Chapter.

4   The most common cause of poor cooling is simply a low system refrigerant charge. If a noticeable drop in cool air output occurs, the following quick check will help you determine if the refrigerant level is low.

Checking the refrigerant charge

Refer to illustrations 13.7 and 13.8

5   Warm the engine up to normal operating temperature.

6   Place the air conditioning temperature selector at the coldest setting and the blower at the highest setting. Open the vehicle doors (to make sure the air conditioning system doesn't cycle off as soon as it cools the passenger compartment).

7   With the compressor engaged - the clutch will make an audible click and the center of the clutch will rotate - feel the evaporator inlet and outlet lines at the firewall **(see illustration).** The inlet (small diameter) line should feel warm and the outlet (large diameter) line should feel cold. If so, the system is properly charged.

8   Place a thermometer in the dashboard vent nearest the evaporator and operate the system until the indicated temperature is around 40 to 45-degrees F **(see illustration).** If the ambient (outside) air temperature is very high, say 110-degrees F, the duct air temperature may be as high as 60-degrees F, but generally the air conditioning is 30 to 40-degrees F cooler than the ambient air. **Note:** *Humidity of the ambient air also affects the cooling capacity of the system. Higher ambient humidity lowers the effectiveness of the air conditioning system.*

Adding refrigerant

Refer to illustrations 13.9 and 13.12

9   Buy an automotive charging kit at an auto parts store **(see illustration).** A charging kit includes a can of refrigerant, a tap valve and a short section of hose that can be attached between the tap valve and the system low side service valve. **Note:** *Leak detection kits with refrigerant dye, a UV light and special glasses*

13.8 Insert a thermometer into the center register, turn on the air conditioning system and wait for it to cool; depending on the humidity, the air should be 30 to 40-degrees cooler than the ambient air

13.9 A basic charging kit for R-134a systems is available at most auto parts stores - it must say R-134a (not R-12) and so should the can of refrigerant

13.12 Add R-134a to the low-side port only - the procedure is easier if you wrap the can with a warm, wet towel to prevent icing

are also available at most automotive supply stores. This kit can help you pinpoint leaks in your air conditioning system. **Caution:** *There are two types of refrigerant used in automotive systems; R-12 - which has been widely used on earlier models, and the more environmentally-friendly R-134a used in all models covered by this manual. These two refrigerants (and their appropriate refrigerant oils) are not compatible and must never be mixed or components will be damaged. Use only R-134a refrigerant in the models covered by this manual.*

10   Hook up the charging kit by following the manufacturer's instructions. **Warning:** *DO NOT hook the charging kit hose to the system high side! The fittings on the charging kit are designed to fit only on the low side of the system.*

11   Back off the valve handle on the charging kit and screw the kit onto the refrigerant can, making sure first that the O-ring or rubber seal inside the threaded portion of the kit is in place. **Warning:** *Wear protective eyewear when dealing with pressurized refrigerant cans.*

12   Remove the dust cap from the low-side charging connection and attach the quick-connect fitting on the kit hose **(see illustration)**.

13   Warm up the engine and turn on the air conditioner. Keep the charging kit hose away from the fan and other moving parts. **Note:** *The compressor needs to be running in order to charge the system. However, if your system is very low on refrigerant, the compressor may turn off or not come on at all. If this happens, disconnect the A/C pressure switch electrical connector and use a jumper wire between the two terminals in the connector. This will keep the compressor running.*

14   Turn the valve handle on the kit until the stem pierces the can, then back the handle out to release the refrigerant. You should be able to hear the rush of gas. Add refrigerant to the low side of the system until both the receiver-drier surface and the evaporator inlet pipe feel about the same temperature. Allow stabilization time between each addition.

15   If you have an accurate thermometer, place it in the center air conditioning vent and

note the temperature of the air coming out of the vent. A fully-charged system which is working correctly should cool down to about 40-degrees F. Generally, an air conditioning system will put out air that is 30 to 40-degrees F cooler than the ambient air. For example, if the ambient (outside) air temperature is very high (over 100-degrees F), the temperature of air coming out of the registers should be 60 to 70-degrees F.

16   When the can is empty, turn the valve handle to the closed position and release the connection from the low-side port. Replace the dust cap. **Caution:** *Never add more than one can of refrigerant to the system. If more refrigerant than that is required, the system should be evacuated and leak tested.*

17   Remove the charging kit from the can and store the kit for future use with the piercing valve in the UP position, to prevent inadvertently piercing the can on the next use.

Heating systems

18   If the carpet under the heater core is damp, or if antifreeze vapor or steam is coming through the vents, the heater core is leaking. Remove it (see Section 12) and install a new unit (most radiator shops will not repair a leaking heater core).

19   If the air coming out of the heater vents isn't hot, the problem could stem from any of the following causes:

a)   *The thermostat is stuck open, preventing the engine coolant from warming up enough to carry heat to the heater core. Replace the thermostat (see Section 3).*

b)   *There is a blockage in the system, preventing the flow of coolant through the heater core. Feel both heater hoses at the firewall. They should be hot. If one of them is cold, there is an obstruction in one of the hoses or in the heater core, or the heater control valve is shut. Detach the hoses and back flush the heater core with a water hose. If the heater core is clear but circulation is impeded, remove the two hoses and flush them out with a water hose.*

c)   *If flushing fails to remove the blockage from the heater core, the core must be replaced (see Section 12).*

Eliminating air conditioning odors

20   Unpleasant odors that often develop in air conditioning systems are caused by the growth of a fungus, usually on the surface of the evaporator core. The warm, humid environment there is a perfect breeding ground for mildew to develop.

21   The evaporator core on most vehicles is difficult to access, and factory dealerships have a lengthy, expensive process for eliminating the fungus by opening up the evaporator case and using a powerful disinfectant and rinse on the core until the fungus is gone. You can service your own system at home, but it takes something much stronger than basic household germ-killers or deodorizers.

22   Aerosol disinfectants for automotive air conditioning systems are available in most auto parts stores, but remember when shopping for them that the most effective treatments are also the most expensive. The basic procedure for using these sprays is to start by running the system in the RECIRC mode for ten minutes with the blower on its highest speed. Use the highest heat mode to dry out the system and keep the compressor from engaging by disconnecting the wiring connector at the compressor (see Section 14).

23   Make sure that the disinfectant can comes with a long spray hose. Point the nozzle through the air recirculation door so that it protrudes inside the evaporator housing, then spray according to the manufacturer's recommendations. Try to cover the whole surface of the evaporator core, by aiming the spray up, down and sideways. Follow the manufacturer's recommendations for the length of spray and waiting time between applications.

24   Once the evaporator has been cleaned, the best way to prevent the mildew from coming back again is to make sure your evaporator housing drain tube is clear **(see illustration 13.1)**.

---

14   Air conditioning compressor - removal and installation

Warning: *The air conditioning system is under high pressure. Do not loosen any hose fittings or remove any components until after the system has been discharged. Air conditioning refrigerant must be properly discharged into an EPA-approved recovery/recycling unit at a dealer service department or an automotive air conditioning repair facility. Always wear eye protection when disconnecting air conditioning system fittings.*
Note: *The receiver-drier should be replaced whenever the compressor is replaced.*

Removal

Refer to illustration 14.5

1   Have the air conditioning system refrigerant discharged and recovered by an air conditioning technician.

14.5 Air conditioning compressor mounting details:

1   *Clutch electrical connector*
2   *Bottom mounting bolt (other mounting bolts are located near the top side of the compressor)*
3   *Line fitting mounting bolt*

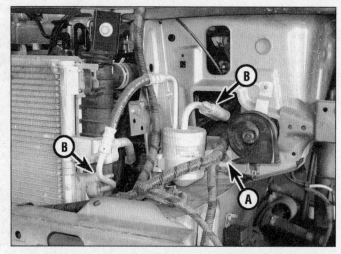

15.3 The receiver/drier is mounted behind the left headlight (bumper cover removed for clarity)

A   *Clamp mounting bolt*   B   *Refrigerant line fitting nuts*

2   Disconnect the cable from the negative battery terminal (see Chapter 5).
3   Remove the drivebelt (see Chapter 1).
4   Set the parking brake, block the rear wheels and raise the front of the vehicle, supporting it securely on jackstands.
5   Disconnect the compressor clutch electrical connector **(see illustration)**.
6   Remove the bolt for the line fitting, then pull the fitting from the compressor and discard the seals. Seal all open connections to avoid contamination.
7   Remove the compressor mounting bolts.
8   Remove the compressor while noting the position of the mounting spacer.

Installation

9   The clutch may have to be transferred from the old compressor to the new unit.
10   If a new or rebuilt compressor is being installed, follow the directions supplied with the compressor to properly adjust the oil level before installing it.
11   Installation is the reverse of removal, using new seals where the hose fitting attaches to the compressor.
12   Reconnect the battery (see Chapter 5).
13   Have the system evacuated, recharged and leak-tested by the shop that discharged it.

15   Air conditioning receiver-drier - removal and installation

Warning: *The air conditioning system is under high pressure. Do not loosen any hose fittings or remove any components until after the system has been discharged. Air conditioning refrigerant must be properly discharged into an EPA-approved recovery/recycling unit at a dealer service department or an automotive air conditioning repair facility. Always wear*

eye protection when disconnecting air conditioning system fittings.
Caution: *When replacing entire components, additional refrigerant oil must be added to them. Be sure to read the label on the container before adding any oil to the system; make sure it is compatible with the R-134a system.*
1   Have the refrigerant discharged and recovered by an air conditioning technician.

2007 and earlier models

Refer to illustration 15.3
2   Refer to Chapter 12 and remove the left headlight housing.
3   Remove the receiver-drier nuts from the condenser and from the refrigerant lines **(see illustration)**. Seal all open ports to avoid contamination. **Caution:** *Hold the condenser fitting with a large pair of pliers to avoid damage to the condenser.*
4   Remove the clamp mounting bolt and lift out the receiver-drier.

2008 and later models

5   Remove the front bumper cover (see Chapter 11).
6   On 2010 and later models, remove the air baffle fasteners and remove the baffles from around the front of the condenser.
7   Remove the mounting clamp bolt at the top of the receiver-drier.
8   Remove the mounting bolt from the bottom of the receiver-drier, then lift the receiver-drier directly up and out.

All models

9   Installation is the reverse of removal. Be sure to install new O-ring seals onto the receiver-drier fittings. Apply a thin layer of refrigerant oil to the seals before installing

them. If a new receiver-drier is being installed, add 1 ounce (30 ml) of fresh R-134a refrigerant oil. On 2008 and later models, tighten the receiver-drier mounting bolt (bottom) to the torque listed in this Chapter's Specifications and tighten the upper clamp bolt securely.
10   Have the system evacuated, charged and leak tested by the shop that discharged it.

16   Air conditioning condenser - removal and installation

Warning: *The air conditioning system is under high pressure. Do not loosen any hose fittings or remove any components until after the system has been discharged. Air conditioning refrigerant must be properly discharged into an EPA-approved recovery/recycling unit at a dealer service department or an automotive air conditioning repair facility. Always wear eye protection when disconnecting air conditioning system fittings.*
Caution: *When replacing entire components, additional refrigerant oil must be added to them. Be sure to read the label on the container before adding any oil to the system; make sure it is compatible with the R-134a system.*
Note: *The receiver-drier should be replaced whenever the condenser is replaced.*

Removal

Refer to illustration 16.5
1   Have the refrigerant discharged and recovered by an air conditioning technician.
2   Disconnect the cable from the negative battery terminal (see Chapter 5).
3   Remove the front bumper cover (see Chapter 11).
4   Follow Steps 6 through 14 in Section 4.

5   Disconnect the refrigerant lines from the condenser **(see illustration)**. Cap all fittings on the condenser and lines to prevent entry of dirt or moisture.

6   Detach the top fasteners of the air conditioning condenser from the radiator (see Section 6, Step 2).

7   Carefully lift the condenser to release the lower retaining tabs and separate it from the radiator. Be careful not to damage the fins on either unit while removing it from the engine compartment.

Installation

8   Installation is the reverse of removal. Tighten the line fitting mounting bolts to the torque listed in this Chapter's Specifications. If a new condenser is being installed, add 1 ounce (30 ml) of fresh refrigerant oil. Assemble all connections with new O-rings, lightly lubricated with R-134a refrigerant oil. **Note:** *If the receiver-drier and condenser are being replaced together, add 2 ounces (60 ml) of fresh refrigerant oil.*

9   Have the system evacuated, charged and leak tested by the shop that discharged it

---

17   Thermal expansion valve (TXV)

Warning: *The air conditioning system is under high pressure. DO NOT loosen any fittings or remove any components until after the system has been discharged. Air conditioning refrigerant must be properly discharged into an EPA-approved container at a dealer service department or an automotive air conditioning repair facility. Always wear eye protection when disconnecting air conditioning system fittings.*

16.5 Location of the refrigerant line fittings for the condenser (2005 model shown - other models similar)

1   Have the air conditioning system discharged by an automotive air conditioning technician (see Warning above).

2   Detach the fuel and brake lines and move them aside for access to the Thermal Expansion Valve (TXV).

3   Remove the refrigerant line fitting nut and detach the line fitting from the TXV.

4   Remove the TXV-to-firewall mounting bolts, then remove the valve. **Caution:** *Always discard the old sealing O-rings. If re-used, leaks may occur.* **Note:** *Plug all open lines and fittings to prevent contamination of the air conditioning system.*

5   Installation is the reverse of removal. Use new O-rings or seals lubricated with refrigerant oil as necessary.

Chapter 4
Fuel and exhaust systems

Contents

Specifications

Fuel system pressure (approximate)

2009 and earlier models	56 to 62 psi (384 to 425 kPa)
2010 through 2012 models	43 to 58 psi (296 to 400 kPa)
2013 and later 3.6L (direct injection, low pressure side)	
Key on, engine off	50 to 100 psi (345 to 689 kPa)
Engine running	43 to 58 psi (296 to 400 kPa)
Fuel pressure leak down (one minute after turning key to OFF)	No more than 5 psi (35 kPa) drop

Torque specifications

Note: *One foot-pound (ft-lb) of torque is equivalent to 12 inch-pounds (in-lbs) of torque. Torque values below approximately 15 ft-lbs are expressed in inch-pounds, since most foot-pound torque wrenches are not accurate at these smaller values.*

Fuel feed line fuel pressure sensor (2010 and later models)	80 in-lbs
Fuel rail pressure sensor (2010 and later models)	24
High-pressure fuel line fittings	
2.4L engines (2010 and later models)	22
3.6L engines (2013 and later models)	21
High-pressure fuel line bracket bolts (2013 and later 3.6L engines)	28
Fuel rail mounting bolts	
2009 and earlier models	89 in-lbs
2010 and later models	18

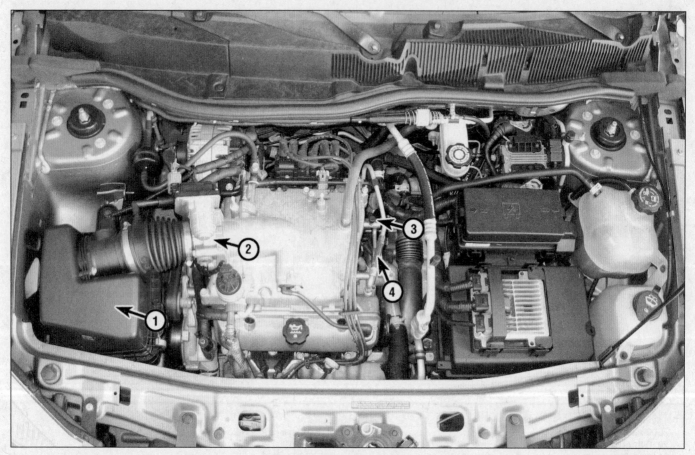

Fuel system details - 3.4L V6

1	Air filter housing	3	Fuel pressure test port
2	Throttle body	4	Fuel line

1 General information

Fuel system warnings

Gasoline is extremely flammable and repairing fuel system components can be dangerous. Consider your automotive repair knowledge and experience before attempting repairs which may be better suited for a professional mechanic.

* *Don't smoke or allow open flames or bare light bulbs near the work area*
* *Don't work in a garage with a gas-type appliance (water heater, clothes dryer)*
* *Use fuel-resistant gloves. If any fuel spills on your skin, wash it off immediately with soap and water*
* *Clean up spills immediately*
* *Do not store fuel-soaked rags where they could ignite*
* *Prior to disconnecting any fuel line, you must relieve the fuel pressure (see Section 3)*
* *Wear safety glasses*
* *Have a proper fire extinguisher on hand*

Fuel system

The fuel system consists of the fuel tank, electric fuel pump/fuel level sending unit (located in the fuel tank), fuel rail and fuel injectors. On 3.0L models, 2.4L models and 2013 and later 3.6L models, fuel pressure is increased greatly by a camshaft-driven high-pressure fuel pump, which supplies high-pressure fuel to the direct fuel injectors (which spray fuel directly into the combustion chambers). The fuel injection system is a multi-port system; multi-port fuel injection uses timed impulses to inject the fuel directly into the intake port of each cylinder. The Powertrain Control Module (PCM) controls the injectors. The PCM monitors various engine parameters and delivers the exact amount of fuel required into the intake ports.

Fuel is circulated from the fuel pump to the fuel rail through fuel lines running along the underside of the vehicle. Various sections of the fuel line are either rigid metal or nylon, or flexible fuel hose. The various sections of the fuel hose are connected either by quick-connect fittings or threaded metal fittings.

Exhaust system

The exhaust system consists of the exhaust manifold(s), catalytic converter(s), muffler(s), tailpipe and all connecting pipes, flanges and clamps. The catalytic converters are an emission control device added to the exhaust system to reduce pollutants.

2 Troubleshooting

Fuel pump

Refer to illustration 2.2

1   The low-pressure fuel pump is located inside the fuel tank. Sit inside the vehicle with the windows closed, turn the ignition key to On (not Start) and listen for the sound of the fuel pump as it's briefly activated. You will only hear the sound for a second or two, but that sound tells you that the pump is working. Alternatively, have an assistant listen at the fuel filler cap.

2   If the pump does not come on, check the fuses in the underhood fuse and relay box **(see illustration)**. On 2009 and earlier models there is also a fuel pump relay; check the underside of the fuse box cover for the location. If the fuses (and, on 2009 and earlier models, the relay) are okay, check the wiring back to the fuel pump. If the fuses and wiring are okay, the fuel pump is probably defective, but the problem could also lie in the Powertrain Control Module (PCM) or, on 2010 and later models, the Fuel Pump Flow Control Module (FPFCM).

If the pump runs continuously with the ignition key in the On position, the PCM or FPFCM is probably defective. Have the circuit checked by a professional mechanic, because if the PCM is defective, the new one will have to be programmed with a special proprietary scan tool.

Note: *On 2009 and earlier models, the fuel pump fuse is fuse no. 22. On 2010 and later models, check the engine compartment fuse box panel cover label for the correct fuse identifications.*

Fuel injection system

Refer to illustration 2.9

Note: *The following procedure is based on the assumption that the fuel pump is working and the fuel pressure is adequate (see Section 4).*

3   Check all electrical connectors that are related to the system. Check the ground wire connections for tightness.

4   Verify that the battery is fully charged (see Chapter 5).

5   Inspect the air filter element (see Chapter 1).

6   Check all fuses related to the fuel system (see Chapter 12).

7   Check the air induction system between the throttle body and the intake manifold for air leaks. Also inspect the condition of all vacuum hoses connected to the intake manifold and to the throttle body.

8   Remove the air intake duct from the throttle body and look for dirt, carbon, varnish, or other residue in the throttle body, particularly around the throttle plate. If it's dirty, clean it with carb cleaner, a toothbrush and a clean shop towel.

9   With the engine running, place an automotive stethoscope against each injector, one at a time, and listen for a clicking sound that indicates operation **(see illustration)**. **Warning:** *Stay clear of the drivebelt and any rotating or hot components.*

10   If you can hear the injectors operating,

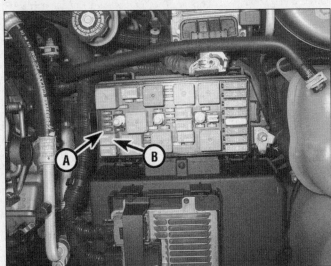

2.2 The fuel pump fuse (A) is located in the engine compartment fuse box; (B) is the fuel pump relay (2009 and earlier models)

2.9 An automotive stethoscope is used to listen to the fuel injectors in operation

but the engine is misfiring, the electrical circuits are functioning correctly, but the injectors might be dirty or clogged. Try a commercial injector cleaning product (available at auto parts stores). If cleaning the injectors doesn't help, replace the injectors.

11   If an injector is not operating (it makes no sound), disconnect the injector electrical connector and measure the resistance across the injector terminals with an ohmmeter. Compare this measurement to the other injectors. If the resistance of the non-operational injector is quite different from the other injectors, replace it.

12   If the injector is not operating, but the resistance reading is within the range of resistance of the other injectors, the PCM or the circuit between the PCM and the injector might be faulty.

3   Fuel pressure relief procedure

Warning: *Gasoline is extremely flammable.* See **Fuel system warnings** in Section 1.

1   Remove the fuel filler cap to relieve any pressure built-up in the fuel tank.

2009 and earlier models

2   Disconnect the cable from the negative terminal of the battery (see Chapter 5).

3   Locate the fuel pressure test port on the fuel rail, then unscrew the cap **(see illustration 4.1b)**.

4   Surround and cover the test port with shop rags, then depress the Schrader valve inside the test port with a small screwdriver until the pressure in the fuel system is relieved. Properly dispose of the rags.

2010 and later models

Warning: *The fuel system on these models operates at very high pressures (in excess of 2,000 psi) and can cause injury. Do not attempt to work on the fuel system until you are absolutely sure the fuel pressure has been relieved.*

Caution: *The fuel delivery system is made up of a low-pressure system and a high-pressure system. Once the pressure on the low-pressure side of the system has been relieved (as described here), wait at least two hours before loosening any fuel line fittings in the engine compartment.*

5   Open the fuse and relay panel inside the engine compartment and locate the fuel pump fuse. Pull the fuel pump fuse, start the engine and allow it to run until it stalls (it might not even start). Once it has stalled (or has failed to start), crank the starter for three more seconds, then turn off the ignition key.

Note: *Check the engine compartment fuse box panel cover label for the correct fuse identifications.*

Warning: *After removing the fuel pump fuse, verify that the fuel pump does not run when the ignition key is turned to the ON position, and that the engine stalls or won't start.*

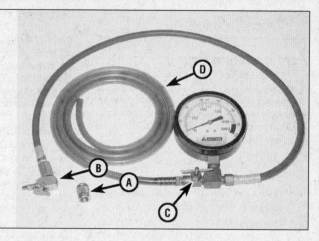

4.1a A typical fuel pressure gauge setup:

A   *Screw-on adapter for the quick-release fitting*
B   *Quick-release fitting for connecting the gauge hose to the test port*
C   *Bleeder valve (optional)*
D   *Bleeder hose (optional)*

6   The fuel pressure is now relieved. Disconnect the cable from the negative terminal of the battery before working on any fuel system component.

Warning: *This procedure merely relieves the pressure that the engine needs to run. But remember that fuel is still present in the system components, and take precautions accordingly before disconnecting any of them.*

4   Fuel pressure - check

Refer to illustrations 4.1a and 4.1b

Warning: *Gasoline is extremely flammable.* See **Fuel system warnings** in Section 1.

Note: *The following procedure assumes that the fuel pump is receiving voltage and runs.*

1   To check the fuel pressure, locate the fuel pressure test port on the fuel rail, unscrew the cap and connect a fuel pressure gauge **(see illustrations)**.

2   Start the engine and allow it to idle. Note the gauge reading as soon as the pressure stabilizes, and compare it with the pressure listed in this Chapter's Specifications.

3   If the fuel pressure is not within specifications, check the following:

a)   *Check for a restriction in the fuel system (kinked fuel line, plugged fuel pump inlet strainer or clogged fuel filter). If no restrictions are found, replace the fuel pump module (see Section 7).*
b)   *If the fuel pressure is higher than specified, replace the fuel pump module (see Section 7).*

4   Turn off the engine. Fuel pressure should not fall more than 8 psi over five minutes. If it does, the problem could be a leaky fuel injector, fuel line leak, or faulty fuel pump module.

5   Disconnect the fuel pressure gauge. Wipe up any spilled gasoline.

5   Fuel lines and fittings - general information and disconnection

Warning: *Gasoline is extremely flammable.* See **Fuel system warnings** in Section 1.

1   Relieve the fuel pressure before servicing fuel lines or fittings (see Section 3), then disconnect the cable from the negative battery terminal (see Chapter 5) before proceeding.

2   The fuel supply line connects the fuel pump in the fuel tank to the fuel rail on the engine. The Evaporative Emission (EVAP) system lines connect the fuel tank to the EVAP canister and connect the canister to the intake manifold.

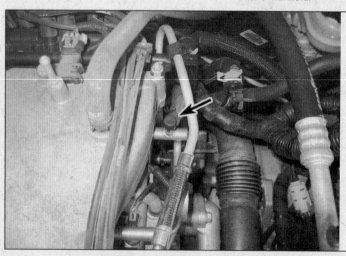

4.1b The fuel pressure test port is located on the fuel rail

Disconnecting Fuel Line Fittings

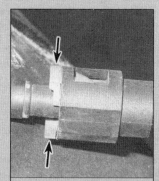

Two-tab type fitting; depress both tabs with your fingers, then pull the fuel line and the fitting apart

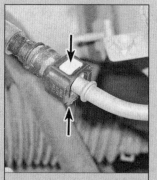

On this type of fitting, depress the two buttons on opposite sides of the fitting, then pull it off the fuel line

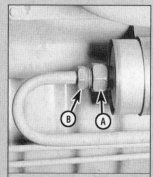

Threaded fuel line fitting; hold the stationary portion of the line or component (A) while loosening the tube nut (B) with a flare-nut wrench

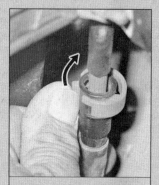

Plastic collar-type fitting; rotate the outer part of the fitting

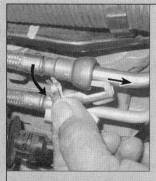

Metal collar quick-connect fitting; pull the end of the retainer off the fuel line, and disengage the other end from the female side of the fitting . . .

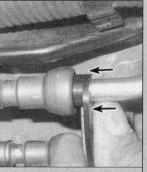

. . . insert a fuel line separator tool into the female side of the fitting, push it into the fitting until it releases the locking tabs inside the fitting, and pull the two halves of the fitting apart

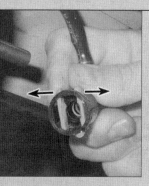

Hairpin-type clip; spread the two legs of the clip apart . . .

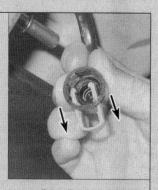

. . . pull the clip out and detach the coupling from the component (fitting detached for clarity)

Spring-lock coupling; remove the safety cover . . .

. . . install a coupling release tool and close the clamshell halves of the tool around the coupling . . .

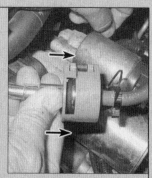

. . . push the tool into the fitting, then pull the two lines apart

3   Whenever you're working under the vehicle, be sure to inspect all fuel and evaporative emission lines for leaks, kinks, dents and other damage. Always replace a damaged fuel or EVAP line immediately.

4   If you find signs of dirt in the lines during disassembly, disconnect all lines and blow them out with compressed air. Inspect the fuel strainer on the fuel pump pick-up unit for damage and deterioration.

Steel tubing

5   It is critical that the fuel lines be replaced with lines of equivalent type and specification.

6   Some steel fuel lines have threaded fittings. When loosening these fittings, hold the stationary fitting with a wrench while turning the tube nut.

Plastic tubing

7   When replacing fuel system plastic tubing, use only original equipment replacement plastic tubing. **Caution:** *When removing or installing plastic fuel line tubing, be careful not to bend or twist it too much, which can damage it. Also, plastic fuel tubing is NOT heat resistant, so keep it away from excessive heat.*

Flexible hoses

8   When replacing fuel system flexible hoses, use only original equipment replacements.

9   Don't route fuel hoses (or metal lines) within four inches of the exhaust system or within ten inches of the catalytic converter. Make sure that no rubber hoses are installed directly against the vehicle, particularly in places where there is any vibration. If allowed to touch some vibrating part of the vehicle, a hose can easily become chafed and it might start leaking. A good rule of thumb is to maintain a minimum of 1/4-inch clearance around a hose (or metal line) to prevent contact with the vehicle underbody.

6   Exhaust system servicing - general information

Refer to illustration 6.1

Warning: *Allow exhaust system components to cool before inspection or repair. Also, when working under the vehicle, make sure it is securely supported on jackstands.*

1   The exhaust system consists of the exhaust manifolds, catalytic converter, muffler, tailpipe and all connecting pipes, flanges and clamps. The exhaust system is isolated from the vehicle body and from chassis components by a series of rubber hangers **(see illustration)**. Periodically inspect these hangers for cracks or other signs of deterioration, replacing them as necessary.

2   Conduct regular inspections of the exhaust system to keep it safe and quiet. Look for any damaged or bent parts, open seams, holes, loose connections, excessive corrosion or other defects which could allow exhaust fumes to enter the vehicle. Do not repair deteriorated exhaust system components; replace them with new parts.

3   If the exhaust system components are extremely corroded, or rusted together, a cutting torch is the most convenient tool for removal. Consult a properly-equipped repair shop. If a cutting torch is not available, you can use a hacksaw, or if you have compressed air, there are special pneumatic cutting chisels that can also be used. Wear safety goggles to protect your eyes from metal chips and wear work gloves to protect your hands.

4   Here are some simple guidelines to follow when repairing the exhaust system:

a) *Work from the back to the front when removing exhaust system components.*

b) *Apply penetrating oil to the exhaust system component fasteners to make them easier to remove.*

c) *Use new gaskets, hangers and clamps.*

d) *Apply anti-seize compound to the threads of all exhaust system fasteners during reassembly.*

e) *Be sure to allow sufficient clearance between newly installed parts and all points on the underbody to avoid overheating the floor pan and possibly damaging the interior carpet and insulation. Pay particularly close attention to the catalytic converter and heat shield.*

7   Fuel pump/fuel level sensor module - removal and installation

Warning: *Gasoline is extremely flammable. See* Fuel system warnings *in Section 1.*

Caution: *The secondary fuel tank module should be removed before attempting to remove the primary module.*

Note: *The primary fuel tank module is the one that contains the main fuel pump.*

1   Detach the cable from the negative battery terminal (see Chapter 5). Relieve the fuel system pressure (see Section 3).

2   Refer to Section 8 and remove the fuel tank from the vehicle.

3   Spray the tangs of the module lock ring with penetrating oil.

Secondary fuel tank module

Refer to illustrations 7.5 and 7.6

4   Disconnect the EVAP hose quick-connect fitting on the top of the secondary fuel tank module.

5   Use a large pair of water pump pliers to remove the fuel tank module lock ring **(see illustration)**. If it won't break loose, tap it with a hammer and a brass punch (to avoid sparks).

6   Disconnect the electrical connector from the unit, then push down on the tab near the base of the module to release the suction port tube, then carefully pull the module out of the tank **(see illustration)**.

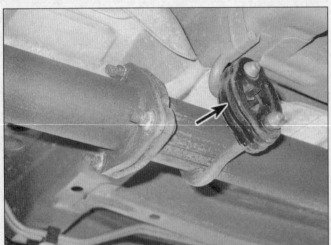

6.1 A typical exhaust system hanger. Inspect regularly and replace at the first sign of damage or deterioration

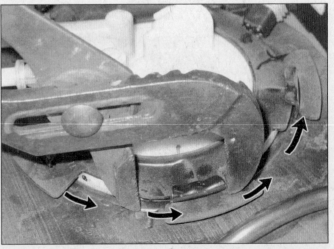

7.5 Use a large pair of water pump pliers to loosen and unscrew the fuel tank module locknut

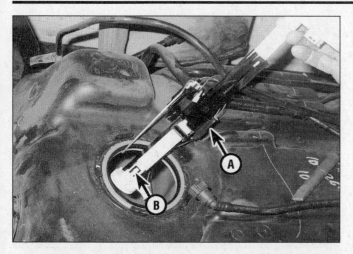

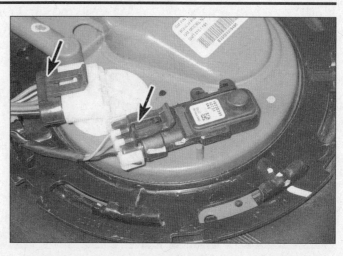

7.6 Disconnect the electrical connector (A), then push down on the tab near the base of the module to release the suction port tube (B)

7.8 Disconnect the electrical connectors from the module and fuel tank pressure sensor

7.10 The lines must be released from the grooves on top of the tank

7.11 Lift the fuel tank module carefully to avoid damaging the fuel level sensor

7   Inspect the O-ring and replace it if it shows any sign of deterioration. Installation is the reverse of removal.

Primary fuel tank module

Refer to illustrations 7.8, 7.10, 7.11 and 7.12

8   Disconnect all the electrical connectors from the module and pressure sensor **(see illustration)**.

9   Use a large pair of water pump pliers to remove the fuel tank module lock ring **(see illustration 6.5)**. If it won't break loose, tap it with a hammer and a brass punch (to avoid sparks).

10   Release the lines from the top of the fuel tank **(see illustration)**.

11   Carefully remove the module from the fuel tank **(see illustration)**.

12   Inspect the O-ring and replace it if it shows any sign of deterioration **(see illustration)**. Installation is the reverse of removal.

8   Fuel tank - removal and installation

Refer to illustrations 8.8 and 8.11

Warning 1: *Gasoline is extremely flammable. See* Fuel system warnings *in Section 1.*

Warning 2: *The following procedure is much easier to perform if the fuel tank is empty.*

1   Remove the fuel tank filler cap to relieve fuel tank pressure.

2   Relieve the fuel system pressure (see Section 3).

3   Disconnect the cable from the negative battery terminal (see Chapter 5).

4   Raise the rear of the vehicle and support it securely on jackstands.

5   Remove the rear section of the exhaust system.

6   On AWD models, remove the driveshaft (see Chapter 8).

7   Carefully clean around all of the vent hoses, squeeze the tabs and pull the connections apart.

7.12 The module's O-ring must be undamaged - if it shows any sign of deterioration, replace it

8.8 The fuel tank fill hose uses a hose clamp - all other fittings are the quick-connect type

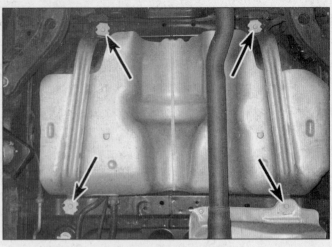

8.11 Fuel tank strap bolts

8   Disconnect the filler hose from the tank **(see illustration)**. Also detach the fuel hose(s) at the fitting(s) near the tank, and the EVAP hoses from the canister. **Warning:** *Be sure to catch fuel spillage with rags. Dispose of the fuel-soaked rags in an approved container.*

9   Disconnect the wiring harnesses from the fuel tank.

10   Support the fuel tank securely.

11   Remove the fuel tank retaining strap bolts **(see illustration)** and remove the straps.

12   Carefully lower the fuel tank.

13   Installation is the reverse of removal. Be sure to tighten the fuel tank strap bolts securely.

14   Reconnect the cable to the negative battery terminal (see Chapter 5), then start the engine and check for fuel leaks.

9   Air filter housing - removal and installation

Air intake duct

Refer to illustration 9.1

1   On 2.4L engine models, disconnect the PCV hose from the resonator only and on V6 engine models, disconnect the PCV fresh air hose from the valve cover **(see illustration)**.

2   Loosen the clamps at the air filter housing and the throttle body and lift off the duct.

3   Installation is the reverse of removal.

Air filter housing

Refer to illustration 9.5

4   Loosen the clamp and detach the duct from the air filter housing.

5   Disconnect the electrical connector from the MAF sensor **(see illustration)**.

6   Remove the air filter housing mounting fastener and lift off the housing.

7   Installation is the reverse of removal.

10   Throttle body - removal and installation

Refer to illustration 10.3

Warning: *Wait until the engine is completely cool before beginning this procedure.*

Caution: *The manufacturer recommends that an idle-learn procedure be done any time the throttle body is removed. This is done to ensure that the idle is stable and that no Diagnostic Trouble Codes (DTCs) occur. Refer to Chapter 5, Section 1 for information on this procedure.*

9.1 Air intake duct details

1 *PCV fresh air hose*	2 *Hose clamp*

9.5 Air filter housing details

1   *MAF sensor electrical connector*
2   *Air filter housing mounting fastener*

10.3 Throttle body details

1   Electrical connector
*2   Throttle body
     mounting fasteners*

1   Disconnect the cable from the negative battery terminal (see Chapter 5).

2   Remove the air intake duct (see Section 9). It can simply be disconnected from the throttle body and moved aside.

3   Disconnect the electrical connector from the throttle body **(see illustration)**.

4   On 3.4L V6 models, remove the fasteners securing the heater hose bracket to the throttle body.

5   Remove the throttle body mounting fasteners and carefully lift off the throttle body. Discard the gasket; it should be replaced with a new one.

6   Cover the intake manifold opening with a clean shop towel.

7   Installation is the reverse of removal. Be sure to use a new gasket and tighten the throttle body fasteners to the torque listed in this Chapter's Specifications.

11 Fuel rail and injectors - removal and installation

Warning: *Gasoline is extremely flammable. See* Fuel system warnings *in Section 1.*

1   Relieve the fuel system pressure (see Section 3).

2009 and earlier models
Removal

Refer to illustrations 11.4 and 11.9

2   Disconnect the cable from the negative battery terminal (see Chapter 5).

3   Remove the upper intake manifold (see Chapter 2B or 2C).

4   Disconnect the fuel supply line at the fuel rail **(see illustration)**.

5   Disconnect the electrical connector for the main fuel injector harness.

6   Disconnect the electrical connectors for the coolant temperature and camshaft position sensors (see Chapter 6).

7   Disconnect the electrical connector from each injector.

8   Remove the fuel rail mounting fasteners, then remove both fuel rails and their injectors as an assembly.

9   Release the fuel injector retaining clips **(see illustration)**. Remove each injector from its bore in the fuel rail. Remove and discard the upper O-ring and the lower O-ring. Repeat this procedure for each injector. **Note:** *Even if you only removed the fuel rail assembly to replace a single injector or a leaking O-ring, it's a good idea to remove all of the injectors from the fuel rail and replace all of the O-rings at the same time.*

Installation

10   Coat the new upper O-rings with clean engine oil and slide them into place on each of the fuel injectors. Coat the new lower O-rings with clean engine oil and install them on the lower ends of the injectors.

11   Coat the outside surface of each upper O-ring with clean engine oil, then insert each injector into its bore in the fuel rail. Install the isolators with new seals on the injectors, if equipped.

12   Install the injectors and fuel rail assembly on the intake manifold. Tighten the fuel rail mounting fasteners securely.

13   The remainder of installation is the reverse of removal.

14   Reconnect the cable to the negative battery terminal (see Chapter 5).

15   Turn the ignition switch to ON (but don't operate the starter).This activates the fuel pump for about two seconds, which builds up fuel pressure in the fuel lines and the fuel rail. Repeat this step two or three times, then check the fuel lines, fuel rails and injectors for fuel leaks.

11.4 Location of the fuel supply line fitting at the fuel rail

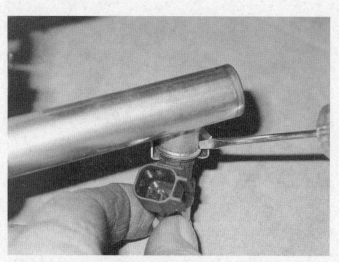

11.9 Carefully pry off each injector retaining clip with a small screwdriver, then pull the injector out of the fuel rail

2010 and later models

Warning: *This procedure can cause fuel to drain from the fuel system. Read the **Warning** in Section 3.*

Note: *To remove the injectors, four special service tools are recommended: an injector bore cleaning tool, seal installer and sizer, slide hammer and injector remover. Check with a dealer or auto parts store concerning availability and pricing of these tools before beginning this procedure.*

14   Relieve the fuel system pressure (see Section 3).

15   Remove the fuel rail feed pipe.

16   Remove the intake manifold (see Chapter 2A or 2C).

17   Carefully remove the foam fuel injector insulator (if equipped), fuel injector electrical connectors and any remaining electrical connections, wiring clips and retainers.

18   Remove the fuel rail using the special injector removal tool #EN-49248 to pull the injectors out of the cylinder head.

19   Remove the retainer that secures each fuel injector to the fuel rail and pull out the injector.

20   Installation is the reverse of removal. New O-rings (correctly sized with the special sizing tool) should be lubricated with clean engine oil prior to installation. Install only into a clean injector bore. If necessary, use the special bore cleaning tool to clean the bore.

21   Start the engine and verify there are no fuel leaks.

12   High-pressure fuel pump (2010 and later models) - removal and installation

Warning: *The fuel system on these models operates at very high pressures (in excess of 2,000 psi) and can cause injury. Do not attempt to work on the fuel system until you are absolutely sure the fuel pressure has been relieved.*

Warning: *This procedure can cause fuel to drain from the fuel system. Read the **Warning** in Section 3.*

Note: *Attached to the rear of the cylinder head and driven by the camshaft, the pump receives fuel from the in-tank fuel pump, and then boosts the fuel pressure to levels high enough for direct injection into the combustion chamber. Integrated to the pump is an electronic fuel pressure regulator, which is controlled by the Powertrain Control Module (PCM). Pressure is regulated between 600 psi and 2176 psi.*

1   Relieve the fuel system pressure (see Section 3).

2   Remove the electrical connector from the high-pressure fuel pump.

3   Disconnect the intake air ducting to the throttle body and move the ducting aside for access to the high-pressure pump.

4   On 3.0L models and 2013 and later 3.6L models, remove the thermostat housing (see Chapter 3).

5   Loosen the fuel line fitting at the injector rail and the inlet/outlet lines at the high-pressure pump.

6   Disconnect the flexible fuel line to the high pressure pump and move it aside. Remove the metal fuel lines from the high pressure pump.

7   Remove and discard the two fuel pump mounting bolts and remove the pump.

Note: *The pump roller lifter should remain in the cylinder, if it comes out slide roller lifter back into the housing.*

8   Installation is the reverse of removal. Start the engine and check for fuel leaks.

13   Fuel Pump Flow Control Module (FPFCM) – removal and installation

1   Disconnect the cable from the negative battery terminal (see Chapter 5).

2   Open the liftgate and remove spare tire cover and the left-side trim panel.

3   Disconnect the electrical connector from the FPFCM.

4   Remove the FPFCM mounting nuts and remove the module.

Note: *If the module is being replaced, it must be programmed using a factory scan tool and the proper calibration software.*

5   Installation is the reverse of removal.

Chapter 5
Engine electrical systems

Contents

1 General information and precautions

General information
Ignition system

The electronic ignition system consists of the Crankshaft Position (CKP) sensor, the Camshaft Position (CMP) sensor, the Knock Sensor (KS), the Powertrain Control Module (PCM), the ignition switch, the battery, the individual ignition coils or a coil pack, and the spark plugs. For more information on the CKP, CMP and KS sensors, as well as the PCM, refer to Chapter 6.

Charging system

The charging system includes the alternator (with an integral voltage regulator), the Powertrain Control Module (PCM), the Body Control Module (BCM), a charge indicator light on the dash, the battery, a fuse or fusible link and the wiring connecting all of these components. The charging system supplies electrical power for the ignition system, the lights, the radio, etc. The alternator is driven by a drivebelt.

Starting system

The starting system consists of the battery, the ignition switch, the starter relay, the Powertrain Control Module (PCM), the Body Control Module (BCM), the Transmission Range (TR) switch, the starter motor and solenoid assembly, and the wiring connecting all of the components.

Precautions

Always observe the following precautions when working on the electrical system:

a) *Be extremely careful when servicing engine electrical components. They are easily damaged if checked, connected or handled improperly.*

b) *Never leave the ignition switched on for long periods of time when the engine is not running.*

c) *Never disconnect the battery cables while the engine is running.*

d) *Maintain correct polarity when connecting battery cables from another vehicle during jump starting - see the "Booster battery (jump) starting" Section at the front of this manual.*

e) *Always disconnect the cable from the negative battery terminal before working on the electrical system, but read the battery disconnection procedure first (see Section 3).*

It's also a good idea to review the safety-related information regarding the engine electrical systems located in the *Safety first!* Section at the front of this manual, before beginning any operation included in this Chapter.

Engine electrical system components - 3.4L V6

1	Alternator	4	Battery (under battery cover)
2	Coil pack	5	Starter motor
3	Fuse/relay block		(side of engine block)

2   Troubleshooting

Ignition system

1   If a malfunction occurs in the ignition system, do not immediately assume that any particular part is causing the problem. First, check the following items:

  a) *Make sure that the cable clamps at the battery terminals are clean and tight.*

  b) *Test the condition of the battery (see Steps 21 through 24). If it doesn't pass all the tests, replace it.*

  c) *Check the ignition coil or coil pack connections.*

  d) *Check any relevant fuses in the engine compartment fuse and relay box (see Chapter 12). If they're burned, determine the cause and repair the circuit.*

Check

Refer to illustration 2.3

Warning: *Because of the high voltage generated by the ignition system, use extreme care when performing a procedure involving ignition components.*

Note 1: *The ignition system components on*

these vehicles are difficult to diagnose. In the event of ignition system failure that you can't diagnose, have the vehicle tested at a dealer service department or other qualified auto repair facility.

Note 2: *You'll need a spark tester for the following test. Spark testers are available at most auto supply stores.*

2   If the engine turns over but won't start, verify that there is sufficient ignition voltage to fire the spark plugs as follows.

3   On models with a coil-over-plug type ignition system, remove a coil and install the tester between the boot at the lower end of the coil and the spark plug **(see illustration)**. On models with spark plug wires, discon-

nect a spark plug wire from a spark plug and install the tester between the spark plug wire boot and the spark plug.

4   Crank the engine and note whether or not the tester flashes. **Caution:** *Do NOT crank the engine or allow it to run for more than five seconds; running the engine for more than five seconds may set a Diagnostic Trouble Code (DTC) for a cylinder misfire.*

Models with a coil-over-plug type ignition system

5   If the tester flashes during cranking, the coil is delivering sufficient voltage to the spark plug to fire it. Repeat this test for each cylinder to verify that the other coils are OK.

6   If the tester doesn't flash, remove a coil

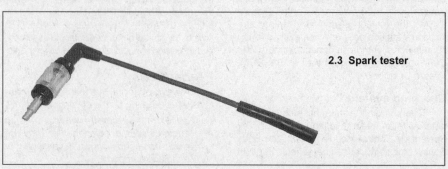

2.3 Spark tester

from another cylinder and swap it for the one being tested. If the tester now flashes, you know that the original coil is bad. If the tester still doesn't flash, the PCM or wiring harness is probably defective. Have the PCM checked out by a dealer service department or other qualified repair shop (testing the PCM is beyond the scope of the do-it-yourselfer because it requires expensive special tools).

7   If the tester flashes during cranking but a misfire code (related to the cylinder being tested) has been stored, the spark plug could be fouled or defective.

Models with spark plug wires

8   If the tester flashes during cranking, sufficient voltage is reaching the spark plug to fire it.

9   Repeat this test on the remaining cylinders.

10   Proceed on this basis until you have verified that there's a good spark from each spark plug wire. If there is, then you have verified that the coils in the coil pack are functioning correctly and that the spark plug wires are OK.

11   If there is no spark from a spark plug wire, then either the coil is bad, the plug wire is bad or a connection at one end of the plug wire is loose. Assuming that you're using new plug wires or known good wires, then the coil is probably defective. Also inspect the coil pack electrical connector. Make sure that it's clean, tight and in good condition.

12   If all the coils are firing correctly, but the engine misfires, then one or more of the plugs might be fouled. Remove and check the spark plugs or install new ones (see Chapter 1).

13   No further testing of the ignition system is possible without special tools. If the problem persists, have the ignition system tested by a dealer service department or other qualified repair shop.

Charging system

14   If a malfunction occurs in the charging system, do not automatically assume the alternator is causing the problem. First check the following items:

a) *Check the drivebelt tension and condition, as described in Chapter 1. Replace it if it's worn or deteriorated.*

b) *Make sure the alternator mounting bolts are tight.*

c) *Inspect the alternator wiring harness and the connectors at the alternator and voltage regulator. They must be in good condition, tight and have no corrosion.*

d) *Check the fusible link (if equipped) or main fuse in the underhood fuse/relay box. If it is burned, determine the cause, repair the circuit and replace the link or fuse (the vehicle will not start and/or the accessories will not work if the fusible link or main fuse is blown).*

e) *Start the engine and check the alternator for abnormal noises (a shrieking or squealing sound indicates a bad bearing).*

f) *Check the battery. Make sure it's fully charged and in good condition (one bad cell in a battery can cause overcharging by the alternator).*

g) *Disconnect the battery cables (negative first, then positive). Inspect the battery posts and the cable clamps for corrosion. Clean them thoroughly if necessary (see Chapter 1). Reconnect the cables (positive first, negative last).*

Alternator - check

15   Use a voltmeter to check the battery voltage with the engine off. It should be at least 12.6 volts **(see illustration 2.21)**.

16   Start the engine and check the battery voltage again. It should now be approximately 13.5 to 15 volts.

17   If the voltage reading is more or less than the specified charging voltage, the voltage regulator is probably defective, which will require replacement of the alternator (the voltage regulator is not replaceable separately). Remove the alternator and have it bench tested (most auto parts stores will do this for you).

18   The charging system (battery) light on the instrument cluster lights up when the ignition key is turned to ON, but it should go out when the engine starts.

19   If the charging system light stays on after the engine has been started, there is a problem with the charging system. Before replacing the alternator, check the battery condition, alternator belt tension and electrical cable connections.

20   If replacing the alternator doesn't restore voltage to the specified range, have the charging system tested by a dealer service department or other qualified repair shop.

Battery - check

Refer to illustrations 2.21 and 2.23

21   Check the battery state of charge. Visually inspect the indicator eye on the top of the battery (if equipped with one); if the indicator eye is black in color, charge the battery as described in Chapter 1. Next perform an open circuit voltage test using a digital voltmeter. **Note:** *The battery's surface charge must be removed before accurate voltage measurements can be made. Turn on the high beams for ten seconds, then turn them off and let the vehicle stand for two minutes.* With the engine and all accessories Off, touch the negative probe of the voltmeter to the negative terminal of the battery and the positive probe to the positive terminal of the battery **(see illustration)**. The battery voltage should be 12.6 volts or slightly above. If the battery is less than the specified voltage, charge the battery before proceeding to the next test. Do not proceed with the battery load test unless the battery charge is correct.

22   Disconnect the negative battery cable, then the positive cable from the battery.

23   Perform a battery load test. An accurate check of the battery condition can only be performed with a load tester **(see illustration)**. This test evaluates the ability of

2.21 To test the open circuit voltage of the battery, touch the black probe of the voltmeter to the negative terminal and the red probe to the positive terminal of the battery; a fully charged battery should be at least 12.6 volts

2.23 Connect a battery load tester to the battery and check the battery condition under load following the tool manufacturer's instructions

the battery to operate the starter and other accessories during periods of high current draw. Connect the load tester to the battery terminals. Load test the battery according to the tool manufacturer's instructions. This tool increases the load demand (current draw) on the battery.

24   Maintain the load on the battery for 15 seconds and observe that the battery voltage does not drop below 9.6 volts. If the battery condition is weak or defective, the tool will indicate this condition immediately. **Note:** *Cold temperatures will cause the minimum voltage reading to drop slightly. Follow the chart given in the manufacturer's instructions to compensate for cold climates. Minimum load voltage for freezing temperatures (32 degrees F) should be approximately 9.1 volts.*

Starting system

The starter rotates, but the engine doesn't

25   Remove the starter (see Section 7), check the overrunning clutch and bench test the starter to make sure the drive mechanism extends fully for proper engagement with the flywheel ring gear. If it doesn't, replace the starter.

26   Check the flywheel ring gear for missing teeth and other damage. With the ignition turned off, rotate the flywheel so you can check the entire ring gear.

The starter is noisy

27   If the solenoid is making a chattering noise, first check the battery (see Steps 21 through 24). If the battery is okay, check the cables and connections.

28   If you hear a grinding, crashing metallic sound when you turn the key to Start, check for loose starter mounting bolts. If they're tight, remove the starter and inspect the teeth on the starter pinion gear and flywheel ring gear. Look for missing or damaged teeth.

29   If the starter sounds fine when you first turn the key to Start, but then stops rotating the engine and emits a zinging sound, the problem is probably a defective starter drive that's not staying engaged with the ring gear. Replace the starter.

The starter rotates slowly

30   Check the battery (see Steps 21 through 24).

31   If the battery is okay, verify all connections (at the battery, the starter solenoid and motor) are clean, corrosion-free and tight. Make sure the cables aren't frayed or damaged.

32   Check that the starter mounting bolts are tight so it grounds properly. Also check the pinion gear and flywheel ring gear for evidence of a mechanical bind (galling, deformed gear teeth or other damage).

The starter does not rotate at all

33   Check the battery (see Steps 21 through 24).

34   If the battery is okay, verify all connections (at the battery, the starter solenoid and motor) are clean, corrosion-free and tight. Make sure the cables aren't frayed or damaged.

35   Check all of the fuses in the underhood fuse/relay box.

36   Check that the starter mounting bolts are tight so it grounds properly.

37   Check for voltage at the starter solenoid "S" terminal when the ignition key is turned to the start position. If voltage is present, replace the starter/solenoid assembly. If no voltage is present, the problem could be the starter relay, the Transmission Range (TR) switch (see Chapter 6) or clutch start switch (see Chapter 8), or with an electrical connector somewhere in the circuit (see the wiring diagrams at the end of Chapter 12). Also, on many modern vehicles, the Powertrain Control Module (PCM) and the Body Control Module (BCM) control the voltage signal to the starter solenoid; on such vehicles a special scan tool is required for diagnosis.

3   Battery - disconnection

Warning: *If the vehicle is equipped with OnStar, make absolutely sure the ignition key is in the Off position and Retained Accessory Power (RAP) has been depleted before disconnecting the cable from the negative battery terminal. Also, never remove the OnStar fuse with the ignition key in any position other than Off. If these precautions are not taken, the OnStar system's back-up battery will be activated, and remain activated, until it goes dead. If this happens, the OnStar system will not function as it should in the event that the main vehicle battery power is cut off (as might happen during a collision).*
Caution: *Always disconnect the cable from the negative battery terminal FIRST and hook it up LAST or the battery may be shorted by the tool being used to loosen the cable clamps.*

Some systems on the vehicle require battery power to be available at all times, either to maintain continuous operation (alarm system, power door locks, etc.), or to maintain control unit memory (radio station presets, Powertrain Control Module and other control units). When the battery is disconnected, the power that maintains these systems is cut. So, before you disconnect the battery, please note that on a vehicle with power door locks, it's a wise precaution to remove the key from the ignition and to keep it with you, so that it does not get locked inside if the power door locks should engage accidentally when the battery is reconnected!

Devices known as "memory-savers" can be used to avoid some of these problems. Precise details vary according to the device used. The typical memory saver is plugged into the cigarette lighter and is connected to a spare battery. Then the vehicle battery can be disconnected from the electrical system. The memory saver will provide sufficient current to maintain audio unit security codes, PCM memory, etc. and will provide power to always hot circuits such as the clock and radio memory circuits. **Warning 1:** *Some memory savers deliver a considerable amount of current in order to keep vehicle systems operational after the main battery is disconnected. If you're using a memory saver, make sure that the circuit concerned is actually open before servicing it.* **Warning 2:** *If you're going to work near any of the airbag system components, the battery MUST be disconnected and a memory saver must NOT be used. If a memory saver is used, power will be supplied to the airbag, which means that it could accidentally deploy and cause serious personal injury.*

To disconnect the battery for service procedures requiring power to be cut from the vehicle, first open the driver's door to disable Retained Accessory Power (RAP), then loosen the cable end bolt and disconnect the cable from the negative battery terminal. Isolate the cable end to prevent it from coming into accidental contact with the battery terminal.

4   Battery - removal and installation

Refer to illustrations 4.2 and 4.3

1   On 2010 and later models, detach the Powertrain Control Module (PCM) from the battery cover (see Chapter 6).

2   Remove the fasteners securing the battery cover, then remove the cover and set it aside **(see illustration)**.

3   Disconnect the cable from the negative battery terminal first, then disconnect the cable from the positive battery terminal **(see illustration)**.

4   Remove the battery hold-down clamp.

5   Lift out the battery. Be careful - it's heavy. **Note:** *Battery straps and handlers are available at most auto parts stores for reasonable prices. They make it easier to remove and carry the battery.*

6   If you are replacing the battery, make sure you get one that's identical, with the same dimensions, amperage rating, cold cranking rating, etc.

7   Installation is the reverse of removal. Be sure to connect the positive cable first and the negative cable last.

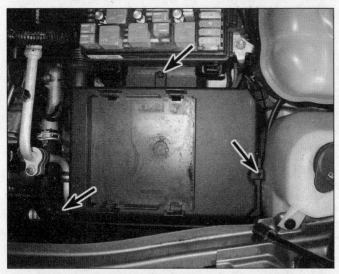

4.2 Battery cover mounting fasteners (3.4L V6 engine shown)

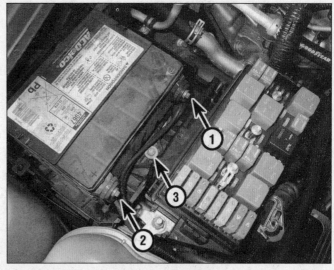

4.3 Battery details:

1	Negative battery cable	3	Battery hold-down clamp
2	Positive battery cable		

5   Battery cables - replacement

1   When removing the cables, always disconnect the cable from the negative battery terminal first and hook it up last, or you might accidentally short out the battery with the tool you're using to loosen the cable clamps. Even if you're only replacing the cable for the positive terminal, be sure to disconnect the negative cable from the battery first.

2   Disconnect the old cables from the battery, then trace each of them to their opposite ends and disconnect them. Be sure to note the routing of each cable before disconnecting it to ensure correct installation.

3   If you are replacing any of the old cables, take them with you when buying new cables. It is vitally important that you replace the cables with identical parts.

4   Clean the threads of the solenoid or ground connection with a wire brush to remove rust and corrosion. Apply a light coat of battery terminal corrosion inhibitor or petroleum jelly to the threads to prevent future corrosion.

5   Attach the cable to the solenoid or ground connection and tighten the mounting nut/bolt securely.

6   Before connecting a new cable to the battery, make sure that it reaches the battery post without having to be stretched.

7   Connect the cable to the positive battery terminal first, *then* connect the ground cable to the negative battery terminal.

6   Ignition coils or coil pack - replacement

1   Disconnect the cable from the negative battery terminal (see Section 3).

2   Remove the engine cover.

3.4L V6 models

Refer to illustration 6.3

3   Label and disconnect the spark plug wires from the coil pack **(see illustration)**.

4   Disconnect the electrical connectors, then remove the mounting fasteners from the coil pack.

5   Remove the coil pack.

All models except 3.4L V6 models

6   Remove the air cleaner outlet duct if it will interfere with the coil(s) you're removing.

7   Disconnect the electrical connector and mounting fastener from the ignition coil.

8   Remove the ignition coil from the spark plug.

All models

9   Apply a little silicone dielectric compound to the inside of each boot before installing it.

10   Installation is the reverse of removal.

7   Alternator - removal and installation

Refer to illustrations 7.5 and 7.7

1   Disconnect the cable from the negative battery terminal (see Section 3).

2   Remove the engine cover.

3   On 2.4L engine models, remove the inlet air duct (see Chapter 4).

4   Remove the drivebelt (see Chapter 1).

5   Disconnect the battery cable and the other wiring connections from the alter-

6.3 Ignition coil pack mounting details (3.4L V6 engine)

A	Spark plug wires
B	Electrical connector
C	Front mounting bolts

7.5 Disconnect the wiring from the back of the alternator (control circuit electrical connector not visible)

7.7 Alternator front mounting bolt (two rear bolts on the firewall side not visible) (3.4L V6 engine shown)

nator **(see illustration)**.

6   On 3.0L and 3.6L models, remove the idler pulley mounting bolt.

7   Remove the alternator mounting bolts and maneuver it out of the vehicle **(see illustration)**.

8   If you're replacing the alternator, take the old one with you when purchasing the replacement unit. Make sure that the new/rebuilt unit looks identical to the old alternator. Look at the electrical terminals on the backside of the alternator. They should be the same in number, size and location as the terminals on the

old alternator. Finally, look at the identification numbers. They will be stamped into the housing or printed on a tag attached to the housing. Make sure that the I.D. numbers are the same on both alternators.

9   Many new/rebuilt alternators DO NOT have a pulley installed, so you might have to swap the pulley from the old unit to the new/rebuilt one. When buying an alternator, find out the store's policy regarding pulley swaps. Some stores perform this service free of charge. If your local auto parts store doesn't offer this service, you'll have to purchase a puller for removing the

pulley and do it yourself.

10   Installation is the reverse of removal. Be sure to tighten the alternator mounting bolts securely.

11   Reconnect the cable to the negative terminal of the battery. Check the charging voltage (see Section 2) to verify that the alternator is operating correctly.

8   Starter motor - removal and installation

Refer to illustration 8.5

1   Detach the cable from the negative terminal of the battery (see Section 3).

2   Raise the vehicle and support it securely on jackstands.

3   On 2.4L engine models, remove the power brake booster pump fasteners and secure the pump out of the way.

4   If you're working on 3.0L and 3.6L models, remove the front catalytic converter (see Chapter 6).

5   Disconnect the wiring from the starter **(see illustration)**.

6   Remove the torque converter cover, then unscrew the starter mounting bolts and remove the starter.

7   Installation is the reverse of removal. Be sure to tighten the starter mounting bolts securely, then reconnect the cable to the negative terminal of the battery (see Section 4).

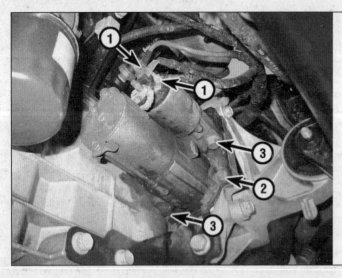

8.5 Starter motor details:

1   *Starter motor wiring*
2   *Torque converter cover mounting fastener*
3   *Starter motor mounting bolts*

Chapter 6
Emissions and engine control systems

Contents

Specifications

Torque specifications

Ft-lbs

Knock sensor
2.4L four-cylinder engine	17
3.0L V6 engine	18
3.4L V6 engine	18
3.6L V6 engine	
2009 and earlier models	17
2013 and later models	18

1   General information

To prevent pollution of the atmosphere from incompletely burned and evaporating gases, and to maintain good driveability and fuel economy, a number of emission control systems are incorporated. They include the:

Catalytic converter

A catalytic converter is an emission control device in the exhaust system that reduces certain pollutants in the exhaust gas stream. There are two types of converters: oxidation converters and reduction converters.

Oxidation converters contain a monolithic substrate (a ceramic honeycomb) coated with the semi-precious metals platinum and palladium. An oxidation catalyst reduces unburned hydrocarbons (HC) and carbon monoxide (CO) by adding oxygen to the exhaust stream as it passes through the substrate, which, in the presence of high temperature and the catalyst materials, converts the HC and CO to water vapor (H_2O) and carbon dioxide (CO_2).

Reduction converters contain a monolithic substrate coated with platinum and rhodium. A reduction catalyst reduces oxides of nitrogen (NOx) by removing oxygen, which in the presence of high temperature and the catalyst material produces nitrogen (N) and carbon dioxide (CO_2).

Catalytic converters that combine both types of catalysts in one assembly are known as "three-way catalysts" or TWCs. A TWC can reduce all three pollutants.

Evaporative Emissions Control (EVAP) system

The Evaporative Emissions Control (EVAP) system prevents fuel system vapors (which contain unburned hydrocarbons) from escaping into the atmosphere. On warm days, vapors trapped inside the fuel tank expand until the pressure reaches a certain threshold. Then the fuel vapors are routed from the fuel tank through the fuel vapor vent valve and the fuel vapor control valve to the EVAP canister, where they're stored temporarily until the next time the vehicle is operated. When the conditions are right (engine warmed up, vehicle up to speed, moderate or heavy load on the engine, etc.) the PCM opens the canister purge valve, which allows fuel vapors to be drawn from the canister into the intake manifold. Once in the intake manifold, the fuel vapors mix with incoming air before being drawn through the intake ports into the combustion chambers where they're burned up with the rest of the air/fuel mixture. The EVAP system is complex and virtually impossible to troubleshoot without the right tools and training.

Exhaust Gas Recirculation (EGR) system

The EGR system reduces oxides of nitrogen by recirculating exhaust gases from the exhaust manifold, through the EGR valve and intake manifold, then back to the combustion chambers, where it mixes with the incoming air/fuel mixture before being consumed. These recirculated exhaust gases dilute the incoming air/fuel mixture, which cools the combustion chambers, thereby reducing NOx emissions.

The EGR system consists of the Powertrain Control Module (PCM), the EGR valve, the EGR valve position sensor and various other information sensors that the PCM uses to determine when to open the EGR valve. The degree to which the EGR valve is opened is referred to as "EGR valve lift." The PCM is programmed to produce the ideal EGR valve lift for varying operating conditions. The EGR valve position sensor, which is an integral part of the EGR valve, detects the amount of EGR valve lift and sends this information to the PCM. The PCM then compares it with the appropriate EGR valve lift for the operating conditions. The PCM increases current flow to the EGR valve to increase valve lift and reduces the current to reduce the amount of lift. If EGR flow is inappropriate to the operating conditions (idle, cold engine, etc.) the PCM simply cuts the current to the EGR valve and the valve closes.

Secondary Air Injection (AIR) system

Some models are equipped with a secondary air injection (AIR) system. The secondary air injection system is used to reduce tailpipe emissions on initial engine start-up. The system uses an electric motor/pump assembly, relay, vacuum valve/solenoid, air shut-off valve, check valves and tubing to inject fresh air directly into the exhaust manifolds. The fresh air (oxygen) reacts with the exhaust gas in the catalytic converter to reduce HC and CO levels. The air pump and solenoid are controlled by the PCM through the AIR relay. During initial start-up, the PCM energizes the AIR relay, the relay supplies battery voltage to the air pump and the vacuum valve/solenoid, engine vacuum is applied to the air shut-off valve which opens and allows air to flow through the tubing into the exhaust manifolds. The PCM will operate the air pump until closed loop operation is reached (approximately four minutes). During normal operation, the check valves prevent exhaust backflow into the system.

Powertrain Control Module (PCM)

The Powertrain Control Module (PCM) is the brain of the engine management system. It also controls a wide variety of other vehicle systems. In order to program the new PCM, the dealer needs the vehicle as well as the new PCM. If you're planning to replace the PCM with a new one, there is no point in trying to do so at home because you won't be able to program it yourself.

Positive Crankcase Ventilation (PCV) system

The Positive Crankcase Ventilation (PCV) system reduces hydrocarbon emissions by scavenging crankcase vapors, which are rich in unburned hydrocarbons. A PCV valve or orifice regulates the flow of gases into the intake manifold in proportion to the amount of intake vacuum available.

The PCV system generally consists of the fresh air inlet hose, the PCV valve or orifice and the crankcase ventilation hose (or PCV hose). The fresh air inlet hose connects the air intake duct to a pipe on the valve cover. The crankcase ventilation hose (or PCV hose) connects the PCV valve or orifice in the valve cover to the intake manifold.

Emissions and engine control system locations (3.4L V6)

1   Accelerator Pedal Position (APP) sensor (top of the accelerator pedal)
2   Output shaft speed sensor (at the back of the transmission near the firewall)
3   Engine Coolant Temperature (ECT) sensor (on the left side of the rear cylinder head)
4   Powertrain Control Module (PCM)
5   Transmission range switch (under the battery box and tray)
6   Manifold Absolute Pressure (MAP) sensor
7   Knock sensors (on the sides of the engine block)
8   Oxygen sensors (on exhaust manifold and exhaust pipe)
9   EVAP purge control solenoid valve
10   Exhaust Gas Recirculation (EGR) valve
11   Mass Air Flow (MAF) sensor/Intake Air Temperature (IAT) sensor
12   Camshaft Position (CMP) sensor (front of the engine)
13   Crankshaft Position (CKP) sensor (side of the engine block, by the firewall)

Information Sensors

Accelerator Pedal Position (APP) sensor - as you press the accelerator pedal, the APP sensor alters its voltage signal to the PCM in proportion to the angle of the pedal, and the PCM commands a motor inside the throttle body to open or close the throttle plate accordingly

Camshaft Position (CMP) sensor - produces a signal that the PCM uses to identify the number 1 cylinder and to time the firing sequence of the fuel injectors

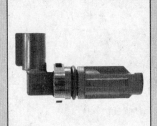

Crankshaft Position (CKP) sensor - produces a signal that the PCM uses to calculate engine speed and crankshaft position, which enables it to synchronize ignition timing with fuel injector timing, and to detect misfires

Engine Coolant Temperature (ECT) sensor - a thermistor (temperature-sensitive variable resistor) that sends a voltage signal to the PCM, which uses this data to determine the temperature of the engine coolant

Fuel tank pressure sensor - measures the fuel tank pressure and controls fuel tank pressure by signaling the EVAP system to purge the fuel tank vapors when the pressure becomes excessive

Intake Air Temperature (IAT) sensor - monitors the temperature of the air entering the engine and sends a signal to the PCM to determine injector pulse-width (the duration of each injector's on-time) and to adjust spark timing (to prevent spark knock)

Knock sensor - a piezoelectric crystal that oscillates in proportion to engine vibration which produces a voltage output that is monitored by the PCM. This retards the ignition timing when the oscillation exceeds a certain threshold

Manifold Absolute Pressure (MAP) sensor - monitors the pressure or vacuum inside the intake manifold. The PCM uses this data to determine engine load so that it can alter the ignition advance and fuel enrichment

Mass Air Flow (MAF) sensor - measures the amount of intake air drawn into the engine. It uses a hot-wire sensing element to measure the amount of air entering the engine

Oxygen sensors - generates a small variable voltage signal in proportion to the difference between the oxygen content in the exhaust stream and the oxygen content in the ambient air. The PCM uses this information to maintain the proper air/fuel ratio. A second oxygen sensor monitors the efficiency of the catalytic converter

Throttle Position (TP) sensor - a potentiometer that generates a voltage signal that varies in relation to the opening angle of the throttle plate inside the throttle body. Works with the PCM and other sensors to calculate injector pulse width (the duration of each injector's on-time)

Photos courtesy of Wells Manufacturing, except APP and MAF sensors.

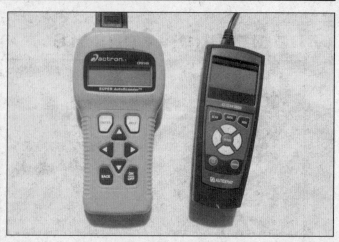

2.4a   Simple code readers are an economical way to extract trouble codes when the CHECK ENGINE light comes on

2.4b   Hand-held scan tools like these can extract computer codes and also perform diagnostics

2   On Board Diagnosis (OBD) system

OBD system general description

1   All models are equipped with the second generation OBD-II system. This system consists of an on-board computer known as the Powertrain Control Module (PCM), and information sensors, which monitor various functions of the engine and send data to the PCM. This system incorporates a series of diagnostic monitors that detect and identify fuel injection and emissions control system faults and store the information in the computer memory. This system also tests sensors and output actuators, diagnoses drive cycles, freezes data and clears codes.

2   The PCM is the brain of the electronically controlled fuel and emissions system. It receives data from a number of sensors and other electronic components (switches, relays, etc.). Based on the information it receives, the PCM generates output signals to control various relays, solenoids (fuel injectors) and other actuators. The PCM is specifically calibrated to optimize the emissions, fuel economy and driveability of the vehicle.

3   It isn't a good idea to attempt diagnosis or replacement of the PCM or emission control components at home while the vehicle is under warranty. Because of a federally-mandated warranty which covers the emissions system components and because any owner-induced damage to the PCM, the sensors and/or the control devices may void this warranty, take the vehicle to a dealer service department if the PCM or a system component malfunctions.

Scan tool information

Refer to illustrations 2.4a and 2.4b

4   Because extracting the Diagnostic Trouble Codes (DTCs) from an engine management system is now the first step in troubleshooting many computer-controlled systems and components, a code reader, at the very least, will be required **(see illustration)**. More powerful scan tools can also perform many of the diagnostics once associated with expensive factory scan tools **(see illustration)**. If you're planning to obtain a generic scan tool for your vehicle, make sure that it's compatible with OBD-II systems. If you don't plan to purchase a code reader or scan tool and don't have access to one, you can have the codes extracted by a dealer service department or an independent repair shop. **Note:** *Some auto parts stores even provide this service.*

3   Obtaining and clearing Diagnostic Trouble Codes (DTCs)

All models covered by this manual are equipped with on-board diagnostics. When the PCM recognizes a malfunction in a monitored emission or engine control system, component or circuit, it turns on the Malfunction Indicator Light (MIL) on the dash. The PCM will continue to display the MIL until the problem is fixed and the Diagnostic Trouble Code (DTC) is cleared from the PCM's memory. You'll need a scan tool to access any DTCs stored in the PCM.

Before outputting any DTCs stored in the PCM, thoroughly inspect ALL electrical connectors and hoses. Make sure that all electrical connections are tight, clean and free of corrosion. And make sure that all hoses are correctly connected, fit tightly and are in good condition (no cracks or tears).

Accessing the DTCs

Refer to illustration 3.1

1   The Diagnostic Trouble Codes (DTCs) can only be accessed with a code reader or scan tool. Professional scan tools are expensive, but relatively inexpensive generic code readers or scan tools **(see illustrations 2.4a and 2.4b)** are available at most auto parts stores. Simply plug the connector of the scan tool into the diagnostic connector **(see illustration)**. Then follow the instructions included with the scan tool to extract the DTCs.

2   Once you have outputted all of the stored DTCs, look them up on the accompanying DTC chart.

3   After troubleshooting the source of each DTC, make any necessary repairs or replace the defective component(s).

Clearing the DTCs

4   Clear the DTCs with the code reader or scan tool in accordance with the instructions provided by the tool's manufacturer.

Diagnostic Trouble Codes

5   The accompanying tables are a list of the Diagnostic Trouble Codes (DTCs) that can be accessed by a do-it-yourselfer working at home (there are many, many more DTCs available to professional mechanics with proprietary scan tools and software, but those codes cannot be accessed by a generic scan tool). If, after you have checked and repaired the connectors, wire harness and vacuum hoses (if applicable) for an emission-related system, component or circuit, the problem persists, have the vehicle checked by a dealer service department or other qualified repair shop.

3.1   The Data Link Connector (DLC) is located under the lower edge of the dash, to the left of the steering column

OBD-II trouble codes

Code	Probable cause
P0008	Engine position system performance (bank 1)
P0009	Engine position system performance (bank 2)
P0010	Intake camshaft position actuator circuit open (bank 1)
P0011	"A" Camshaft position - timing over-advanced (bank 1)
P0013	"B" Camshaft position - actuator circuit malfunction (bank 1)
P0014	"B" Camshaft position - timing over-advanced (bank 1)
P0016	Crankshaft position/camshaft position, bank 1, sensor A - correlation
P0017	Crankshaft position/camshaft position, bank 1, sensor B - correlation
P0018	Crankshaft position/camshaft position, bank 2, sensor A - correlation
P0019	Crankshaft position/camshaft position, bank 2, sensor B - correlation
P0020	Intake camshaft position actuator circuit open (bank 2)
P0021	Intake camshaft position-timing over-advanced (bank 2)
P0023	"B" Camshaft position - actuator circuit (bank 2)
P0024	"B" Camshaft position - timing over-advanced or system performance problem (bank 2)
P0030	HO2S heater control circuit (bank 1, sensor 1)
P0031	HO2S heater control circuit low (bank 1, sensor 1)
P0032	HO2S heater control circuit high (bank 1, sensor 1)
P0036	HO2S heater control circuit (bank 1 sensor 2)
P0037	HO2S heater control circuit low (bank 1, sensor 2)
P0038	HO2S heater control circuit high (bank 1, sensor 2)
P0050	HO2S heater control circuit (bank 2, sensor 1)
P0051	HO2S heater control circuit low (bank 2, sensor 1)
P0052	HO2S heater control circuit high (bank 2, sensor 1)
P0053	HO2S heater resistance (bank 1, sensor 1)
P0054	HO2S heater resistance (bank 1, sensor 2)

OBD-II trouble codes (continued)

Code	Probable cause
P0056	HO2S heater control circuit malfunction (bank 2, sensor 2)
P0057	HO2S heater control circuit low (bank 2, sensor 2)
P0058	HO2S heater control circuit high (bank 2, sensor 2)
P0068	Throttle Position (TP) sensor inconsistent with Mass Air Flow (MAF) sensor
P0089	Fuel pressure regulator performance
P0100	Mass air flow or volume air flow circuit malfunction
P0101	Mass air flow or volume air flow circuit, range or performance problem
P0102	Mass air flow or volume air flow circuit, low input
P0103	Mass air flow or volume air flow circuit, high input
P0106	Manifold absolute pressure or barometric pressure circuit, range or performance problem
P0107	Manifold absolute pressure or barometric pressure circuit, low input
P0108	Manifold absolute pressure or barometric pressure circuit, high input
P0111	Intake air temperature circuit, range or performance problem
P0112	Intake air temperature circuit, low input
P0113	Intake air temperature circuit, high input
P0115	Engine coolant temperature circuit
P0116	Engine coolant temperature circuit range/performance problem
P0117	Engine coolant temperature circuit, low input
P0118	Engine coolant temperature circuit, high input
P0119	Engine coolant temperature circuit, intermittent
P0120	Throttle position or pedal position sensor/switch circuit malfunction
P0121	Throttle position or pedal position sensor/switch circuit, range or performance problem
P0122	Throttle position or pedal position sensor/switch circuit, low input
P0123	Throttle position or pedal position sensor/switch circuit, high input
P0125	Insufficient coolant temperature for closed loop fuel control

Code	Probable cause
P0128	Coolant thermostat (coolant temperature below thermostat regulating temperature)
P0130	O2 sensor circuit malfunction (bank 1, sensor 1)
P0131	O2 sensor circuit, low voltage (bank 1, sensor 1)
P0132	O2 sensor circuit, high voltage (bank 1, sensor 1)
P0133	O2 sensor circuit, slow response (bank 1, sensor 1)
P0134	O2 sensor circuit - no activity detected (bank 1, sensor 1)
P0135	O2 sensor heater circuit malfunction (bank 1, sensor 1)
P0137	O2 sensor circuit, low voltage (bank 1, sensor 2)
P0138	O2 sensor circuit, high voltage (bank 1, sensor 2)
P0139	O2 sensor circuit, slow response (bank 1, sensor 2)
P0140	O2 sensor circuit - no activity detected (bank 1, sensor 2)
P0141	O2 sensor heater circuit malfunction (bank 1, sensor 2)
P0151	O2 sensor circuit, low voltage (bank 2, sensor 1)
P0152	O2 sensor circuit, high voltage (bank 2, sensor 1)
P0153	O2 sensor circuit, slow response (bank 2, sensor 1)
P0154	O2 sensor circuit - no activity detected (bank 2, sensor 1)
P0155	O2 sensor heater circuit malfunction (bank 2, sensor 1)
P0157	O2 sensor circuit, low voltage (bank 2, sensor 2)
P0158	O2 sensor circuit, high voltage (bank 2, sensor 2)
P0160	O2 sensor circuit - no activity detected (bank 2, sensor 2)
P0161	O2 sensor heater circuit malfunction (bank 2, sensor 2)
P0171	System too lean (bank 1)
P0172	System too rich (bank 1)
P0191	Fuel rail pressure sensor performance
P0192	Fuel rail pressure sensor circuit, low voltage
P0193	Fuel rail pressure sensor circuit, high voltage
P0201	Injector circuit malfunction - cylinder no. 1
P0202	Injector circuit malfunction - cylinder no. 2

OBD-II trouble codes (continued)

Code	Probable cause
P0203	Injector circuit malfunction - cylinder no. 3
P0204	Injector circuit malfunction - cylinder no. 4
P0205	Injector circuit malfunction - cylinder no. 5
P0206	Injector circuit malfunction - cylinder no. 6
P0218	Transmission overheating condition
P0220	Throttle position or pedal position sensor/switch B circuit malfunction
P0221	Throttle position or pedal position sensor/switch B, range or performance problem
P0222	Throttle position or pedal position sensor/switch B circuit, low input
P0223	Throttle position or pedal position sensor/switch B circuit, high input
P0230	Fuel pump primary circuit malfunction
P0261	Cylinder no. 1 injector circuit, low
P0262	Cylinder no. 1 injector circuit, high
P0264	Cylinder no. 2 injector circuit, low
P0265	Cylinder no. 2 injector circuit, high
P0267	Cylinder no. 3 injector circuit, low
P0268	Cylinder no. 3 injector circuit, high
P0270	Cylinder no. 4 injector circuit, low
P0271	Cylinder no. 4 injector circuit, high
P0273	Cylinder no. 5 injector circuit, low
P0274	Cylinder no. 5 injector circuit, high
P0276	Cylinder no. 6 injector circuit, low
P0277	Cylinder no. 6 injector circuit, high
P0300	Random/multiple cylinder misfire detected
P0301	Cylinder no. 1 misfire detected
P0302	Cylinder no. 2 misfire detected
P0303	Cylinder no. 3 misfire detected
P0304	Cylinder no. 4 misfire detected
P0305	Cylinder no. 5 misfire detected

Code	Probable cause
P0306	Cylinder no. 6 misfire detected
P0315	Crankshaft position system - variation not learned
P0324	Knock control system error
P0325	Knock sensor no. 1 circuit malfunction (bank 1 or single sensor)
P0326	Knock sensor no. 1 circuit, range or performance problem (bank 1 or single sensor)
P0327	Knock sensor no. 1 circuit, low input (bank 1 or single sensor)
P0328	Knock sensor no. 1 circuit, high input (bank 1 or single sensor)
P0330	Knock sensor no. 2 circuit malfunction (bank 2)
P0331	Knock sensor no. 2 circuit, range or performance problem (bank 2)
P0332	Knock sensor no. 2 circuit, low input (bank 2)
P0333	Knock sensor no. 2 circuit, high input (bank 2)
P0335	Crankshaft position sensor "A" - circuit malfunction
P0336	Crankshaft position sensor "A" - range or performance problem
P0338	Crankshaft position sensor "A" - high input
P0340	Camshaft position sensor "A" - circuit malfunction (bank 1)
P0341	Camshaft position sensor "A" - range or performance problem (bank 1)
P0342	Camshaft position sensor "A" - low input (bank 1)
P0343	Camshaft position sensor "A" - high input (bank 1)
P0346	Camshaft position sensor "A" - range/performance problem (bank 2)
P0347	Camshaft position sensor "A" - low input (bank 2)
P0348	Camshaft position sensor "A" - range/performance problem (bank 2)
P0351	Ignition coil 1 primary or secondary circuit malfunction
P0352	Ignition coil 2 primary or secondary circuit malfunction
P0353	Ignition coil 3 primary or secondary circuit malfunction
P0354	Ignition coil 4 primary or secondary circuit malfunction
P0355	Ignition coil 5 primary or secondary circuit malfunction
P0356	Ignition coil 6 primary or secondary circuit malfunction
P0366	Camshaft position sensor "B" - range/performance problem (bank 1)
P0367	Camshaft position sensor "B" - low input (bank 1)

OBD-II trouble codes (continued)

Code	Probable cause
P0368	Camshaft position sensor "B" circuit high input (bank 1)
P0391	Camshaft position sensor "B" - range/performance problem (bank 2)
P0392	Camshaft position sensor "B" - low input (bank 2)
P0393	Camshaft position sensor "B" - high input (bank 2)
P0401	Exhaust gas recirculation - insufficient flow detected
P0403	Exhaust gas recirculation - circuit malfunction
P0404	Exhaust gas recirculation - range or performance problem
P0405	Exhaust gas recirculation valve position sensor A - circuit low
P0406	Exhaust gas recirculation valve position sensor A - circuit high
P0411	Secondary air Injection system incorrect air flow detected
P0412	Secondary air injection valve control circuit
P0418	Secondary air injection pump relay control circuit
P0420	Catalyst system efficiency below threshold (bank 1)
P0430	Catalyst system efficiency below threshold (bank 2)
P0442	Evaporative emission control system, small leak detected
P0443	Evaporative emission control system, purge control valve circuit malfunction
P0446	Evaporative emission control system, vent control circuit malfunction
P0449	Evaporative emission control system, vent valve/solenoid circuit malfunction
P0450	Evaporative emission control system, pressure sensor malfunction
P0451	Evaporative emission control system, pressure sensor range or performance problem
P0452	Evaporative emission control system, pressure sensor low input
P0453	Evaporative emission control system, pressure sensor high input
P0454	Evaporative emission control system, pressure sensor intermittent
P0455	Evaporative emission (EVAP) control system leak detected (no purge flow or large leak)
P0458	Evaporative emission control system, purge control valve - circuit low
P0459	Evaporative emission control system, purge control valve - circuit high
P0461	Fuel level sensor circuit, range or performance problem
P0462	Fuel level sensor circuit, low input

Code	Probable cause
P0463	Fuel level sensor circuit, high input
P0464	Fuel level sensor circuit, intermittent
P0480	Cooling fan no. 1, control circuit malfunction
P0481	Cooling fan no. 2, control circuit malfunction
P0496	Evaporative emission system - high purge flow
P0497	Evaporative emission system - low purge flow
P0498	Evaporative emission system, vent control - circuit low
P0499	Evaporative emission system, vent control - circuit high
P0506	Idle control system, rpm lower than expected
P0507	Idle control system, rpm higher than expected
P0513	Incorrect immobilizer key
P0520	Engine oil pressure sensor/switch circuit malfunction
P0532	A/C refrigerant pressure sensor, low input
P0533	A/C refrigerant pressure sensor, high input
P0556	Brake booster pressure sensor performance problem
P0557	Brake booster pressure sensor circuit - low voltage
P0558	Brake booster pressure sensor circuit - high voltage
P0562	System voltage low
P0563	System voltage high
P0571	Cruise control/brake switch A, circuit malfunction
P0572	Cruise control/brake switch A, circuit low
P0573	Cruise control/brake switch A, circuit high
P0575	Cruise control system - input circuit malfunction
P0601	Internal control module, memory check sum error
P0602	Control module, programming error
P0603	Internal control module, keep alive memory (KAM) error
P0604	Internal control module, random access memory (RAM) error
P0606	PCM processor fault
P0607	Control module performance

OBD-II trouble codes (continued)

Code	Probable cause
P0615	Starter relay - circuit malfunction
P0616	Starter relay - circuit low
P0617	Starter relay - circuit high
P0621	Alternator L terminal circuit malfunction
P0622	Alternator F terminal circuit malfunction
P0625	Alternator field terminal - circuit low
P0626	Alternator field terminal - circuit high
P0627	Fuel pump control - circuit open
P0628	Fuel pump control - circuit low
P0629	Fuel pump control - circuit high
P0633	Immobilizer key not programmed - ECM
P0634	ECM/TCM - internal temperature too high
P0638	Throttle actuator control range/performance problem (bank 1)
P0641	Sensor reference voltage A - circuit open
P0642	Engine control module (ECM), knock control - defective
P0643	Sensor reference voltage A - circuit high
P0644	Driver display, serial communication - circuit malfunction
P0645	A/C clutch relay control circuit
P0650	Malfunction indicator lamp (MIL), control circuit malfunction
P0651	Sensor reference voltage B - circuit open
P0653	Sensor reference voltage B - circuit high
P0667	ECM/TCM internal temperature sensor - circuit range/performance problem
P0668	ECM/TCM internal temperature sensor - circuit low
P0669	ECM/TCM internal temperature sensor - circuit high
P0685	ECM power relay, control - circuit open
P0686	ECM power relay control - circuit low
P0687	Engine, control relay - short to ground
P0689	ECM power relay sense - circuit low

Code	Probable cause
P0690	ECM power relay sense - circuit high
P0691	Engine coolant blower motor 1 - short to ground
P0700	Transmission control system malfunction
P0703	Torque converter/brake switch B, circuit malfunction
P0705	Transmission range sensor, circuit malfunction (PRNDL input)
P0711	Transmission fluid temperature sensor circuit, range or performance problem
P0712	Transmission fluid temperature sensor circuit, low input
P0713	Transmission fluid temperature sensor circuit, high input
P0717	Input/turbine speed sensor circuit, no signal
P0722	Output speed sensor circuit, no signal
P0727	Engine speed input circuit, no signal
P0730	Incorrect gear ratio
P0731	Incorrect gear ratio, first gear
P0732	Incorrect gear ratio, second gear
P0733	Incorrect gear ratio, third gear
P0734	Incorrect gear ratio, fourth gear
P0735	Incorrect gear ratio, fifth gear
P0736	Incorrect gear ratio, reverse gear
P0741	Torque converter clutch, circuit performance problem or stuck in Off position
P0742	Torque converter clutch circuit, stuck in On position
P0762	Shift solenoid C, stuck in On position
P0856	Traction Control Torque Request Circuit
P0962	Pressure control (PC) solenoid A - control circuit low
P0963	Pressure control (PC) solenoid A - control circuit high
P0966	Pressure control (PC) solenoid B - control circuit low
P0967	Pressure control (PC) solenoid B - control circuit high
P0970	Pressure control (PC) solenoid C - control circuit low
P0971	Pressure control (PC) solenoid C - control circuit high
P0973	Shift solenoid (SS) A - control circuit low

OBD-II trouble codes (continued)

Code	Probable cause
P0974	Shift solenoid (SS) A - control circuit high
P0976	Shift solenoid (SS) B - control circuit low
P0977	Shift solenoid (SS) B - control circuit high
P0979	Shift solenoid (SS) C - control circuit low
P0980	Shift solenoid (SS) C - control circuit high
P0982	Shift solenoid (SS) D - control circuit low
P0983	Shift solenoid (SS) D - control circuit high
P0985	Shift solenoid (SS) E - control circuit low
P0986	Shift solenoid (SS) E - control circuit high

4   Accelerator Pedal Position (APP) sensor - replacement

Refer to illustration 4.1

1   Disconnect the electrical connector from the APP sensor **(see illustration)**.
2   Remove the APP sensor module mounting bolts and remove the sensor module. **Note:** *The top bolt must be left in the sensor until it's removed from the vehicle. Make sure to place it in its hole before installing the sensor.*
3   Installation is the reverse of removal.

4.1 The APP sensor is located at the top of the accelerator pedal. To unplug the connector, slide the white lock (A) out, then depress the tab on the side (B) and pull the connector off

5   Camshaft Position (CMP) sensor - replacement

3.4L V6 models

Refer to illustration 5.2

1   Remove the air intake duct (see Chapter 4, Section 9).
2   Disconnect the CMP sensor electrical connector **(see illustration)**.
3   Remove the mounting bolt and CMP sensor.
4   Installation is the reverse of removal.

3.0L and 3.6L V6 models

Refer to illustration 5.5

5   The 3.0L and 3.6L V6 engines have four camshaft sensors **(see illustration)**.
6   If you're replacing a CMP sensor on the right side, remove the air filter housing (see Chapter 4).
7   On 2013 and later 3.6L models, if you're replacing the CMP sensor on the left side (front cylinder bank), remove the front engine mount (see Chapter 2C).
8   Disconnect the CMP sensor electrical connector.
9   Remove the CMP sensor mounting bolt and the CMP sensor.
10   Installation is the reverse of removal.

2.4L models

11   The 2.4L four-cylinder engine has two camshaft sensors, the intake camshaft position sensor, located just below the fuel pump and the exhaust camshaft sensor located at the corner of the cylinder head above the exhaust manifold.

12   To remove the intake camshaft position sensor, remove the fuel pump cover (see Chapter 4).
13   Disconnect the electrical connector to the sensor(s).
14   Remove the mounting bolt and the sensor from the cylinder head.
15   Installation is the reverse of removal.

6   Camshaft Position Actuator Solenoid Valve (3.0L and 3.6L V6 models) - replacement

Refer to illustration 6.1

1   The 3.6L V6 engine has four camshaft position actuator solenoid valves **(see illustration)**.
2   If you're working on a right side solenoid valve, remove the air filter housing (see Chapter 4).
3   On 2013 and later 3.6L models, if you're replacing the solenoid valve on the left side (front cylinder bank), remove the front engine mount (see Chapter 2C).
4   Disconnect the solenoid valve electrical connector.
5   Remove the solenoid valve mounting bolt and the solenoid valve.
6   Installation is the reverse of removal.

7   Crankshaft Position (CKP) sensor - replacement

3.4L V6 models

Refer to illustration 7.2

1   Raise the front of the vehicle and place it securely on jackstands.

5.2 On 3.4L V6 models, the camshaft position sensor is located at the front of the engine

5.5 On 3.0L and 3.6L V6 models, the camshaft position sensors are located on the front of the cylinder heads, just below the valve cover (left side [front bank cylinder head] intake sensor shown, right side [rear bank cylinder head] sensors similar)

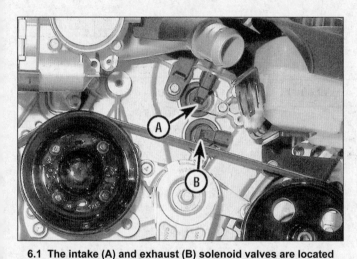

6.1 The intake (A) and exhaust (B) solenoid valves are located next to each other, on the front of the cylinder heads, just below the camshaft position sensors (left side [front bank cylinder head] shown, right side [rear bank cylinder head] similar)

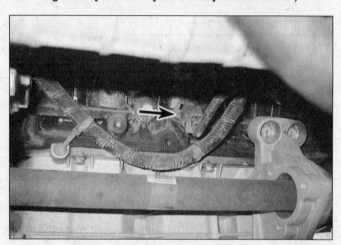

7.2 The CKP sensor on 3.4L V6 models is located on the firewall side of the engine block

2   Disconnect the wiring from the CKP sensor **(see illustration).**
3   Unscrew the mounting fastener and remove the CKP sensor.
4   Installation is the reverse of removal.

3.0L and 3.6L V6 models

Refer to illustration 7.7

5   Raise the front of the vehicle and place it securely on jackstands.
6   If you're working on 2009 and earlier models with AWD, remove the transfer case (see Chapter 7B). If you're working on 2010 and later models, remove the right side catalytic converter (see Section 16).
7   Disconnect the wiring from the CKP sensor **(see illustration).**
8   Unscrew the mounting fastener and remove the CKP sensor.
9   Installation is the reverse of removal.

2.4L models

10   The 2.4L four-cylinder engine the crankshaft sensor is located just below the oil filter housing.

11   Disconnect the electrical connector to the sensor.
12   Remove the mounting bolt and the sensor from the cylinder block.
13   Installation is the reverse of removal.

7.7 The CKP sensor on 3.6L V6 models is located at the rear of the engine block on the firewall side

8.3 On 3.4L V6 engines, the ECT sensor is located on the left (driver's) side of the rear cylinder head

9.2 The MAF/IAT sensor is located on the air cleaner cover

10.4 Two knock sensors are used on the engine - one on each side of the engine block

8   Engine Coolant Temperature (ECT) sensor - replacement

Refer to illustration 8.3

Warning: *Wait until the engine has cooled completely before beginning this procedure.*

1   Remove the engine cover.

2   Drain the engine coolant (see Chapter 1). **Note:** *It is possible to replace the sensor without draining the coolant. If you don't drain the coolant, some coolant will run out of the coolant crossover when you remove the ECT sensor, so install the new sensor as quickly as possible.*

3   Disconnect the electrical connector from the ECT sensor **(see illustration)**. **Note:** *The ECT sensor on 3.0L and 3.6L V6 models is located at the front of the engine, on the side of the engine block, just below the valve cover. On 2.4L four-cylinder models the sensor is located below the exhaust manifold towards the end of the engine.*

4   Unscrew the ECT sensor with a wrench (most deep sockets won't fit over the ECT sensor).

5   When installing the ECT sensor, apply some sealant to the threads

6   Installation is the reverse of removal. Refill the cooling system (or, if you didn't drain the coolant, check the coolant level, adding as necessary) (see Chapter 1).

9   Mass Air Flow (MAF) sensor/ Intake Air Temperature (IAT) sensor - replacement

Refer to illustration 9.2

Note: *The IAT sensor is an integral component of the Mass Air Flow (MAF) sensor.*

1   Remove the air intake duct (see Chapter 4, Section 9).

2   Disconnect the electrical connector from the MAF sensor **(see illustration)**.

3   Remove the fasteners and detach the

sensor from the air filter housing.

4   Installation is the reverse of removal.

10   Knock sensor - replacement

Refer to illustration 10.4

1   Disconnect the cable from the negative battery terminal (see Chapter 5).

2   Raise the vehicle and support it securely on jackstands. On 3.0L models, remove the starter motor (see Chapter 5).

Note: *On 2013 and later 3.6L models, the left-side (front cylinder bank) knock sensor is located between the oil filter and the starter motor.*

3   If you're working on a 3.0L or 3.6L V6 model and are removing a knock sensor from the rear of the engine, remove the right catalytic converter (see Section 16).

4   Disconnect the knock sensor electrical connector **(see illustration)**.

5   Unscrew the knock sensor retaining bolt and detach the sensor from the engine block.

6   Installation is the reverse of removal. Be sure to tighten the knock sensor bolt to the torque listed in this Chapter's Specifications.

11   Manifold Absolute Pressure (MAP) sensor - replacement

1   Remove the engine cover.

2.4L models

2   Remove the air cleaner outlet duct from the throttle body (see Chapter 4).

3   Disconnect the electrical connector to the sensor.

4   Remove the mounting bolt and the sensor from the intake manifold.

5   Installation is the reverse of removal.

3.4L models

Refer to illustration 11.6

6   Disconnect the electrical connector from the MAP sensor **(see illustration)**. Remove the bolt and hold-down bracket and detach the sensor from the upper intake manifold.

7   Installation is the reverse of removal. Be sure to use a new O-ring and tighten the hold-down bolt securely.

2013 and later 3.6L models

8   Remove the engine cover then disconnect the electrical connector from the MAP sensor located on top of the intake manifold.

11.6 The Manifold Absolute Pressure (MAP) sensor is located on the upper intake manifold (3.4L V6 engine)

12.4 On 3.4L V6 engines, the upstream oxygen sensor is located in the rear exhaust manifold. On 3.6L V6 engines, there are two upstream oxygen sensors, one in each exhaust manifold

12.9 The downstream oxygen sensor(s) are located in the exhaust pipe, just after the catalytic converter(s)

9   Remove the sensor mounting bolt and pull the sensor from the manifold.
10   Installation is the reverse of removal. Be sure to use a new O-ring and tighten the hold-down bolt securely.

12 Oxygen sensors - replacement

Note: *Because it is installed in the exhaust manifold or pipe, both of which contract when cool, an oxygen sensor might be very difficult to loosen when the engine is cold. Rather than risk damage to the sensor or its mounting threads, start and run the engine for a minute or two, then shut it off. Be careful not to burn yourself during the following procedure.*
1   Be particularly careful when servicing an oxygen sensor:
  a) *Oxygen sensors have a permanently attached pigtail and an electrical connector that cannot be removed. Damaging or removing the pigtail or electrical connector will render the sensor useless.*
  b) *Keep grease, dirt and other contaminants away from the electrical connector and the louvered end of the sensor.*
  c) *Do not use cleaning solvents of any kind on an oxygen sensor.*
  d) *Oxygen sensors are extremely delicate. Do not drop a sensor or handle it roughly.*

  e) *Make sure that the silicone boot on the sensor is installed in the correct position. Otherwise, the boot might melt and it might prevent the sensor from operating correctly.*

Replacement
Upstream oxygen sensors

Refer to illustration 12.4
Note: *These sensors are either in the exhaust pipe or screwed directly into the exhaust manifold(s). The 2.4L four-cylinder and 3.4L V6 engines have two oxygen sensors; one in the rear exhaust manifold and one in the exhaust pipe after the catalytic converter. The 3.0L and 3.6L V6 have four oxygen sensors; one in each exhaust manifold, and one just after each catalytic converter.*
2   If you're working on a 2.4L four-cylinder, 3.0L V6 or 3.6L V6 model, remove the engine cover.
3   If you're working on a 3.4L V6 model, raise the vehicle and support it securely on jackstands.
4   Disconnect the upstream oxygen sensor electrical connector **(see illustration)**. Unclip it from any retainers.
5   Remove the upstream oxygen sensor.
6   If you're going to install the old sensor, apply anti-seize compound to the threads of the sensor to facilitate future removal. If you're going to install a new oxygen sensor, it's not necessary to apply anti-seize compound to the threads; the threads on new sensors

already have anti-seize compound on them.
7   Installation is the reverse of removal. Be sure to tighten the oxygen sensor securely.

Downstream oxygen sensors

Refer to illustration 12.9
8   Raise the vehicle and support it securely on jackstands.
9   Locate the downstream oxygen sensor **(see illustration)**, then trace the lead up to the electrical connector and disconnect the connector.
10   Unscrew the downstream oxygen sensor.
11   If you're going to install the old sensor, apply anti-seize compound to the threads of the sensor to facilitate future removal. If you're going to install a new oxygen sensor, it's not necessary to apply anti-seize compound to the threads. The threads on new sensors already have anti-seize compound on them.
12   Installation is the reverse of removal. Be sure to tighten the oxygen sensor securely.

13 Transmission range switch - removal and installation

Note: *This switch is sometimes referred to as the Park/Neutral Position (PNP) switch.*

Removal

Refer to illustrations 13.3a, 13.3b and 13.6
1   Place the shift lever in Neutral.
2   Refer to Chapter 5 and remove the battery.

13.3a Battery box fasteners

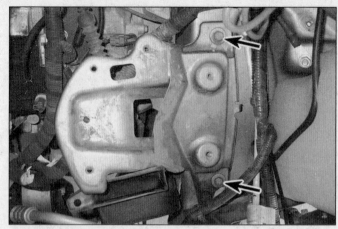

13.3b Battery tray upper mounting fasteners (lower fastener not visible in photo)

13.6 Hold the lever with a wrench while unscrewing the lever nut

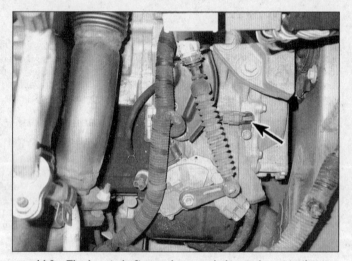

14.3a The input shaft speed sensor is located next to the transmission range switch

3   Remove the battery (see Chapter 5) battery box and tray **(see illustrations).**
4   Pry the end of the shift cable off of the ball stud on the transaxle range switch. Carefully use two screwdrivers to do this without twisting the end of the cable.
5   Disconnect the electrical connector from the switch.
6   Remove the control lever nut and remove the lever **(see illustration)**.
7   Remove the mounting fasteners and lift off the switch.

Installation

8   Make sure that the transaxle is in Neutral.
9   Put the new switch in place, aligning the flats on the switch with the flats on the shaft.
10   Install the switch, but leave the mounting fasteners snug until you verify that the engine will only start in Park and Neutral.
11   Install the lever.
12   The remainder of installation is the reverse of removal.
13   Verify that the engine will only start in

Park or Neutral.
14   Tighten the switch bolts securely.

14   Transmission speed sensors - replacement

Refer to illustrations 14.3a and 14.3b
1   Refer to Chapter 5 and remove the battery.
2   If you're removing the input shaft speed sensor, remove the battery (see Chapter 5), battery box and tray **(see illustrations 13.3a and 13.b)**. If you're removing the output shaft speed sensor, raise the front of the vehicle and support it securely on jackstands.
3   Remove the sensor mounting bolt and remove the sensor along with its O-ring **(see illustrations).**
4   Installation is the reverse of removal. Be sure to use a new O-ring and to tighten the sensor mounting bolt securely.

15   Powertrain Control Module (PCM) - removal and installation

Refer to illustration 15.2
Caution: *To avoid electrostatic discharge damage to the PCM, handle the PCM only by its case. Do not touch the electrical terminals during removal and installation. If available, ground yourself to the vehicle with an anti-static ground strap, available at computer supply stores.*
Note: *This procedure applies only to disconnecting, removing and installing the PCM that is already installed in your vehicle. If the PCM is defective and has to be replaced, it must be programmed with new software and calibrations. This procedure requires the use of GM's TECH-2 scan tool and GM's latest PCM-programming software, so you WILL NOT BE ABLE TO REPLACE THE PCM AT HOME.*
1   Disconnect the cable from the negative battery terminal (see Chapter 5).

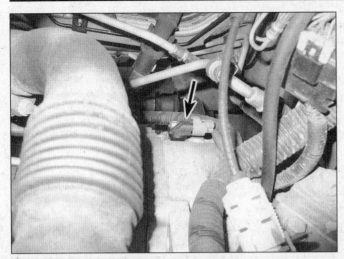

14.3b The output shaft speed sensor is located at the back of the transaxle near the firewall

15.2 The PCM is located on top of the battery cover

2   Disconnect the wiring harnesses from the PCM **(see illustration).**

3   Release the retaining tabs and slide the PCM forward to remove it.

4   Remove the PCM from its housing.

5   Installation is the reverse of removal.

16   Catalytic converter - replacement

Warning: *Replace catalytic converters only after enough time has elapsed after driving the vehicle to allow the system components to cool completely. Also, when working under the vehicle, make sure it is securely supported on jackstands.*

Note 1: *The following is a generalized procedure for most catalytic converters. The specifics vary among the engines used in these vehicles.*

Note 2: *Many exhaust specialist shops are able to replace catalytic converters at a lower cost than what you might pay for a new one from a dealer.*

1   Raise the vehicle and place it securely

on jackstands.

2   Disconnect the electrical connector from the related oxygen sensor and remove the sensor (see Section 12).

3   Support the other sections of the exhaust system as necessary.

4   Remove the retaining nuts from the front and rear catalyst mounting flanges.

5   Remove the catalyst and pull off the gaskets.

6   Installation is the reverse of removal. Be sure to replace any rusted or damaged fasteners along with the gaskets.

17   Evaporative Emissions Control (EVAP) system - component replacement

EVAP purge control solenoid valve

Refer to illustration 17.2

1   Remove the engine cover.

2   Disconnect the electrical connector from the valve **(see illustration).**

3   Disconnect the hose. See Chapter 4 for information on quick-connect fittings.

4   Unscrew the bolt and remove the purge valve from the upper intake manifold.

5   Installation is the reverse of removal.

EVAP canister

Refer to illustration 17.7

6   Raise the vehicle and support it securely on jackstands.

7   Disconnect the hoses from the canister **(see illustration).** See Chapter 4 for information on quick-connect fittings.

8   Disconnect the electrical connector from the canister vent solenoid valve.

9   Remove the fasteners securing the canister, and remove the canister.

10   Installation is the reverse of removal.

17.2 The EVAP purge control solenoid is located on the top of the upper intake manifold (3.4L V6 model shown, 3.6L V6 similar)

17.7 The EVAP canister is located under the left side of the vehicle

18　Exhaust Gas Recirculation (EGR) valve (3.4L V6 models) - replacement

Refer to illustration 18.1

1　Disconnect the electrical connector from the EGR valve **(see illustration)**.
2　Disconnect the pipe on the EGR valve.
3　Remove the valve mounting fasteners and lift off the EGR valve.
4　Remove the gasket and clean both gasket surfaces.
5　Installation is the reverse of removal. Be sure to use a new EGR valve gasket, and tighten the EGR valve mounting nuts securely.

19　Positive Crankcase Ventilation (PCV) valve - replacement

3.4L V6 models

Refer to illustration 19.1

1　The PCV valve on the 3.4L V6 engine is part of the front valve cover **(see illustration)** and therefore not removable from the valve cover. The valve cover and PCV valve must be replaced as an assembly.

3.0L and 3.6L V6 models

Note: *The PCV valve on this engine is actually a fixed orifice which regulates the flow of blow-by gases into the intake manifold.*

2　Remove the engine cover.
3　The 3.6L engine has a PCV valve in both the front and rear valve covers. Disconnect the hose from the PCV valve.
4　Using a pair of pliers, carefully grip the top of the valve, and with a twisting motion pull the valve out of the valve cover.
5　Installation is the reverse of removal. Apply a small amount of RTV sealant at the mating points of PCV valve and the valve cover.

2.4L models

6　On 2.4L models the PCV is not removable from the valve cover. If hose to the PCV end or the PCV is damaged the valve cover must be replaced.

18.1　The EGR valve is located at the front of the upper intake manifold

19.1　The PCV valve on the 3.4L V6 engine is part of the front valve cover

Notes

Notes

Chapter 7 Part A
Automatic transaxle

Contents

Specifications

General

Fluid type and capacity	See Chapter 1

Torque specifications

	Ft-lbs
Engine/transaxle subframe bolts*	114
Torque converter-to-driveplate bolts	45
Transaxle-to-engine mounting bolts	56
Transaxle mount fasteners	
Five-speed transaxle	
Left-side mount	
To chassis	27
To transaxle	42
Front mount	
Through-bolt	81
To frame bolts	
AWD models	37
FWD models	42
Rear mount	
Through-bolt	81
To frame bolts	
AWD models	37
FWD models	42
Mount bracket-to-transaxle	81
Six-speed transaxle	
Left side mount	
To frame bolts	20
To transmission bolts	20
Front mount	
Through-bolt	
2009 and earlier models	81
2010 and later models	74
To frame bolts	
AWD models	37
FWD models	42
Rear mount	
Through-bolt	
2009 and earlier models	81
2010 and later models	74
To frame bolts	
AWD models	37
FWD models	42
Mount bracket-to-transaxle	
Large bolts	81
Small bolt(s)	37
Mount through-bolts	81
Mount-to-frame fasteners	37
Mount-to-engine fasteners	37

*These bolts must be replaced with **new** ones whenever they have been loosened or removed.*

1 General information

Information on the automatic transaxle is included in this Part of Chapter 7. Information on the transfer case used on all-wheel drive models can be found in Part B.

Because of the complexity of the automatic transaxles and the specialized equipment necessary to perform most service operations, this Chapter contains only those procedures related to general diagnosis, routine maintenance, adjustment and removal and installation.

If the transaxle requires major repair work, it should be left to a dealer service department or an automotive or transmission repair shop. Once properly diagnosed you can, however, remove and install the transaxle yourself and save the expense, even if the repair work is done by a transmission shop. Keep in mind, however, that transaxle removal is difficult on these models. Transmission shops are generally equipped with vehicle hoists and other specialized equipment that is necessary for transaxle removal.

2 Diagnosis - general

1   Automatic transaxle malfunctions may be caused by five general conditions:

a) *Poor engine performance*
b) *Improper adjustments*
c) *Hydraulic malfunctions*
d) *Mechanical malfunctions*
e) *Malfunctions in the computer or its signal network*

2   Diagnosis of these problems should always begin with a check of the easily repaired items: fluid level and condition (see Chapter 1), shift cable adjustment and shift lever installation. Next, perform a road test to determine if the problem has been corrected or if more diagnosis is necessary. If the problem persists after the preliminary tests and corrections are completed, additional diagnosis should be performed by a dealer service department or other qualified transmission repair shop. Refer to the *Troubleshooting* Section at the front of this manual for information on symptoms of transaxle problems.

Preliminary checks

3   Drive the vehicle to warm the transaxle to normal operating temperature.
4   Check the fluid level as described in Chapter 1:

a) *If the fluid level is unusually low, add enough fluid to bring the level within the designated area of the dipstick, then check for external leaks (see following).*
b) *If the fluid level is abnormally high, drain off the excess, then check the drained fluid for contamination by coolant. The presence of engine coolant in the automatic transmission fluid indicates that a failure has occurred in the internal*

radiator oil cooler walls that separate the coolant from the transmission fluid (see Chapter 3).

c) *If the fluid is foaming, drain it and refill the transaxle, then check for coolant in the fluid, or a high fluid level.*

5   Check the engine idle speed. **Note:** *If the engine is malfunctioning, do not proceed with the preliminary checks until it has been repaired and runs normally.*
6   Check and adjust the shift cable, if necessary (see Section 4).
7   If hard shifting is experienced, inspect the shift cable under the steering column and at the manual lever on the transaxle (see Section 4).

Fluid leak diagnosis

8   Most fluid leaks are easy to locate visually. Repair usually consists of replacing a seal or gasket. If a leak is difficult to find, the following procedure may help.
9   Identify the fluid. Make sure it's transmission fluid and not engine oil or brake fluid (automatic transmission fluid is a deep red color).
10   Try to pinpoint the source of the leak. Drive the vehicle several miles, then park it over a large sheet of cardboard. After a minute or two, you should be able to locate the leak by determining the source of the fluid dripping onto the cardboard.
11   Make a careful visual inspection of the suspected component and the area immediately around it. Pay particular attention to gasket mating surfaces. A mirror is often helpful for finding leaks in areas that are hard to see.
12   If the leak still cannot be found, clean the suspected area thoroughly with a degreaser or solvent, then dry it thoroughly.
13   Drive the vehicle for several miles at normal operating temperature and varying speeds. After driving the vehicle, visually inspect the suspected component again.
14   Once the leak has been located, the cause must be determined before it can be properly repaired. If a gasket is replaced but the sealing flange is bent, the new gasket will not stop the leak. The bent flange must be straightened.
15   Before attempting to repair a leak, check to make sure that the following conditions are corrected or they may cause another leak. **Note:** *Some of the following conditions cannot be fixed without highly specialized tools and expertise. Such problems must be referred to a qualified transmission shop or a dealer service department.*

Gasket leaks

16   Check the pan periodically. Make sure the bolts are tight, no bolts are missing, the gasket is in good condition and the pan is flat (dents in the pan may indicate damage to the valve body inside).
17   If the pan gasket is leaking, the fluid level or the fluid pressure may be too high, the vent may be plugged, the pan bolts may be too tight, the pan sealing flange may be warped,

the sealing surface of the transaxle housing may be damaged, the gasket may be damaged or the transaxle casting may be cracked or porous. If sealant instead of gasket material has been used to form a seal between the pan and the transaxle housing, it may be the wrong type of sealant.

Seal leaks

18   If a transaxle seal is leaking, the fluid level or pressure may be too high, the vent may be plugged, the seal bore may be damaged, the seal itself may be damaged or improperly installed, the surface of the shaft protruding through the seal may be damaged or a loose bearing may be causing excessive shaft movement.
19   Make sure the dipstick tube seal is in good condition and the tube is properly seated. Periodically check the area around the sensors for leakage. If transmission fluid is evident, check the seals for damage.

Case leaks

20   If the case itself appears to be leaking, the casting is porous and will have to be repaired or replaced.
21   Make sure the oil cooler hose fittings are tight and in good condition.

Fluid comes out vent pipe or fill tube

22   If this condition occurs, the possible causes are: the transaxle is overfilled, there is coolant in the fluid, the case is porous, the dipstick is incorrect, the vent is plugged or the drain-back holes are plugged.

3 Shift lever - replacement

Warning: *These models are equipped with airbags. Always disable the airbag system before working in the vicinity of any airbag system component to avoid the possibility of accidental deployment of the airbag(s), which could cause personal injury (see Chapter 12).*
1   Disconnect the cable from the negative battery terminal (see Chapter 5).
2   Turn the key to the Run position, then put the shifter into Neutral.
3   Use a screwdriver to carefully push the trim collar at the base of the shift knob downward.
4   Loosen (but don't remove) the set screw, then lift off the knob.

2005 through 2006 models

5   Remove the console (see Chapter 11).
6   Release the ignition interlock cable from the bracket by pinching the tabs on the clip, then pry the end from the shift lever.

2007 and later models

7   Refer to Chapter 11 and remove the instrument panel center trim panel, the console top panel and the right console side panel.

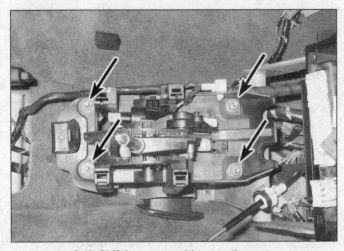

3.10 Shift lever assembly mounting nuts

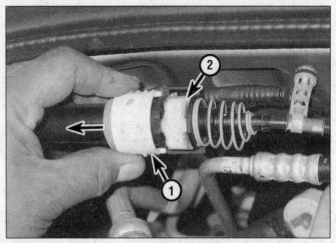

4.3 Slide the white plastic cover (1) on the shift cable junction to expose the lock (2)

All models

Refer to illustration 3.10

8   Pry the end of the shift cable from the shift lever, then pinch the tabs on the cable to release it from the bracket.

9   Disconnect the wiring from the shift assembly.

10   Remove the mounting nuts and lift the shifter assembly up **(see illustration)**.

11   Installation is the reverse of removal.

---

4   Shift cable - replacement and adjustment

Warning: *These models are equipped with airbags. Always disable the airbag system before working in the vicinity of any airbag system component to avoid the possibility of accidental deployment of the airbag(s), which could cause personal injury (see Chapter 12).*

Replacement

2005 and 2006 models

Refer to illustrations 4.3, 4.7, 4.12, 4.13a and 4.13b

Note: *These vehicles use a two-piece cable. If one half is damaged, it can be replaced independently. The procedure below is for detaching the two pieces and removing both of them. If both parts need to be replaced, you can ignore the Steps regarding separation of the cable halves.*

Separation of cable halves

1   Make sure that the shift lever is in Park.

2   Unclip the shift cable from the retainer on the firewall.

3   Retract the coupling cover **(see illustration)**, then pull up the lock.

4   Turn the cable to access the small clip, then pull the clip out to separate the two cable halves.

Rear cable section

5   Remove the console (see Chapter 11).

6   Make sure that the shift lever is in Park.

7   Pry the end of the shift cable from the shift lever, then pinch the tabs on the cable body to release it from the bracket **(see illustration)**.

8   Pull the firewall grommet out. Detach the cable from the cowl and firewall.

9   Note the cable routing, then pull it out.

10   Installation is the reverse of removal. Proceed to adjust the cable.

Front cable section

11   Remove the PCM (see Chapter 6), the battery and the battery tray (see Chapter 5).

12   Remove the covers from the underhood fuse box, then unclip it and move it aside **(see illustration)**. Remove the other interfering components under the fuse box.

13   Carefully pry the cable end from the transaxle shift lever. Pinch the tabs of the

4.7 Pry the cable end off of the ballstud at the shift lever

4.12 Remove the two fuse block covers, then use a screwdriver to release these clips - you can then position the fuse block aside

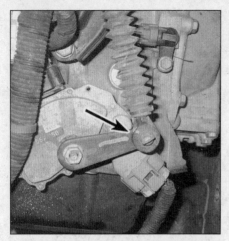

4.13a Pry this end of the cable off of the transaxle shift lever

4.13b The cable housing is detached from the transaxle bracket by pinching the retainer

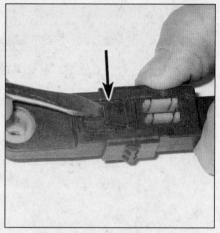

4.32 Release the lock tab with a screwdriver, then slide it to release the adjuster

cable to disconnect it from the transaxle bracket **(see illustrations)**.
14   Detach the cable from the retainers on the firewall and remove it.
15   Installation is the reverse of removal. Proceed to adjust the cable.

2007 and later models
16   Remove the PCM (see Chapter 6), the battery and the battery tray (see Chapter 5).
17   Make sure that the shift lever is in Park.
18   Carefully pry the cable end from the transaxle shift lever. Pinch the tabs of the cable to disconnect it from the transaxle bracket.
19   Use a screwdriver to carefully push the trim collar at the base of the shift knob downward. Loosen (but don't remove) the set screw, then lift off the knob.
20   Remove the instrument panel center trim panel, the console top panel and the right console side panel (see Chapter 11).

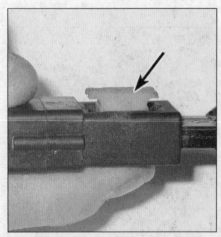

4.33 Push the white plastic adjuster to release it - when the cable has adjusted itself, push it to the locked position and slide the cover back into place

21   Pry the end of the shift cable from the shift lever, then pinch the tabs on the cable to release it from the bracket.
22   Note how the cable is routed, then pull the firewall grommet out and extract the cable.
23   Installation is the reverse of removal. Proceed to adjust the cable.

Adjustment

2005 and 2006 models
Note: *This procedure is performed at the junction of the two cable halves.*
24   Place the shift lever in Park.
25   Slide back the white cover on the coupling at the junction of the two cables **(see illustration 4.3)**.
26   Pull up the center tabs of the lock button.
27   Let go of the rear cable and allow the mechanism to adjust itself.
28   Slide the white cover back, then push the clear-colored lock button down in place.
29   Let go of the white cover and ensure that it covers the lock button. If it doesn't, do the adjustment over again.
30   Check the transmission for correct operation in all ranges. If there are any problems, perform the adjustment again.

2007 and later models
Refer to illustrations 4.32 and 4.33
Note: *This procedure is performed at the transaxle end of the cable.*
31   Place the shift lever in Park.
32   Access the transaxle shift lever. Press the tab, then slide the adjustment lock cover (on the bottom of the cable) toward the end of the cable **(see illustration)**. This uncovers the adjustment mechanism.
33   Pinch the adjustment tabs to release the lock, then lift the adjustment lock **(see illustration)**.
34   Verify that both the transaxle and the

shift lever are in Park.
35   Press the adjustment lock into place. Slide the cover back into place over the lock.
36   Check the transmission for correct operation of all gears. If there are any problems, perform the adjustment again.

---

5   Brake Transmission Shift Interlock (BTSI) system - description, component replacement and adjustment

Warning: *These models are equipped with airbags. Always disable the airbag system before working in the vicinity of any airbag system component to avoid the possibility of accidental deployment of the airbag(s), which could cause personal injury (see Chapter 12).*

Description

1   This system prevents the transmission from being shifted out of Park unless the key is turned On and the brake pedal is pressed. When the vehicle is started, the BTSI is energized, locking the shift lever in Park; when the brake pedal is pressed, the solenoid is de-energized, so the shift lever can be moved to another gear. 2005 and 2006 models also use a cable attached to the ignition switch that ensures that the shift lever can't be moved when the key is turned Off; this function is controlled electrically on later models.

BTSI solenoid replacement

Refer to illustration 5.4
2   Remove the center shifter console panels as necessary for access to the BTSI solenoid (see Chapter 11).
3   Unplug the BTSI solenoid electrical connector.
4   Disconnect the solenoid from the shift lever assembly **(see illustration)**.
5   Installation is the reverse of removal.

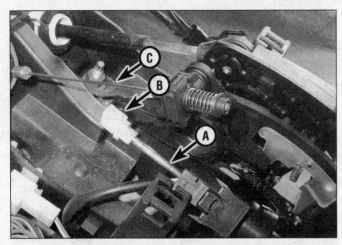

5.4 Brake Transaxle Shift Interlock system details

5.9 Pinch the tabs of the cable retainer to release it

A   *Solenoid shaft*
B   *Solenoid rod end*
C   *BTSI system cable to ignition switch (2005 and 2006 models only)*

BTSI cable replacement

Refer to illustrations 5.9 and 5.10

Note: *This procedure applies only to 2005 and 2006 models.*

6   Disconnect the cable from the negative terminal of the battery (see Chapter 5).
7   Remove the console and the steering column covers (see Chapter 11).
8   Make sure the ignition key is in the key off position.
9   Squeeze the retainer clip tabs and detach the cable from the shifter base **(see illustration)**. Disconnect the end of the cable from the shift lever **(see illustration 5.4)**.
10   Detach the cable from the ignition switch lock cylinder housing **(see illustration)**.
11   Remove the cable, noting how it's routed.
12   Use a screwdriver to push in the black tab on the ignition switch end of the cable. This will release the end fitting, which must be replaced with a new one.

5.10 Cut this strap, then release the tabs and detach the end of the BTSI cable from the ignition lock cylinder housing

13   Installation is the reverse of removal. Use a new end fitting in the forward end of the cable and attach the cable to the ignition switch module.
14   Snap the cable housing to its bracket on the shift lever, then align the cable end fitting to the lever pin and push forward. **Note:** *Handle only the end fitting, not the cable, when installing it. Make sure it's installed straight and that it snaps into place.*
15   Check the operation of the cable as follows:
a) *With the shift lever in Park and the key in Lock, make sure the shifter lever cannot be moved to another position and the key can be removed.*
b) *With the key in Run and the shift lever in Neutral, make sure the key cannot be turned to Lock.*

BTSI cable adjustment

16   Pull out the adjustment lock tab at the shift lever end of the cable.
17   Put the shift lever into Park.
18   Push in the lock tab, then refer to Step 15 to check the adjustment.

6   Driveaxle oil seals - replacement

Note: *This procedure does not apply to the transfer case seals on the right side of AWD models.*

1   Oil leaks frequently occur due to wear of the driveaxle oil seals. Replacement of these seals is relatively easy since the repairs can be performed without removing the transaxle from the vehicle.
2   The seals are located on the sides of the transaxle where the driveaxles are attached. If leakage of a seal is suspected, raise the vehicle and support it securely on jackstands. If the seal is leaking, fluid will be found on the sides of the transaxle.

3   Remove the driveaxle (see Chapter 8).
4   Note how deep the seal is installed, then use a screwdriver to carefully pry the seal from its bore. If it can't be removed with a screwdriver, a special seal removal tool (available at most auto supply stores) will be needed.
5   Compare the old seal to the new one to be sure it's correct.
6   Coat the inside and outside diameters of the new seal with transaxle fluid.
7   Using a seal installation tool or a large socket, install the new seal. Drive it into the bore squarely and make sure it's seated to its original depth.
8   Install the driveaxle (see Chapter 8).
9   Installation is the reverse of removal. Check the transaxle fluid level, adding if necessary (see Chapter 1).

7   Automatic transaxle - removal and installation

Removal

Refer to illustrations 7.16 and 7.19

1   Remove the battery and the battery tray (see Chapter 5).
2   Remove the shift cable bracket. Disconnect the shift cable (see Section 4).
3   Disconnect all wiring from the transaxle and transfer case (if so equipped). This includes the ground wires, the transaxle range switch, the transaxle speed sensors and the transaxle control module.
4   Detach the fuel line retainer from the transaxle. Disconnect the transaxle and transfer case vent tubes.
5   Use wire to tie the radiator, fan module and condenser to the support structure so they can remain with the vehicle when the transaxle is lowered.
6   Install an engine support fixture and connect the chains solidly to the lifting brackets

7.16 The transaxle cooler lines are secured by this stud and nut

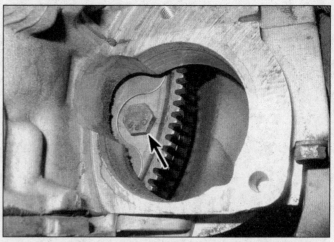

7.19 The starter must be removed on 3.4L models for access to the torque converter bolts; on 3.6L engines, there is also a plate on the transaxle side that can be removed

on top of the engine. **Note:** *Engine support fixtures are available at most equipment rental yards.*

7　　Remove the uppermost transaxle-to-engine mounting bolts.

8　　With the weight of the engine and transaxle supported by the fixture, remove the driver-side transaxle mount.

9　　Loosen the front wheel lug nuts and the driveaxle/hub nuts, then raise the vehicle and support it securely on jackstands.

10　Drain the transaxle fluid (see Chapter 1).

11　Remove the front wheels and the inner fender splash shields (see Chapter 11).

12　Remove the pinch bolt from the lower steering shaft. Throw the bolt away - it will have to be replaced with a new one. Disconnect the steering shaft from the steering gear.

13　Refer to Chapter 10 and disconnect the outer tie rod ends and the lower balljoints from the steering knuckles.

14　Disconnect the stabilizer bar links from the front stabilizer bar.

15　Remove the lower air deflector from the front bumper cover.

16　Disconnect the transaxle fluid cooler lines from the transaxle **(see illustration)**. Seal the openings to prevent contamination.

17　On 3.4L models, remove the transaxle-to-engine brace.

18　Remove the starter (see Chapter 5). On 3.6L models, also remove the flywheel inspection cover. Make matchmarks on the torque converter and flywheel so they can be assembled in the same position.

19　Remove the torque converter bolts **(see illustration)**.

20　Disconnect the front and rear engine mounts.

21　Remove the driveaxles and the intermediate shaft (see Chapter 8).

22　On AWD models, remove the driveshaft (see Chapter 8) and the transfer case mounting bracket.

23　On 3.6L AWD models, remove the transfer case (see Chapter 7B).

24　Remove the subframe (see Chapter 10).

25　Support the transaxle with a jack; preferably one designed for this purpose. Transmission/transaxle adapters are available for most heavy duty floor jacks. If the jack is equipped with safety chains or straps, use them to secure the transaxle to the jack.

26　Remove the lowermost transaxle-to-engine bolts, then pull the transaxle away from the engine until it can be lowered using the jack.

27　Lower the transaxle until it can be safely set on the ground.

Installation

28　Installation of the transaxle is a reversal of the removal procedure, but note the following points:

a) *As the torque converter is reinstalled, ensure that the drive tangs at the center of the torque converter hub engage with the recesses in the automatic transaxle fluid pump inner gear. This can be confirmed by turning the torque converter while pushing it towards the transaxle. If it isn't fully engaged, it will "clunk" into place. Align the matchmarks you made on the driveplate and torque converter.*

b) *Replace the steering shaft pinch bolt with a new one and tighten it to the torque listed in Chapter 10.*

c) *Install all of the driveplate-to-torque converter nuts before tightening any of them.*

d) *Tighten the driveplate-to-torque converter bolts to the specified torque.*

e) *Tighten the transaxle mounting bolts to the specified torque.*

f) *Replace all O-rings discarded previously.*

g) *Tighten the new subframe bolts to the torque values listed in this Chapter's Specifications.*

h) *Tighten the wheel lug nuts to the torque listed in the Chapter 1 Specifications.*

i) *Fill the transaxle with the correct type and amount of automatic transmission fluid as described in Chapter 1.*

j) *On completion, adjust the shift cable as described in Section 4.*

8　Automatic transaxle overhaul - general information

In the event of a problem occurring, it will be necessary to establish whether the fault is electrical, mechanical or hydraulic in nature, before repair work can be contemplated. Diagnosis requires detailed knowledge of the transaxle's operation and construction, as well as access to specialized test equipment, and so is deemed to be beyond the scope of this manual. It is therefore essential that problems with the automatic transaxle are referred to a dealer service department or other qualified repair facility for assessment.

Note that a faulty transaxle should not be removed before the vehicle has been diagnosed by a knowledgeable technician equipped with the proper tools, as troubleshooting must be performed with the transaxle installed in the vehicle.

9　Transaxle mounts - replacement

5 speed transaxle
Left mount

Refer to illustration 9.3

1　　Remove the battery and battery tray (see Chapter 5)

2　　Install an engine support fixture and connect the chains solidly to the lifting brackets on top of the engine. **Note:** *Engine support fixtures are available at most equipment rental yards.* Alternatively, the transaxle can be supported from below with a jack and a block of wood.

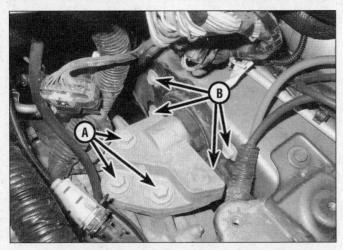

9.3 Left transaxle mount details (five-speed transaxle)

A   Mount-to-transaxle bolts   B   Mount-to-chassis bolts

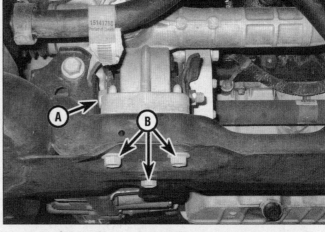

9.9 Rear transaxle mount details (five-speed transaxle)

A   Through-bolt   B   Mount-to-frame bolts

3   Unbolt the mount from the engine and the chassis, then remove it **(see illustration)**.
4   Installation is the reverse of removal. Tighten all bolts to the torque listed in this Chapter's Specifications.

Rear mount

Refer to illustration 9.9

5   Install an engine support fixture and connect the chains solidly to the lifting brackets on top of the engine. **Note:** *Engine support fixtures are available at most equipment rental yards.* Alternatively, the engine/transaxle can be supported from below with a jack and a block of wood.
6   On AWD models, loosen the front wheel lug nuts.
7   Raise the vehicle and support it securely on jackstands.
8   On AWD models, remove the front wheels, then disconnect the stabilizer bar from its links (see Chapter 10). Rotate the bar up for access.
9   Remove the through-bolt from the rear mount **(see illustration)**.
10   Unbolt the mount from the frame and remove it.
11   Installation is the reverse of removal. Tighten all bolts to the torque listed in this Chapter's Specifications.

Front mount

Refer to illustration 9.14

12   Install an engine support fixture and connect the chains solidly to the lifting brackets on top of the engine. **Note:** *Engine support fixtures are available at most equipment rental yards.* Alternatively, the engine/transaxle can be supported from below with a jack and a block of wood.
13   Raise the vehicle and support it securely on jackstands.
14   Remove the mount through-bolt **(see illustration)**.

15   Unbolt the mount from the transaxle and remove it.
16   Installation is the reverse of removal. Tighten all bolts to the torque listed in this Chapter's Specifications.

6 speed transaxles

Left mount

17   Loosen the left front wheel lug nuts. Raise the vehicle and support it securely on jackstands.
18   Remove the left front wheel and the inner fender splash shield (see Chapter 11).
19   Support the transaxle with a floor jack.
20   Remove the mount bolts, then raise the transaxle using the jack.
21   With the transaxle raised, maneuver the mount out.
22   Installation is the reverse of removal. Tighten all bolts to the torque listed in this Chapter's Specifications.

Rear mount

23   Install an engine support fixture and connect the chains solidly to the lifting brackets

on top of the engine. **Note:** *Engine support fixtures are available at most equipment rental yards.* Alternatively, the engine/transaxle can be supported from below with a jack and a block of wood.
24   Remove the rear exhaust manifold (see Chapter 2B).
25   Raise the vehicle and support it securely on jackstands.
26   Remove the mount through-bolt.
27   Unbolt the mount from the frame and remove it.
28   Installation is the reverse of removal. Tighten all bolts to the torque listed in this Chapter's Specifications.

Front mount

29   Raise the vehicle and support it securely on jackstands.
30   Support the transaxle with a floor jack.
31   Remove the mount through-bolt.
32   Unbolt the mount from the transaxle and remove it.
33   Installation is the reverse of removal. Tighten all bolts to the torque listed in this Chapter's Specifications.

9.14 Front transaxle mount details (five-speed transaxle)

A   Through-bolt
B   Mount-to-transaxle bolts

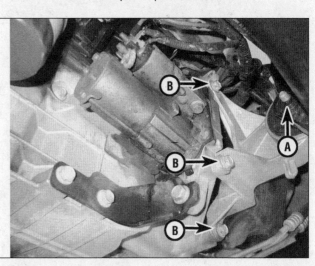

Notes

Chapter 7 Part B
Transfer case

Contents

Specifications

Transfer case fluid type .... See Chapter 1

Torque specifications

Ft-lbs (unless otherwise indicated)

Note: *One foot-pound (ft-lb) of torque is equivalent to 12 inch-pounds (in-lbs) of torque. Torque values below approximately 15 foot-pounds are expressed in inch-pounds, because most foot-pound torque wrenches are not accurate at these smaller values.*

Heat shield bolts, Getrag 760	97 in-lbs
Transfer case-to-transaxle bolts/nuts	
NVG-900	44
Getrag 760	
Large bolt (5 bolts)	37
Large bolt/small bolt combination (6 bolts)	
Large bolt	37
Small bolt	17
Transfer case mounting bracket bolts	
NVG-900	45
Getrag 760	
Small bolts	17
Large bolts	37
Transfer case bracket-to-engine block mounting bolts, NVG-900	44

1 General information

Due to the complexity of the transfer case covered in this manual and the need for specialized equipment to perform most service operations, this Chapter contains only routine maintenance and removal and installation procedures.

If the transfer case requires major repair work, it should be taken to a dealer service department or an automotive or transmission repair shop. You can, however, remove and install the transfer case yourself and save the expense of that labor, even if the repair work is done by a transmission shop.

2 Transfer case rear output shaft oil seal - replacement

Note: *This procedure applies only to NVG-900 transfer cases. The output shaft seal on the Getrag 760 transfer case is not normally serviced while in the vehicle.*

1   Raise the vehicle and support it securely on jackstands.

2   Drain the transfer case lubricant (see Chapter 1).

3   Remove the driveshaft (see Chapter 8).

4   Mark the relative positions of the pinion, nut and flange.

5   Use a beam- or dial-type inch-pound torque wrench to determine the torque required to rotate the pinion. Record it for use later.

6   Count the number of threads visible between the end of the nut and the end of the pinion shaft and record it for use later.

7   Remove the flange mounting nut using a chain wrench to hold the pinion flange while loosening the locknut.

8   Remove the companion flange; a small puller will be required for removal. The puller must be the type that has two bolts that screw into the companion flange.

9   Pry out the seal with a screwdriver or a seal removal tool. Don't damage the seal bore.

10   Lubricate the lips of the new seal with multi-purpose grease and tap it evenly into position with a seal installation tool or a large socket. Make sure it enters the housing squarely and is tapped into its full depth.

11   Align the mating marks made before disassembly and install the companion flange. If necessary, tighten the pinion nut to draw the flange into place.

12   Tighten the nut carefully until the original number of threads are exposed and the marks are aligned.

13   Measure the torque required to rotate the pinion and tighten the nut in small increments until it matches the figure recorded in Step 5.

14   Connect the driveshaft, add the specified lubricant to the transfer case (see Chapter 1) and lower the vehicle.

3   Transfer case driveaxle oil seal (right side) - removal and installation

Note: *This procedure is for NVG-900 transfer cases only. Getrag 760 transfer cases use seals that are not normally serviced with the transfer case in the vehicle.*

1   Break the right front driveaxle/hub nut loose with a socket and large breaker bar.

2   Loosen the right-front wheel lug nuts, raise the vehicle and support it securely on jackstands. Remove the wheel.

3   Remove the right driveaxle and intermediate shaft (see Chapter 8).

4   Remove the exhaust crossover pipe if necessary for access.

5   Remove the stub shaft. If it's necessary to use a puller to extract the stub shaft, remove the retaining ring from the shaft. It must be replaced with a new one.

6   Carefully pry out the oil seal with a seal removal tool or a large screwdriver; make sure you don't scratch the seal bore.

7   Using a seal installer or a large deep socket as a drift, install the new oil seal. Drive

it into the bore squarely and make sure it's completely seated.

8   Lubricate the lip of the new seal with multi-purpose grease, then install a new dust shield.

9   Install the intermediate shaft and driveaxle (see Chapter 8).

10   The remainder of installation is the reverse of removal. Check the transfer case lubricant level and add some, if necessary, to bring it to the appropriate level (see Chapter 1).

4   Transfer case - removal and installation

1   Loosen the right front wheel lug nuts and driveaxle/hub nut. Raise the vehicle and support it securely on jackstands.

2   Drain the transfer case lubricant (see Chapter 1).

3   Remove the driveshaft (see Chapter 8).

4   Remove the right front wheel and driveaxle (see Chapter 8).

NVG-900 transfer case

5   Remove the intermediate shaft (see Chapter 8). Also remove the retaining ring from the inner part of the intermediate shaft stub shaft.

6   Remove the stub shaft using a puller (see Section 3).

7   Remove the transfer case mounting bracket and disconnect the vent hose.

8   Use a floor jack to support the weight of the transfer case, then remove the four mounting bolts.

9   Unbolt the rear transaxle mount from the frame, leaving it attached to the transfer case for now.

10   Pull the transfer case from the transaxle, then turn it to point the driveshaft flange toward the transaxle.

11   Turn the transfer case while lifting it, in order to aim the driveshaft flange toward the ground.

12   Carefully lower the transfer case. The rear mount can now be removed if necessary.

13   Installation is the reverse of removal, noting the following points:

a)   *Install all of the bolts hand tight. After they're in place, tighten them to the torque values listed in this Chapter's Specifications.*

b)   *Tighten the driveshaft fasteners to the torque listed in the Chapter 8 Specifications.*

c)   *Refill the transfer case with the proper type and amount of lubricant (see Chapter 1).*

d)   *Tighten the wheel lug nuts to the torque listed in the Chapter 1 Specifications.*

Getrag 760 transfer case

14   Remove both catalytic converters along with the front exhaust pipe sections (see Chapter 4). Support the rear exhaust pipes with wire.

15   Remove the transfer case mounting bracket.

16   Place a floor jack under the transaxle, then remove the rear transaxle mount and bracket (see Chapter 7A).

17   Remove the transfer case mounting bolts, then remove it from the vehicle.

18   Installation is the reverse of removal, noting the following points:

a)   *Tighten the exhaust system fasteners to the torque listed in the Chapter 4 Specifications.*

b)   *Tighten the driveshaft fasteners to the torque listed in the Chapter 8 Specifications.*

c)   *Tighten the transfer case fasteners to the torque listed in this Chapter's Specifications.*

d)   *Refill the transfer case with the proper type and amount of lubricant (see Chapter 1).*

e)   *Tighten the wheel lug nuts to the torque listed in the Chapter 1 Specifications.*

Chapter 8
Driveline

Contents

Specifications

Torque specifications

Ft-lbs

Driveaxles

Driveaxle/hub nut*

Front 151

Rear (AWD models).. 151

Intermediate shaft bearing bracket fasteners 26

Nut must be replaced

Driveshaft (AWD models)

Center bearing mounting nuts 19

Driveshaft-to-transfer case bolts.................................... 25

Driveshaft-to-rear differential bolts 37

Rear differential assembly (AWD models)

2005 and 2006 models

Rear bracket bushing-to-body bolts 66

Side bracket bushing-to-body bolts.............................. 77

2007 and later models

Rear support bushing bolts 139

Differential front mounting nuts 90

Left rear differential mounting bolts.................................... 21

2.1 Loosen the driveaxle/hub nut with a long breaker bar

2.9 Carefully pry the inner end of the driveaxle from the transaxle

1   General information

The information in this Chapter deals with the components from the rear of the engine to the drive wheels, except for the transaxle (and transfer case on AWD models), which is dealt with in the previous Chapter.

Since nearly all the procedures covered in this Chapter involve working under the vehicle, make sure it's securely supported on sturdy jackstands or on a hoist where the vehicle can be easily raised and lowered.

2   Driveaxles - removal and installation

Front

Removal

Refer to illustrations 2.1 and 2.9

1   Remove the wheel center cover. Break the driveaxle/hub nut loose with a socket and large breaker bar **(see illustration)**. **Note:** *If the opening in the wheel is too small to accommodate your socket, wait until after the wheel is removed to loosen the nut. You can prevent the brake disc from turning by inserting a long punch into a disc cooling vane and letting it come to rest against the caliper mounting bracket.*

2   Loosen the wheel lug nuts, raise the vehicle and support it securely on jackstands. Remove the wheel.

3   Separate the lower control arm from the steering knuckle (see Chapter 10).

4   Refer to Chapter 10 and disconnect the tie-rod end from the steering knuckle.

5   Disconnect the stabilizer bar from the link at each end (see Chapter 10).

6   Remove the driveaxle/hub nut from the axle and discard it.

7   Swing the knuckle/hub assembly out (away from the vehicle) until the end of the driveaxle is free of the hub. **Note:** *If the drive-axle splines stick in the hub, tap on the end of the driveaxle with a plastic hammer.* Support the outer end of the driveaxle with a piece of wire to avoid unnecessary strain on the inner CV joint.

8   If you're removing the right driveaxle, carefully pry the inner CV joint off the interme-diate shaft using a large screwdriver or prybar positioned between the CV joint housing and the intermediate shaft bearing support.

9   If you're removing the left driveaxle, pry the inner CV joint out of the transaxle using a large screwdriver or prybar positioned between the transaxle and the CV joint hous-ing **(see illustration)**. Be careful not to dam-age the differential seal.

10   Support the CV joints and carefully remove the driveaxle from the vehicle.

Installation

11   Pry the old spring clip from the inner end of the driveaxle (left side) or outer end of the intermediate shaft (right side) and install a new one. Lubricate the differential or interme-diate shaft seal with multi-purpose grease and raise the driveaxle into position while support-ing the CV joints. **Note:** *Position the spring clip with the opening facing down; this will ease insertion of the driveaxle and prevent damage to the clip.*

12   Push the splined end of the inner CV joint into the differential side gear (left side) or onto the intermediate shaft (right side) and make sure the spring clip locks in its groove.

13   Apply a light coat of multi-purpose grease to the outer CV joint splines, pull out on the steering knuckle assembly and install the stub axle into the hub.

14   Insert the balljoint stud into the steer-ing knuckle and tighten the bolt to the torque listed in the Chapter 10 Specifications. Install a new cotter pin. Also reconnect the tie-rod ends and the stabilizer bar links (see Chap-ter 10).

15   Install a **new** driveaxle/hub nut. Tighten the hub nut securely, but don't try to tighten it to the actual torque specification until you've lowered the vehicle to the ground. **Note:** *If your socket won't fit through the opening in the wheel, tighten the driveaxle/ hub nut now, using the technique described in Step 1 to prevent the brake disc from turning.*

16   Grasp the inner CV joint housing (not the driveaxle) and pull out to make sure the drive-axle has seated securely in the transaxle or on the intermediate shaft.

17   Install the wheel and lug nuts, then lower the vehicle. Tighten the lug nuts to the torque listed in the Chapter 1 Specifications.

18   Tighten the driveaxle/hub nut to the torque listed in this Chapter's Specifications.

Intermediate shaft

Removal

19   Remove the right driveaxle (see Steps 1 through 10).

20   Remove the O-ring seal from the inter-mediate shaft.

21   Unscrew the bolts from the intermediate shaft bearing retainer.

22   Pull the intermediate shaft from the transaxle.

Installation

23   Lubricate the lips of the transaxle seal with multi-purpose grease. Carefully guide the intermediate shaft into the transaxle side gear, then install the mounting nuts for the bearing support.

24   Tighten the nuts to the torque listed in this Chapter's Specifications. Make sure that the shaft is positively engaged into the transaxle.

25   Put the O-ring on the shaft.

26   Put a new wheel-retaining clip on the shaft.

27   The remainder of installation is the reverse of removal.

3.3 Cut off the boot clamps and discard them

3.4 Mark the relationship of the tri-pod assembly to the outer race

3.5 Use a center punch to place marks on the tri-pod and the driveaxle to ensure that they are properly reassembled

Rear (AWD models)

Removal

Note: *On 2007 and later models, because of the design of the inner seal for the rear drive-axle, the seal will come out of the rear differential at the same time the driveaxle is removed.*

28   Remove the wheel center cover. Break the driveaxle/hub nut loose with a socket and large breaker bar **(see illustration 2.1)**.

29   Block the front wheels to prevent the vehicle from rolling. Loosen the wheel lug nuts, raise the rear of the vehicle and support it securely on jackstands. Remove the wheel.

30   Use a brass punch or a block of wood with a hammer to break the end of the drive-axle loose from the wheel hub. Remove the nut and discard it. **Caution:** *The manufacturer recommends that all suspension nuts and cotter pins be replaced with new ones any time they are removed.*

31   Remove the rear knuckle (see Chapter 10).

32   Pry the driveaxle out of the differential. Remove the retaining ring - it will have to be replaced with a new one.

Installation

33   Apply a light film of grease to the area on the inner CV joint stub shaft where the seal rides, then insert the splined end of the inner CV joint into the differential. Make sure the spring clip locks in its groove. **Note:** *Position the spring clip with the opening facing down; this will ease insertion of the driveaxle and prevent damage to the clip.*

34   Apply a light film of grease to the outer CV joint splines and insert the outer end of the driveaxle into the hub.

35   Reconnect the rear knuckle following the procedure in Chapter 10.

36   Install a *new* driveaxle/hub nut. Tighten the hub nut securely, but don't try to tighten it to the actual torque specification until you've lowered the vehicle to the ground.

37   Install the wheel and lug nuts, then lower the vehicle. Tighten the lug nuts to the torque listed in the Chapter 1 Specifications.

38   Tighten the driveaxle/hub nut to the torque listed in this Chapter's Specifications. Install the wheel center cover.

---

3   Driveaxle boot - replacement

Note 1: *If the CV joints are worn, indicating the need for an overhaul (usually due to torn boots), explore all options before beginning the job. Complete rebuilt driveaxles are available on an exchange basis, which eliminates much time and work.*

Note 2: *Some auto parts stores carry split-type replacement boots, which can be installed without removing the driveaxle from the vehicle. This is a convenient alternative; however, the driveaxle should be removed and the CV joint disassembled and cleaned to ensure the joint is free from contaminants such as moisture and dirt which will accelerate CV joint wear.*

Note 3: *These procedures apply to rear drive-axle boots as well as those used on the front of the vehicle.*

1   Remove the driveaxle from the vehicle

(see Section 2).

2   Mount the driveaxle in a vise. The jaws of the vise should be lined with wood or rags to prevent damage to the driveaxle.

Inner CV joint and boot

Refer to illustrations 3.3, 3.4, 3.5, 3.6a, 3.6b, 3.7, 3.11, 3.14, 3.16, 3.17a, 3.17b, 3.17c, 3.17d and 3.17e

Removal

3   Remove the boot clamps **(see illustration)**.

4   Pull the boot back from the inner CV joint and slide the joint housing off. Be sure to mark the relationship of the tri-pod to the outer race **(see illustration)**.

5   Use a center punch to mark the tri-pod and axleshaft to ensure that they are reassembled properly **(see illustration)**.

6   Spread the ends of the stop-ring apart, slide it towards the center of the shaft, then remove the retainer clip from the end of the axleshaft **(see illustrations)**.

3.6a Spread the ends of the stop-ring apart and slide it towards the center of the shaft . . .

3.6b . . . then slide the tri-pod assembly back and remove the retainer clip

3.7 Drive the tri-pod joint from the axleshaft with a brass punch and hammer - make sure you don't damage the bearing surfaces or the splines on the shaft

3.11 Wrap the splined area of the axleshaft with tape to prevent damage to the boot(s) when installing it

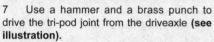

3.14 Pack the outer race with CV joint grease and slide it over the tri-pod assembly - make sure the match marks on the CV joint housing and tri-pod line up

3.16 Equalize the pressure inside the boot by inserting a small, dull screwdriver between the boot and the outer race

14   Apply CV joint grease to the tri-pod assembly, the inside of the joint housing and the inside of the boot **(see illustration)**.
15   Slide the boot into place.
16   Position the CV joint mid-way through its travel, then equalize the pressure in the boot **(see illustration)**.
17   Tighten the boot clamps **(see illustrations)**.
18   Install the driveaxle assembly (see Section 2).

Outer CV joint and boot

Refer to illustrations 3.20, 3.25a, 3.25b and 3.25c

Removal

19   Remove the boot clamps **(see illustration 3.3)**.
20   Strike the edge of the CV joint housing sharply with a soft-face hammer to dislodge the outer CV joint from the axleshaft **(see illustration)**. Remove and discard the bearing retainer clip from the axleshaft.
21   Slide the outer CV joint boot off the axleshaft.

7   Use a hammer and a brass punch to drive the tri-pod joint from the driveaxle **(see illustration)**.
8   Remove the stop-ring from the axleshaft and discard it.

Inspection

9   Clean the old grease from the outer race and the tri-pod bearing assembly. Carefully disassemble each section of the tri-pod assembly, one at a time so as not to mix up the parts, and clean the needle bearings with solvent.
10   Inspect the rollers, tri-pod, bearings and outer race for scoring, pitting or other signs of abnormal wear, which will warrant the replacement of the inner CV joint.

Reassembly

11   Slide the clamps and boot onto the axleshaft. It's a good idea to wrap the axleshaft splines with tape to prevent damaging the boot **(see illustration)**.
12   Install a new stop-ring on the axleshaft, but don't seat it in its groove; position it on the

shaft past the groove.
13   Place the tri-pod on the shaft (making sure the marks are aligned) and install a new bearing retainer clip. Now slide the tri-pod up against the retainer clip and seat the stop-ring in its groove.

3.17a To install new fold-over type clamps, bend the tang down . . .

3.17b . . . and flatten the tabs to hold it in place

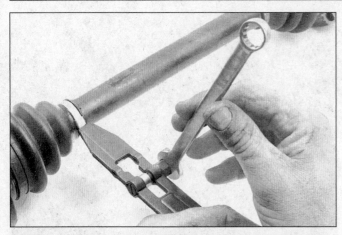

3.17c To install band-type clamps, you'll need a special tool; install the band with its end pointing in the direction of axle rotation and tighten it securely, then pivot the tool up 90-degrees and tap the center of the clip with a center punch . . .

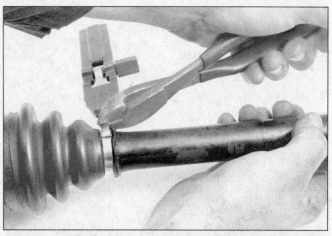

3.17d . . . then bend the end of the clamp back over the clip and cut off the excess

3.17e If you're installing crimp-type boot clamps, you'll need a pair of special crimping pliers (available at most auto parts stores)

3.20 Strike the edge of the CV joint housing sharply with a soft-faced hammer to dislodge the CV joint from the shaft

3.25a Pack the outer CV joint assembly with CV grease . . .

Inspection

22   Thoroughly clean all components with solvent until the old CV grease is completely removed. Inspect the bearing surfaces of the inner tri-pods and housings for cracks, pitting, scoring, and other signs of wear. If any part of the outer CV joint is worn, you must replace the entire driveaxle assembly (inner CV joint, axleshaft and outer CV joint).

Reassembly

23   Slide a new sealing boot clamp and sealing boot onto the axleshaft. It's a good idea to wrap the axleshaft splines with tape to prevent damaging the boot **(see illustration 3.11)**.
24   Place a new bearing retainer clip onto the axleshaft.
25   Place half the grease provided in the sealing boot kit into the outer CV joint assembly housing **(see illustration)**. Put the remaining grease into the sealing boot **(see illustrations)**.
26   Align the splines on the axleshaft with the splines on the outer CV joint assembly and gently drive the CV joint onto the axle-

shaft using a soft-faced hammer until the CV joint is seated to the axleshaft.
27   Position the CV joint mid-way through its travel, then equalize the pressure in the boot

(see illustration 3.16).
28   Tighten the boot clamps **(see illustrations 3.17a through 3.17e)**.
29   Install the driveaxle as outlined in Section 2.

3.25b . . . then apply grease to the inside of the boot . . .

3.25c . . . until the level is up to the end of the axle

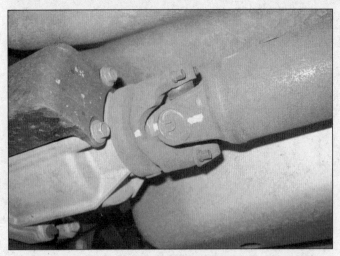

4.2 Mark the relationship of the driveshaft to the differential pinion yoke - also mark the other end of the driveshaft to the transfer case flange

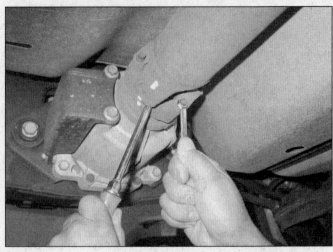

4.3 Immobilize the driveshaft by placing a screwdriver into the universal joint while loosening the bolts

4 Driveshaft (AWD models) - removal and installation

Refer to illustrations 4.2, 4.3, 4.4 and 4.5

Note: *The manufacturer recommends replacing driveshaft fasteners with new ones when installing the driveshaft.*

1   Raise the vehicle and support it securely on jackstands. Place the shift selector lever in Neutral.
2   Use chalk or a scribe to index the relationship of the driveshaft to the differential pinion yoke and the flange at the transfer case **(see illustration)**.
3   Disconnect the rear U-joint of the driveshaft from the rear differential assembly **(see illustration)**.
4   Remove the front constant velocity joint bolts from the transfer case **(see illustration)**.
5   Remove the mounting bolts from the center support bearing **(see illustration).**

6   Remove the driveshaft from the vehicle.
7   Installation is the reverse of removal. Make sure the marks you made previously are aligned. Tighten the fasteners to the torques listed in this Chapter's Specifications.

5 Driveshaft universal and constant velocity joints (AWD models) - general information and check

Universal joints

1   Universal joints are mechanical couplings which connect two rotating components that meet each other at different angles.
2   These joints are composed of a yoke on each side connected by a crosspiece called a trunnion. Cups at each end of the trunnion contain needle bearings which provide smooth transfer of the torque load. Snap-rings, either

inside or outside of the bearing cups, hold the assembly together.
3   Wear in the needle roller bearings is characterized by vibration in the driveline, noise during acceleration, and in extreme cases of lack of lubrication, metallic squeaking and ultimately grating and shrieking sounds as the bearings disintegrate.
4   It is easy to check if the needle bearings are worn with the driveshaft in position, by trying to turn the shaft with one hand, the other hand holding the rear differential pinion flange when the rear universal joint is being checked, and the front half coupling when the front universal joint is being checked. Any movement between the driveshaft and the front half couplings, and around the rear half couplings, is indicative of considerable wear. Another method of checking for universal joint wear is to use a prybar inserted into the gap between the universal joint and the driveshaft or flange. Leave the vehicle in gear and try to pry the joint

4.4 Remove the bolts securing the driveshaft to the transfer case

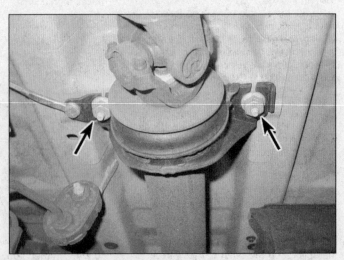

4.5 Remove the nuts securing the center support bearing

both radially and axially. Any looseness should be apparent with this method. A final test for wear is to attempt to lift the shaft and note any movement between the yokes of the joints.

5   If any of the above conditions exist, replace the driveshaft.

Constant velocity joint

6   The constant velocity joint on the forward end of the driveshaft is not serviceable. If the joint wears out or the boot becomes damaged, the front portion of the driveshaft must be replaced.

6   Driveshaft center support bearing (AWD models) - replacement

1   Remove the driveshaft (see Section 4).

2   Mark the relationship of the front portion of the driveshaft to the rear portion of the driveshaft (it's best to make the mark on the slip yoke). Disassemble the U-joint between the front half of the driveshaft and the center support bearing (see Section 6).

3   Remove the snap-ring from the end of the rear half of the driveshaft.

4   If you have access to a hydraulic press (one tall enough to accommodate the shaft) and the necessary fixtures, press the yoke off the driveshaft, then press the driveshaft out of the center support bearing. Reverse this operation to install the new bearing.

5   If you do not have the necessary equipment, take the shaft to an automotive machine shop or other qualified repair facility to have the old bearing pressed off and the new one pressed on.

7   Rear driveaxle oil seals (AWD models) - replacement

1   Raise the rear of the vehicle and support it securely on jackstands. Place the transaxle in Neutral with the parking brake off. Block the front wheels to prevent the vehicle from rolling.

2   Remove the driveaxle(s) (see Section 2). **Note:** *On 2007 and later models, because of the design of the inner seal for the rear driveaxle, the seal will come out of the rear differential at the same time the driveaxle is removed.*

3   Carefully pry out the driveaxle oil seal with a seal removal tool or a large screwdriver. Be careful not to damage or scratch the seal bore.

4   Using a seal installer or a large deep socket as a drift, install the new oil seal. Drive it into the bore squarely and make sure it's completely seated.

5   Lubricate the lip of the new seal with multi-purpose grease, then install the driveaxle. Be careful not to damage the lip of the new seal.

6   Check the differential lubricant level and add some, if necessary, to bring it to the appropriate level (see Chapter 1).

8   Rear differential assembly (AWD models) - removal and installation

Removal

1   Raise the rear of the vehicle and support it securely on jackstands. Block the front wheels to prevent the vehicle from rolling.

2   Remove the driveshaft (see Section 4).

3   Remove the driveaxles (see Section 2).

4   Remove the rear section of the exhaust system.

5   Place a floor jack under the differential assembly and tie the assembly securely using wire or a similar method.

6   Remove the differential rear mounting bolts.

7   Remove both front mounting nuts.

8   Carefully lower the differential just enough to disconnect the electrical connectors

9   Carefully remove the differential.

Installation

10   Installation is the reverse of removal, noting the following points:

a) *Don't tighten any of the mounting fasteners until all of them have been installed.*

b) *Tighten all fasteners to the torque listed in this Chapter's Specifications.*

Notes

Chapter 9   Brakes

Contents

Specifications

General
Brake fluid type .... See Chapter 1

Disc brakes
Brake pad minimum thickness	See Chapter 1
Disc lateral runout limit	0.002 inch
Disc minimum thickness	Cast into disc
Parallelism (thickness variation) limit	0.001 inch

Drum brakes
Maximum drum diameter	Cast into drum
Shoe lining minimum thickness	See Chapter 1

Torque specifications
Ft-lbs (unless otherwise indicated)

Note: *One foot-pound (ft-lb) of torque is equivalent to 12 inch-pounds (in-lbs) of torque. Torque values below approximately 15 foot-pounds are expressed in inch-pounds, because most foot-pound torque wrenches are not accurate at these smaller values.*

ABS hydraulic control unit	
Mounting nuts	180 in-lbs
Mounting bracket nuts	18
Brake hose fitting bolt	
2006 and earlier	32
2007 through 2009	38
2010 and later	
Front	38
Rear	30

Torque specifications (continued)   Ft-lbs (unless otherwise indicated)

Note: *One foot-pound (ft-lb) of torque is equivalent to 12 inch-pounds (in-lbs) of torque. Torque values below approximately 15 foot-pounds are expressed in inch-pounds, because most foot-pound torque wrenches are not accurate at these smaller values.*

Caliper mounting bolts	
2006 and earlier	32
2007 and later*	20
Caliper mounting bracket bolts	
Front*	
2009 and earlier	136
2010 and later	140
Rear*	
2009 and earlier	89
2010 and later	92
Master cylinder mounting nuts	
2009 and earlier models	18
2010 and later models	124 in-lbs
Power brake booster mounting nuts	18
Power brake booster pump bracket mounting bolts	
2.4L models	
2011 and earlier models	
Lower nuts	80 in-lbs
Upper bolt	16
2012 and later models	37
3.0L models	37
Power brake booster pump mounting nuts	
2.4L models	
2011 and earlier models	
Lower bolts (M10)	37
Upper bolt (M8)	16
2012 and later models	18
3.0L models	80 in-lbs
Wheel cylinder	132 in-lbs
Wheel speed sensor mounting bolts	
2012 and earlier models	72 in-lbs
2013 and later models	89 in-lbs
Wheel lug nuts	See Chapter 1

*Use high-strength thread-locking compound during installation

---

1   General information

The vehicles covered by this manual are equipped with hydraulically operated front disc brakes. The rear brakes may be disc or drum type. Both types of brakes are self adjusting and automatically compensate for pad or shoe wear.

Hydraulic system

The hydraulic system consists of two separate circuits that are diagonally split front-to-rear. The master cylinder has separate reservoirs for the two circuits, and, in the event of a leak or failure in one hydraulic circuit, the other circuit will remain operative and a warning indicator will light up on the instrument panel when a substantial amount of brake fluid is lost, showing that a failure has occurred.

Power brake booster

The power brake booster, utilizing engine manifold vacuum and atmospheric pressure to provide assistance to the hydraulically operated brakes, is mounted on the firewall in the engine compartment.

Parking brake

The parking brake operates the rear brakes only, through cable actuation. It's activated by a lever mounted in the center console.

Service

After completing any operation involving disassembly of any part of the brake system, always test drive the vehicle to check for proper braking performance before resuming normal driving. When testing the brakes, perform the tests on a clean, dry, flat surface. Conditions other than these can lead to inaccurate test results.

Test the brakes at various speeds with both light and heavy pedal pressure. The vehicle should stop evenly without pulling to one side or the other.

Tires, vehicle load and wheel alignment are other factors which also affect braking performance.

Precautions

There are some general cautions and warnings involving the brake system on this vehicle:

a) Use only brake fluid conforming to DOT 3 specifications.

b) The brake pads and linings contain fibers which are hazardous to your health if inhaled. Whenever you work on brake system components, clean all parts with brake system cleaner. Do not allow the fine dust to become airborne. Also, wear an approved filtering mask.

c) Safety should be paramount whenever any servicing of the brake components is performed. Do not use parts or fasteners which are not in perfect condition, and be sure that all clearances and torque specifications are adhered to. If you are at all unsure about a certain procedure, seek professional advice. Upon completion of any brake system work, test the brakes carefully in a controlled area before putting the vehicle into normal service. If a problem is suspected in the brake system, don't drive the vehicle until it's fixed.

d) Used brake fluid is considered a hazardous waste and it must be disposed of in accordance with federal, state and local laws. **DO NOT pour it down the sink, into septic tanks or storm drains, or on the ground.**

e) *Clean up any spilled brake fluid immediately and then wash the area with large amounts of water. This is especially true for any finished or painted surfaces.*

2 Anti-lock Brake System (ABS), Traction Control (TCS) and Vehicle stability control (StabiliTrak) systems - general information

1   Anti-lock Brake Systems (ABS) maintain vehicle maneuverability, directional stability, and optimum deceleration under severe braking conditions on most road surfaces. It does so by monitoring the rotational speed of the wheels and controlling the brake line pressure to the wheels during braking. This prevents the wheels from locking up on slippery roads or during hard braking.

Electro-Hydraulic Control Unit

Refer to illustration 2.2

2   The Electro-Hydraulic Control Unit **(see illustration)**, controls hydraulic pressure to the brake calipers by modulating hydraulic pressure to prevent wheel lock-up. It is made up of the Brake Pressure Modulator Valve (BPMV) and the Electronic Brake Control Module (EBCM). Basically, the BPMV reduces pressure in a brake line when the Electronic Brake Control Module (EBCM) detects an abnormal deceleration in the speed of a wheel (via a wheel speed sensor signal). When the speed of the wheel is restored to normal, the modulator once again allows full pressure to the brake. This cycle is repeated as many times as necessary, which results in a pulsing of the brake pedal. **Note:** *The system can't increase brake line pressure above that which is generated by the master cylinder, and it can't apply the brakes by itself (although the control unit **can** apply the brakes for traction control and stability control system functions).*
3   In addition to sensing and processing information received from the brake switch and wheel speed sensors to control the hydraulic line pressure and avoid wheel lock up, the EBCM also continually monitors the system and stores fault codes which indicate specific problems.

Traction Control System

4   Some models are equipped with a Traction Control System (TCS). When wheel slip is detected, the Electronic Brake Control Module (EBCM) will activate the traction control mode. A signal is sent from the EBCM to the PCM (Powertrain Control Module) commanding less torque to the drive wheels. Torque is reduced by retarding the ignition timing and by turning off fuel injectors. Additionally, if the reduction of torque to the wheels doesn't correct the problem, the control unit will apply brake pressure to the drive wheels to eliminate slippage.

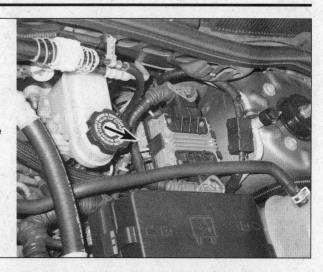

2.2 Location of the ABS Electro-Hydraulic Control Unit (EHCU) (the Transmission Control Module is mounted above it)

Stability Control System

5   This system is designed to assist in correcting over/under steering. This system is integrated with the ABS and TCS systems but also uses additional sensors. It will selectively actuate the brakes, as necessary, to control under- and over-steering conditions.

Wheel speed sensors

6   Generally, there is a wheel speed sensor designated for each wheel. Each sensor generates a signal in the form of a low-voltage electrical current or a frequency when the wheel is turning. A variable signal is generated as a result of a square-toothed ring (tone-ring, exciter-ring, reluctor, etc.) that rotates very close to the sensor. The signal is directly proportional to the wheel speed and is interpreted by an electronic module (computer).
7   The sensors are mounted to the hub and wheel bearing assemblies and the tone-rings are integrated within the assemblies.

Warning lights

8   The ABS system has self-diagnostic capabilities. Each time the vehicle is started, the EBCM runs a self-test. There are warning lights on the instrument panel. During starting, these lights should come on briefly then go out. If a BRAKE warning light stays on, it indicates a problem with the main braking system, such as low fluid level detected or the parking brake is still on. If the light stays on after the parking brake is released, check the brake fluid level in the master cylinder reservoir (see Chapter 1).
9   The ABS light indicates a problem with the ABS system, not the main or basic brake system. If the light stays on, it indicates that there is a problem with the ABS system, but the main system is still working. Take the vehicle to a dealer service department or other qualified repair shop for diagnosis and repair.
10   The TCS/StabiliTrak warning light will flash on and off when the system is activated. It can stay on for as long as 15 minutes while

it completes self-testing. It will also stay on if the system has a problem or if the system has been manually turned off using an on/off button located near the gear shift lever.

Diagnosis and repair

11   If a dashboard warning light comes on and stays on while the vehicle is in operation, the ABS, or other related systems require attention. Information will also be displayed in the Driver Information Center (DIC), as equipped. Although special diagnostic testing and tools are necessary to properly diagnose the system, you can perform a few preliminary checks before taking the vehicle to a dealer service department.

a) *Check the brake fluid level in the reservoir.*
b) *Check the electrical connectors at the EBCM and the hydraulic modulator/motor assembly.*
c) *Check the fuses.*
d) *Follow the wiring harness to each wheel and verify that all connections are secure and that the wiring is undamaged.*

12   If the above preliminary checks do not rectify the problem, the vehicle should be diagnosed by a dealer service department or other qualified repair shop. Due to the complexity of this system, all actual repair work must be done by a qualified automotive technician. **Warning:** *Do NOT try to repair a wheel speed sensor wiring harness. These systems are sensitive to even the smallest changes in resistance. Repairing the harness could alter resistance values and cause the system to malfunction. If the wiring harness is damaged in any way, it must be replaced.* **Caution:** *Make sure the ignition is turned off before unplugging or reattaching any electrical connections.*

Wheel speed sensor - removal and installation

Refer to illustrations 2.16a and 2.16b

13   Loosen the wheel lug nuts, raise the vehicle and support it securely on jackstands. Remove the wheel.

2.16a Front wheel speed sensor location

2.16b Rear wheel speed sensor location

3.5a Before disassembling the brake, wash it thoroughly with brake system cleaner and allow it to dry - position a drain pan under the brake to catch the residue - DO NOT use compressed air to blow off brake dust!

3.5b To make room for the new pads, use a C-clamp to depress the piston(s) into the caliper before removing the caliper - do this a little at a time, keeping an eye on the fluid level in the master cylinder to make sure it doesn't overflow

3.5c Remove the caliper lower mounting bolt (Note: *On 2007 and later model vehicles, hold the caliper slide pin with an open-end wrench while loosening the bolt with another wrench*) . . .

14   Make sure the ignition key is turned to the OFF position.

15   For front sensors, remove the front brake disc (see Section 5). For rear sensors, remove the brake shoes (see Section 6) or, on models with rear disc brakes, the parking brake shoes (see Section 12).

16   Remove the mounting fastener and carefully pull the sensor out from the hub assembly **(see illustrations)**.

17   Follow the wiring harness to the electrical connector and disconnect it. Remove the harness from any brackets that may secure it to other components.

18   Installation is the reverse of the removal procedure. Tighten the mounting fastener to the torque listed in this Chapter's Specifications.

19   Install the wheel and lug nuts, lower the vehicle and tighten the lug nuts to the torque listed in the Chapter 1 Specifications.

3   Disc brake pads - replacement

Refer to illustrations 3.5a through 3.5n

Warning: *Disc brake pads must be replaced on both front or both rear wheels at the same time - never replace the pads on only one wheel. Also, the dust created by the brake system is harmful to your health. Never blow it out with compressed air and don't inhale any of it. An approved filtering mask should be worn when working on the brakes. Do not, under any circumstances, use petroleum-based solvents to clean brake parts. Use brake system cleaner only!*

Note: *This procedure applies to the front and rear brake pads.*

1   Remove the cap from the brake fluid reservoir. Remove about two-thirds of the fluid from the reservoir, then reinstall the cap. **Caution:** *Brake fluid will damage paint. If any fluid is spilled, wash it off immediately with plenty of clean, cold water.*

2   Loosen the front or rear wheel lug nuts, raise the front or rear of the vehicle and support it securely on jackstands. Block the wheels at the opposite end.

3   Remove the wheels. Work on one brake assembly at a time, using the assembled brake for reference if necessary.

4   Inspect the brake disc carefully as outlined in Section 5. If machining is necessary, follow the information in that Section to remove the disc.

5   Follow the accompanying photo sequence for the actual pad replacement procedure **(see illustrations 3.5a through 3.5n)**. Be sure to stay in order and read the caption under each illustration.

3.5d . . . then pivot the caliper up and support it in this position with a piece of wire

3.5e Remove the inner brake pad

3.5f Remove the outer brake pad

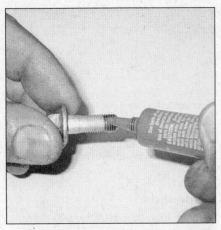

3.5g On 2007 and later model vehicles, clean the caliper mounting bolts of any old thread-locking compound, then apply new high-strength thread-locking compound on the caliper mounting bolt threads and set it aside

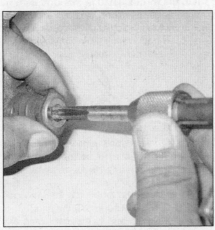

3.5h On 2007 and later models, clean the bolt holes (in the back of the caliper slide pins) of any old thread-locking compound using an appropriate tool (caliper mounting bracket removed for clarity)

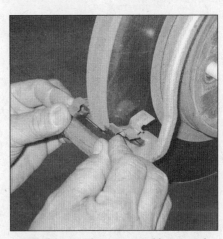

3.5i Remove the upper and lower pad retainers and replace them with new ones. Clean the bracket-to-pad retainer mating surface with a stiff wire brush, then wipe it clean with a rag before installing the new pad retainers

3.5j Install the outer brake pad . . .

3.5k . . . and the inner brake pad

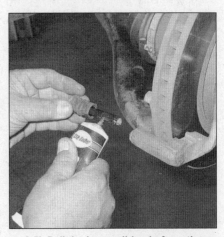

3.5l Pull the lower slide pin from the caliper and clean it. Lubricate it with high-temperature brake grease. Replace damaged or worn slide pins or boots

3.5m Withdraw the upper slide pin and caliper away from the caliper mounting bracket to clean and lubricate it, then place it back into position, making sure the boot seats

3.5n Pivot the caliper back into position over the brake pads (being careful not to damage the upper slide pin boot), then install the lower mounting bolt and tighten it to the torque listed in this Chapter's Specifications. **Note:** *On 2007 and later model vehicles, hold the caliper slide pin with an open-end wrench while tightening the caliper lower mounting bolt*

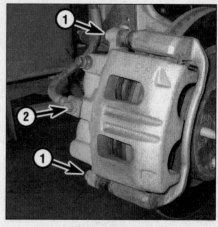

4.2 Brake caliper details:

1   *Caliper mounting bolts*
2   *Brake line banjo fitting bolt*

6   When reinstalling the caliper, be sure to tighten the mounting bolts to the torque listed in this Chapter's Specifications. Tighten the wheel lug nuts to the torque listed in the Chapter 1 Specifications.

7   After the job has been completed, slowly depress the brake pedal about two-thirds of its travel a few times until the pedal is firm. This will bring the pads into contact with the disc. Check the level of the brake fluid, adding some if necessary (see Chapter 1).

8   Break in the new pads by making several low speed stops (from approximately 30 mph). Allow time in between each stop for the discs to cool. About twenty stops should be sufficient to break in the new pads.

9   Confirm that the brakes are fully operational before resuming normal driving.

4   Disc brake caliper - removal and installation

Warning: *Dust created by the brake system is harmful to your health. Never blow it out with compressed air and don't inhale any of it. An approved filtering mask should be worn when working on the brakes. Do not, under any circumstances, use petroleum-based solvents to clean brake parts. Use brake system cleaner only.*
Note: *If replacement is indicated (usually because of fluid leakage), it is recommended that the calipers be replaced, not overhauled. New and factory-rebuilt units are available, which makes this job quite easy. Always replace the calipers in pairs - never replace just one of them.*

Removal

Refer to illustration 4.2

1   Loosen - but don't remove - the lug nuts on the front wheels. Raise the front of the vehicle and place it securely on jackstands. Remove the front wheels.

2   Disconnect the brake line from the caliper and plug it to keep contaminants out of the brake system and to prevent losing any more brake fluid than is necessary **(see illustration)**. **Note:** *If you're simply removing the caliper for access to other components, don't disconnect the hose.*

3   Remove the caliper mounting bolts.

4   Detach the caliper from its mounting bracket. **Caution:** *If the hose is still attached, hang the caliper with a length of wire - don't let it hang by the brake hose.*

Installation

5   Install the caliper by reversing the removal procedure. Remember to replace the copper sealing washers on each side of the brake line fitting with new ones. Tighten the caliper mounting bolts and the brake line banjo fitting bolt to the torque values listed in this Chapter's Specifications.

6   Bleed the brake system (see Section 10).

7   Install the wheels and lug nuts and lower the vehicle. Tighten the wheel lug nuts to the torque listed in the Chapter 1 Specifications.

5   Brake disc - inspection, removal and installation

Warning: *Dust created by the brake system is harmful to your health. Never blow it out with compressed air and don't inhale any of it. An approved filtering mask should be worn when working on the brakes. Do not, under any circumstances, use petroleum-based solvents to clean brake parts. Use brake system cleaner only.*

Inspection

Refer to illustrations 5.4, 5.5a, 5.5b, 5.6a and 5.6b

Note: *The manufacturer may have installed a thin plate (with a small V-notch pointing to one of the wheel studs) between the disc and the hub flange as a method to correct disc runout. Before removing the plate from the hub, make sure that you place a match mark on the stud that the V-notch points to. If the disc is going to be machined or replaced, it's possible that the plate can be removed and discarded. Check the disc runout on the machined or replacement disc with the plate and disc installed. If there is excessive runout, remove the plate and check the disc runout again.*

1   Loosen the wheel lug nuts, raise the vehicle and support it securely on jackstands. Remove the wheel.

2   Remove the brake caliper as outlined in Section 4. It's not necessary to disconnect the brake hose for this procedure. After removing the caliper mounting bolts, suspend the caliper out of the way with a piece of wire. Don't let the caliper hang by the hose and don't stretch or twist the hose.

3   Reinstall three lug nuts (inverted) to hold the disc against the hub. It may be necessary to install washers between the disc and the lug nuts to take up space.

4   Visually check the disc surface for score marks, cracks and other damage. Light

5.4 The brake pads on this vehicle were obviously neglected, as they wore down completely and cut deep grooves into the disc - wear this severe means the disc must be replaced

5.5a Use a dial indicator to check disc runout; if the reading exceeds the limit, the disc must be machined or replaced

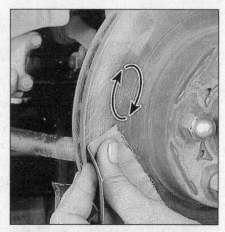

5.5b Using a swirling motion, remove the glaze from the disc surface with sandpaper or emery cloth

scratches and shallow grooves are normal after use and may not always be detrimental to brake operation. Deep score marks or cracks may require disc refinishing by an automotive machine shop or disc replacement **(see illustration)**. Be sure to check both sides of the disc. If pulsating has been noticed during application of the brakes, suspect disc runout. **Note:** *The most common symptoms of damaged or worn brake discs are pulsation in the brake pedal when the brakes are applied or loud grinding noises caused from severely worn brake pads. If these symptoms are extreme, it is very likely that the disc(s) will need to be replaced.*

5 To check disc runout, place a dial indicator at a point about 1/2-inch from the outer edge of the disc **(see illustration)**. Set the indicator to zero and turn the disc. An indicator reading that exceeds 0.003 of an inch could cause pulsation upon brake application and will require disc refinishing by an automotive machine shop or disc replacement. **Note:** *If disc refinishing or replacement is not neces-*

sary, you can deglaze the brake pad surface on the disc with emery cloth or sandpaper (use a swirling motion to ensure a non-directional finish) **(see illustration)**.

6 The disc must not be machined to a thickness less than the specified minimum refinish thickness. The minimum wear (or discard) thickness is cast into either the front or backside of the disc **(see illustration)**. The disc thickness can be checked with a micrometer **(see illustration)**.

Removal and installation

Refer to illustration 5.7

7 Remove the caliper mounting bracket **(see illustration)**. **Note:** *On 2007 and later models, be sure to clean the caliper mounting bracket bolt threads and holes and apply new high-strength thread-locking compound during installation* **(see illustrations 3.5g and 3.5h)**.

8 Mark the disc in relation to the hub so that it can be installed in it's original position

on the hub. Remove the bolt securing the disc to the hub flange and then remove the disc. If the disc appears to be stuck to the hub, use a mallet to knock it loose. **Note:** *Be sure to look for a thin plate that may be installed between the brake disc and hub flange and refer to the* **Note** *at the beginning of this Section.*

9 Clean the hub flange and the inside of the brake disc thoroughly, removing any rust or corrosion, then install the disc onto the hub assembly and tighten the retaining bolt securely.

10 Install the caliper mounting bracket and tighten the bolts to the torque listed in this Chapter's Specifications.

11 Place the caliper over the disc and onto the mounting bracket. Install the caliper mounting bolts and tighten them to the torque listed in this Chapter's Specifications. **Note:** *On 2007 and later models, be sure to clean the caliper mounting bolt threads and holes and apply new high-strength thread-locking compound during installation* **(see illustrations 3.5g and 3.5h)**.

5.6a The minimum wear dimension is cast into the rear of the disc (typical shown)

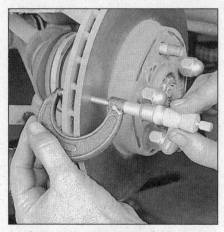

5.6b Use a micrometer to measure disc thickness

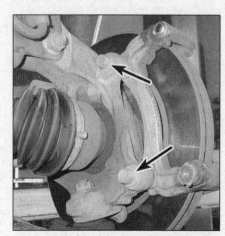

5.7 Caliper bracket mounting bolts

6.4a If the brake drum has never been removed, you'll have to pull off the drum retainer clip - it can be discarded (it's for assembly line purposes - there's no need to install a new one)

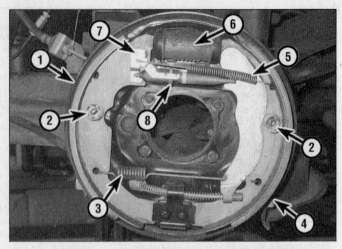

6.4b Rear brake shoe details (left side shown)

1	Leading shoe	5	Upper return spring
2	Shoe retainer and spring	6	Wheel cylinder
3	Lower return spring	7	Adjuster lever
4	Trailing shoe	8	Star-wheel assembly

6.4c Before removing anything, wash the brake assembly with brake system cleaner and allow it to drip dry into a pan - DO NOT USE COMPRESSED AIR TO BLOW BRAKE DUST FROM PARTS!

6.4d Unhook the lower spring and remove it

6.4e The upper spring can now be removed from the adjuster lever - note how it engages

12   Install the wheel and lower the vehicle to the ground. Tighten the wheel lug nuts to the torque listed in the Chapter 1 Specifications. Depress the brake pedal a few times to bring the brake pads into contact with the disc. Bleeding of the system will not be necessary unless the brake hose was disconnected from the caliper. Check the operation of the brakes thoroughly before placing the vehicle into normal service.

---

6   Drum brake shoes - replacement

Refer to illustrations 6.4a through 6.4w
Warning: *Drum brake shoes must be replaced on both wheels at the same time -*

never replace the shoes on only one wheel. Also, the dust created by the brake system is harmful to your health. Never blow it out with compressed air and don't inhale any of it. An approved filtering mask should be worn when working on the brakes. Do not, under any circumstances, use petroleum-based solvents to clean brake parts. Use brake system cleaner only!
1   Loosen the wheel lug nuts, raise the rear of the vehicle and support it securely on jackstands. Block the front wheels to keep the vehicle from rolling.
2   Release the parking brake.
3   Remove the wheel. **Note:** *All four rear brake shoes must be replaced at the same time, but to avoid mixing up parts, work on only one brake assembly at a time.*

4   Follow the accompanying illustrations for the brake shoe replacement procedure **(see illustrations 6.4a through 6.4w)**. Be sure to stay in order and read the caption under each illustration. **Note:** *If the brake drum cannot be easily removed, make sure the parking brake is completely released. If the drum still cannot be pulled off, the brake shoes will have to be retracted. This is done by first removing the plug from the backing plate. With the plug removed, push the lever off the adjuster star wheel with a narrow screwdriver while turning the adjuster wheel with another screwdriver, moving the shoes away from the drum (see illustration 6.4w). The drum should now come off.*
5   Before reinstalling the drum, it should be checked for cracks, score marks, deep

6.4f Remove the adjuster lever

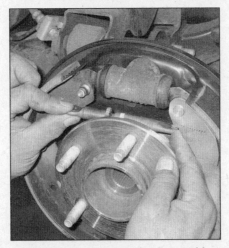
6.4g Remove the star-wheel assembly

6.4h Depress the front shoe retainer and turn it to remove it

6.4i Remove the front brake shoe

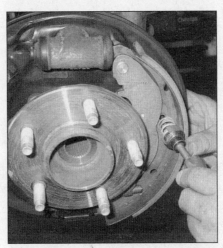
6.4j Remove the rear shoe retainer and allow the shoe to come free

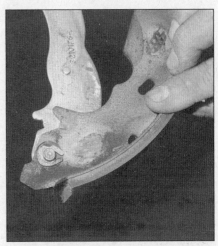
6.4k Pry open the clip to disconnect the parking brake lever from the rear shoe

6.4l Lubricate the backing plate contact points after thoroughly cleaning it

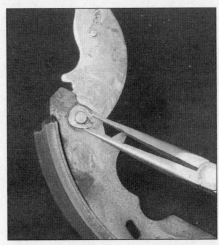
6.4m Crimp the clip to attach the new rear shoe to the parking brake lever

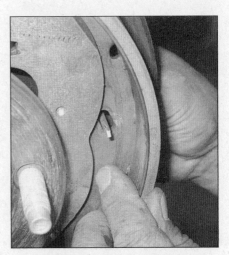
6.4n Install the pin through the backing plate and the rear shoe

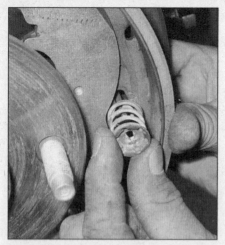

6.4o Install the spring and the retainer to the pin

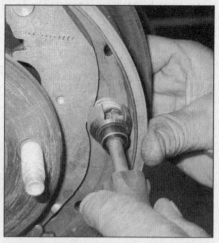

6.4p Use the special tool to compress the spring and engage the retainer

6.4q Perform the same procedure on the front brake shoe

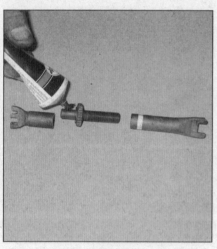

6.4r Clean the star-wheel adjuster and lubricate it lightly

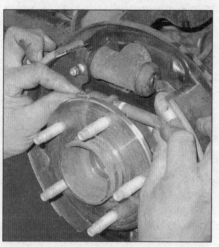

6.4s Install the star-wheel assembly with the star to the front

6.4t Install the adjuster lever

6.4u Install the upper spring, hooking it into the adjuster

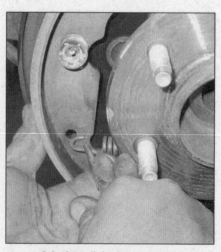

6.4v Install the lower spring

6.4w Adjust the star-wheel to give a very light drag when the brake drum is turned

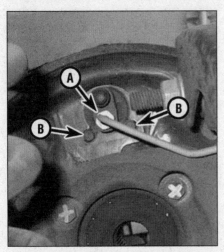

7.5 Brake line fitting (A) and wheel cylinder bolts (B)

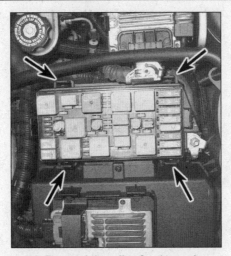

8.2 The retaining clips for the engine compartment fuse/relay box

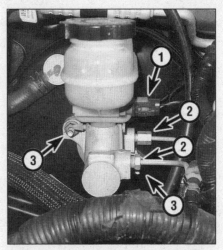

8.5 Master cylinder installation details:

1   *Electrical connector*
2   *Brake line fittings*
3   *Mounting nuts*

scratches and hard spots, which will appear as small discolored areas. If the hard spots cannot be removed with fine emery cloth or if any of the other conditions listed above exist, the drum must be taken to an automotive machine shop to have it resurfaced. **Note:** *Professionals recommend resurfacing the drums each time a brake job is done. Resurfacing will eliminate the possibility of out-of-round drums. If the drums are worn so much that they can't be resurfaced without exceeding the maximum allowable diameter (stamped into the drum), then new drums will be required. At the very least, if you elect not to have the drums resurfaced, remove the glaze from the surface with emery cloth using a swirling motion.*

6   Install the brake drum on the axle flange. Pump the brake pedal a couple of times, then turn the drum and listen for the sound of the shoes rubbing. If they are, turn the adjuster star wheel until the shoes stop rubbing.

7   Install the wheel and lug nuts, then lower the vehicle. Tighten the lug nuts to the torque listed in the Chapter 1 Specifications.

8   Make a number of forward and reverse stops to seat the shoes to the drum and to actuate the auto-adjusting mechanism.

9   Check the operation of the parking brake and adjust, if necessary.

10   Check the operation of the brakes carefully before driving the vehicle.

7   Wheel cylinder - removal and installation

Note: *If replacement is indicated (usually because of fluid leakage or sticky operation), it is recommended that the wheel cylinders be replaced, not overhauled. Always replace the wheel cylinders in pairs - never replace just one of them.*

Removal

Refer to illustration 7.5

1   Block the front wheels to keep the vehicle from rolling. Loosen the rear wheel lug nuts, raise the rear of the vehicle and support it securely on jackstands.

2   Remove the wheel.

3   Remove the brake shoes (see Section 6). **Note:** *An alternative procedure is to spread the top portion of the brake shoes apart enough to remove the wheel cylinder instead of removing the brake shoes entirely.*

4   Remove all dirt and foreign material from around the wheel cylinder.

5   At the rear of the backing plate, unscrew the brake line fitting, using a flare-nut wrench, if available **(see illustration)**. Don't pull the brake line away from the wheel cylinder (it could become kinked).

6   Remove the wheel cylinder mounting bolts and detach the wheel cylinder from the backing plate. Be sure to remove any sealant from the backing plate.

Installation

7   Put the wheel cylinder in position, then connect the brake line fitting finger tight. Install the wheel cylinder mounting bolts and tighten them to the torque listed in this Chapter's Specifications. Tighten the brake line fitting securely.

8   Install the brake shoes, if they were removed (see Section 6).

9   Install the brake drum on the hub flange, then install the wheel and lug nuts.

10   Bleed the brake system as outlined in Section 10.

11   Lower the vehicle to the ground, then tighten the lug nuts to the torque listed in the Chapter 1 Specifications. Check the operation of the brakes carefully before driving the vehicle in traffic.

8   Master cylinder - removal and installation

Removal

Refer to illustrations 8.2 and 8.5

Caution: *Brake fluid will damage paint. Cover all painted surfaces around the work area and be careful not to spill fluid during this procedure.*

1   Disconnect the cable from the negative terminal of the battery (see Chapter 5).

2   Move the engine compartment fuse/relay box aside for clearance as necessary **(see illustration)**.

3   Remove as much fluid as you can from the reservoir with a syringe, such as an old turkey baster. **Warning 1:** *If a baster is used, never again use it for the preparation of food.* **Warning 2:** *Don't depress the brake pedal until the reservoir has been refilled, otherwise air may be introduced into the brake hydraulic system.*

4   Place rags under the fluid fittings and prepare caps or plastic bags to cover the ends of the lines once they are disconnected. Loosen the fittings at the ends of the brake lines where they enter the master cylinder. To prevent rounding off the corners on these nuts, the use of a flare-nut wrench, which wraps around the nut, is preferred. Pull the brake lines slightly away from the master cylinder and plug the ends to prevent contamination.

5   Disconnect the electrical connector at the brake fluid level switch on the master cylinder reservoir (you must pull back the locking tab of the connector before removing it), then remove the nuts attaching the master cylinder to the power booster **(see illustration)**. Pull the master cylinder off the studs and out of the engine compartment. Again, be careful not to spill any fluid as this is done.

6   If necessary, transfer the reservoir to the new master cylinder. Drive out the two retain-

8.8 The best way to bleed air from the master cylinder before installing it is with a pair of bleeder tubes that direct fluid into the reservoir during bleeding (typical unit shown)

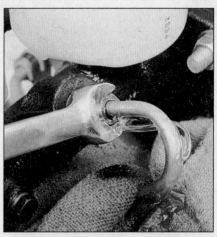

8.17 Loosen the fittings on the master cylinder to bleed it when it is installed on the vehicle (typical shown)

ing pins that hold the reservoir to the master cylinder body and discard them, then pull the reservoir directly up to remove it. **Note:** *Be sure to install new seals and retaining pins when transferring the reservoir.*

Installation

Refer to illustrations 8.8 and 8.17

7   Bench bleed the new master cylinder before installing it. Mount the master cylinder in a vise, with the jaws of the vise clamping on the mounting flange.
8   Attach a pair of master cylinder bleeder tubes to the outlet ports of the master cylinder **(see illustration)**.
9   Fill the reservoir with brake fluid of the recommended type (see Chapter 1).
10   Slowly push the pistons into the master cylinder (a large Phillips screwdriver can be used for this) - air will be expelled from the pressure chambers and into the reservoir. Because the tubes are submerged in fluid, air can't be drawn back into the master cylinder when you release the pistons.
11   Repeat the procedure until no more air bubbles are present.
12   Remove the bleed tubes, one at a time, and install plugs in the open ports to prevent fluid leakage and air from entering. Install the reservoir cap.
13   Install the master cylinder over the studs on the power brake booster and tighten the attaching nuts only finger tight at this time. **Note:** *Be sure to install a new O-ring seal onto the master cylinder.*
14   Thread the brake line fittings into the master cylinder by hand until you know they are started straight. Since the master cylinder is still a bit loose, it can be moved slightly in order for the fittings to thread in easily. Do not strip the threads as the fittings are tightened.
15   Tighten the mounting nuts to the torque listed in this Chapter's Specifications, then tighten the brake line fittings securely.
16   Connect the brake fluid switch electrical connector.
17   Fill the master cylinder reservoir with fluid. To bleed the cylinder on the vehicle, have an assistant depress the brake pedal

and hold the pedal to the floor. Loosen the fitting to allow air and fluid to escape, then close the fitting **(see illustration)**. Repeat this procedure on both fittings until the fluid is clear of air bubbles. **Caution:** *Have plenty of rags on hand to catch the fluid - brake fluid will ruin painted surfaces. After the bleeding procedure is completed, rinse the area under the master cylinder with clean water.* Bleed the remainder of the brake system as described in Section 10.
18   Reinstall the fuse/relay box and reconnect the battery.
19   Test the operation of the brake system carefully before placing the vehicle into normal service. **Warning:** *Do not operate the vehicle if you are in doubt about the effectiveness of the brake system. It is possible for air to become trapped in the anti-lock brake system hydraulic control unit, so, if the pedal continues to feel spongy after repeated bleedings or the BRAKE or ANTI-LOCK light stays on, have the vehicle towed to a dealer service department or other qualified shop to be bled with the aid of a scan tool.*

9   Brake hoses and lines - inspection and replacement

1   About every six months, with the vehicle raised and placed securely on jackstands, the flexible hoses which connect the steel brake lines with the front and rear brake assemblies should be inspected for cracks, chafing of the outer cover, leaks, blisters and other damage. These are important and vulnerable parts of the brake system and inspection should be complete. A light and mirror will be needed for a thorough check. If a hose exhibits any of the above defects, replace it with a new one.

Flexible hoses

Refer to illustrations 9.3a, 9.3b and 9.3c

2   Clean all dirt away from the ends of the hose. Follow the path of the hose and detach any brackets that secure the hose to other components. Also, detach any wiring har-

nesses that may be attached to the bracket hose.
3   To disconnect a brake hose from the brake line, unscrew the metal tube nut with a flare nut wrench, then remove the U-clip from the female fitting at the bracket and remove the hose from the bracket **(see illustrations)**.
4   Disconnect the hose from the caliper, discarding the sealing washers on either side of the fitting.
5   Using new sealing washers, attach the new brake hose to the caliper or wheel cylinder.
6   To reattach a brake hose to the metal line, insert the end of the hose through the frame bracket, make sure the hose isn't twisted, then attach the metal line by tightening the tube nut fitting securely. Install the U-clip at the frame bracket.
7   Carefully check to make sure the suspension or steering components don't make contact with the hose. Have an assistant push down on the vehicle and also turn the steering wheel lock-to-lock during inspection.
8   Bleed the brake system (see Section 10).

Metal brake lines

9   When replacing brake lines, be sure to use the correct parts. Don't use copper tubing for any brake system components. Purchase steel brake lines from a dealer parts department or auto parts store.
10   Prefabricated brake line, with the tube ends already flared and fittings installed, is available at auto parts stores and dealer parts departments. These lines can be bent to the proper shapes using a tubing bender.
11   When installing the new line, make sure it's well supported in the brackets and has plenty of clearance between moving or hot components.
12   After installation, check the master cylinder fluid level and add fluid as necessary. Bleed the brake system as outlined in Section 10 and test the brakes carefully before placing the vehicle into normal operation.

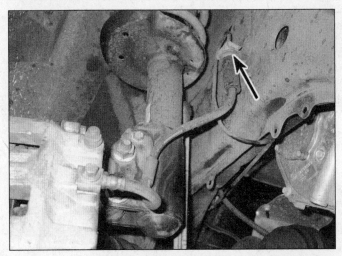

9.3a Front brake hose connection to the steel brake tubing

9.3b Rear brake hose

10 Brake hydraulic system - bleeding

Refer to illustration 10.8

Warning 1: *If air has found its way into the hydraulic control unit, the system must be bled with the use of a scan tool. If the brake pedal feels spongy even after bleeding the brakes, or the ABS light on the instrument panel does not go off, or if you have any doubts whatsoever about the effectiveness of the brake system, have the vehicle towed to a dealer service department or other repair shop equipped with the necessary tools for bleeding the system.*

Warning 2: *Wear eye protection when bleeding the brake system. If the fluid comes in contact with your eyes, immediately rinse them with water and seek medical attention.*

Caution: *Brake fluid will damage paint. Cover all pained surfaces around the work area and be careful not to spill fluid during this procedure.*

Note: *Bleeding the brake system is necessary to remove any air that's trapped in the system when it's opened during removal and installation of a hose, line, caliper or master cylinder.*

1   It will probably be necessary to bleed the system at all four brakes if air has entered the system due to a low fluid level, or if the brake lines have been disconnected at the master cylinder.

2   If a brake line was disconnected only at a wheel, then only that caliper or wheel cylinder must be bled.

3   If a brake line is disconnected at a fitting located between the master cylinder and any of the brakes, that part of the system served by the disconnected line must be bled, beginning with the fitting closest to the master cylinder and then working downstream, bleeding each fitting of each component as described in Step 17 of Section 8. This includes the Brake Pressure Modulator Valve (BPMV) (see Section 2).

4   Remove any residual vacuum (or hydrau-

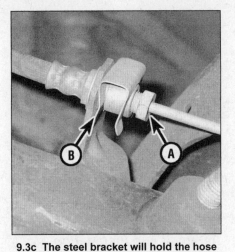

9.3c The steel bracket will hold the hose while the fitting nut (A) is loosened; to remove the hose from the bracket, remove the clip (B)

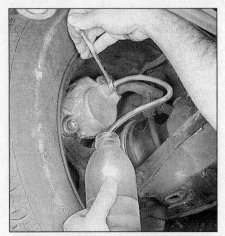

10.8 When bleeding the brakes, a hose is connected to the bleed screw at the caliper or wheel cylinder and submerged in brake fluid - air will be seen as bubbles in the tube and container (all air must be expelled before moving to the next wheel)

lic pressure) from the brake power booster by applying the brake several times with the engine off.

5   Remove the master cylinder reservoir cap and fill the reservoir with brake fluid. Reinstall the cap. **Note:** *Check the fluid level often during the bleeding operation and add fluid as necessary to prevent the fluid level from falling low enough to allow air bubbles into the master cylinder.*

6   Have an assistant on hand, as well as a supply of new brake fluid, an empty clear plastic container, a length of plastic, rubber or vinyl tubing to fit over the bleeder valve and a wrench to open and close the bleeder valve.

7   Working at the right rear wheel, loosen the bleeder screw slightly, then tighten it to a point where it's snug but can still be loosened quickly and easily.

8   Place one end of the tubing over the bleeder screw fitting and submerge the other end in brake fluid in the container **(see illustration)**.

9   Have your assistant slowly depress the

brake pedal and hold it in the depressed position.

10   While the pedal is held depressed, open the bleeder screw just enough to allow a flow of fluid to leave the valve. Watch for air bubbles to exit the submerged end of the tube. When the fluid flow slows after a couple of seconds, tighten the screw and have your assistant release the pedal.

11   Repeat Steps 9 and 10 until no more air is seen leaving the tube, then tighten the bleeder screw and proceed to the left front wheel, the left rear wheel and the right front wheel, in that order, and perform the same procedure. Be sure to check the fluid in the master cylinder reservoir frequently.

12   Never use old brake fluid. It contains moisture that can boil, rendering the brake system inoperative.

13   Refill the master cylinder with fluid at the end of the operation.

14   Check the operation of the brakes. The

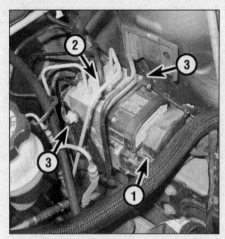

11.15 ABS hydraulic unit mounting
details:

1   *Electrical connector*
2   *Brake line fittings*
3   *Mounting nuts (one hidden from view
    in this photo - vicinity given)*

11.23 This clip secures the brake booster
pushrod to the pedal

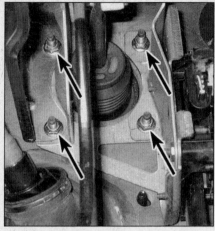

11.24 Brake booster mounting nuts

pedal should feel solid when depressed, with no sponginess. If necessary, repeat the entire process. **Warning:** *Do not operate the vehicle if you are in doubt about the effectiveness of the brake system. It is possible for air to become trapped in the anti-lock brake system hydraulic control unit, so, if the pedal continues to feel spongy after repeated bleedings or the BRAKE or ANTI-LOCK light stays on, have the vehicle towed to a dealer service department or other qualified shop to be bled with the aid of a scan tool.*

11   Power brake booster - removal and installation

Warning: *The ABS hydraulic control unit, if equipped, must be moved aside in order to remove the power brake booster. It's important to note that if the ABS hydraulic control unit is removed, and the hydraulic lines are detached from it (opening the circuit), then air can be introduced into the system. If air has found its way into the hydraulic control unit, the system must be bled with the use of a specialized scan tool. If the brake pedal feels spongy even after bleeding the brakes, or the ABS light or BRAKE warning light on the instrument panel stays on, or if you have any doubts whatsoever about the effectiveness of the brake system, have the vehicle towed to a dealer service department or other qualified repair shop equipped with this tool.*
Note: *The power brake booster is not serviceable. If it has failed, replace it with a new or rebuilt unit.*

Operating check

1   Depress the brake pedal several times with the engine off and make sure that there is no change in the pedal reserve distance.
2   Depress the pedal and start the engine.

If the pedal goes down slightly, operation is normal.

Airtightness check

3   Start the engine and turn it off after one or two minutes. Depress the brake pedal several times slowly. If the pedal goes down farther the first time but gradually rises after the second or third depression, the booster is airtight.
4   Depress the brake pedal while the engine is running, then stop the engine with the pedal depressed. If there is no change in the pedal reserve travel after holding the pedal for 30 seconds, the booster is airtight.

Removal and installation

Refer to illustrations 11.15, 11.23 and 11.24
Caution: *Brake fluid will damage paint. Cover all pained surfaces around the work area and be careful not to spill fluid during this procedure.*
5   The power brake booster is not rebuildable. If a problem develops, it must be replaced with a new one.
6   Disconnect the cable from the negative terminal of the battery (see Chapter 5).
7   Disconnect the vacuum hose check valve where it attaches to the power brake booster.
8   Move the engine compartment fuse/relay box aside for clearance **(see illustration 8.2)**.
9   Refer to Chapter 5 and remove the battery.
10 Remove the battery tray (see Chapter 3, Section 3).
11 When the engine is completely cool, disconnect the hose from the coolant expansion tank. **Warning:** *The engine must be completely cool before disconnecting the hose.*
12 Disconnect the sensor wiring harness from the master cylinder.
13 Remove the master cylinder mounting nuts, carefully pull it from its studs and position it aside. Don't disconnect the brake lines.
14 Disconnect any electrical connectors

from the booster.
15 Disconnect the electrical connector from the ABS hydraulic control unit **(see illustration)**. **Note:** *Move the transmission control module that's mounted above the ABS hydraulic control unit aside, if equipped.*
16 Put rags under the module to catch any spilled brake fluid.
17 Thoroughly clean the areas on the ABS hydraulic control unit where the hydraulic line fittings are connected.
18 Disconnect the hydraulic lines from the ABS hydraulic control unit **(see illustration 11.15)**. Make notes of their locations to help in assembly. Seal all openings to prevent contamination.
19 On 2012 and earlier models, loosen the ABS hydraulic control unit mounting nuts on each side, then pull the unit up from its bracket **(see illustration 11.15)**. On 2013 and later models, carefully label the brake lines to the ABS control unit, then disconnect the lines from the unit and remove the control unit and bracket from the vehicle.
20 On 2012 and earlier models, remove the ABS hydraulic unit mounting bracket nuts.
21 Working in the cab of the vehicle, remove the insulator panel located below the instrument panel on the driver's side (see Chapter 11).
22 On 2013 and later models, remove the mounting fasteners to the communications module and move the module out of the way, then remove the Accelerator Pedal Position (APP) sensor (see Chapter 6). Remove the brake pedal position sensor (see Section 14).
23 Remove the clip and the foam washer that secures the brake booster pushrod to the brake pedal **(see illustration)**.
24 Remove the four brake booster nuts **(see illustration)**.
25 Working in the engine compartment, pull the brake booster forward and toward the middle of the engine compartment to remove it. Remove the foam seal if it didn't come out

12.5 Turn the adjuster until it's at its shortest setting (the threads disappear into the adjuster) (typical shown)

12.6 Remove the adjuster spring (typical shown)

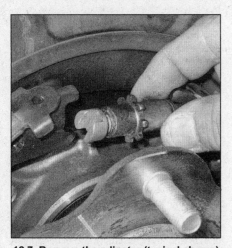

12.7 Remove the adjuster (typical shown)

12.8 Remove the return spring (typical shown)

12.11 Clean the backing plate thoroughly, then lubricate the brake shoe contact areas on the plate with high-temperature grease (some areas are hidden) (typical shown)

with the booster.

26   Installation is the reverse of removal. Before installing the new booster, lubricate the pushrod pin on the brake pedal arm with multi-purpose grease. Connect the pushrod to the brake pedal, then install the foam washer and clip.

27   Install the ABS hydraulic control unit and tighten the mounting fasteners to the torque values in this Chapter's Specifications.

28   Bleed the brakes (see Section 10).

29   Test the operation of the brakes thoroughly before placing the vehicle into normal service. Refer to the **Warning** at the beginning of this Section.

12 Parking brake shoes - replacement

Refer to illustrations 12.5, 12.6, 12.7, 12.8, 12.11, 12.12, 12.15, 12.16 and 12.17

Warning 1: *Dust created by the brake system is hazardous to your health. Never blow it out with compressed air and don't inhale any of*

it. An approved filtering mask should be worn when working on the brakes. Do not, under any circumstances, use petroleum-based solvents to clean brake parts. Use brake system cleaner only!

Warning 2: *Parking brake shoes must be replaced on both wheels at the same time - never replace the shoes on only one wheel.*

Note: *This procedure applies to models equipped with rear disc brakes.*

Note: *On 2007 and later models two design versions were used. Both designs are removed and installed the same but the parts vary slightly in design. Make sure the replacement parts match the parts on your vehicle.*

1   Remove the rear brake discs (see Section 5).

2   Measure the thickness of the lining material on the shoes. Generally, if the lining is 1.0 mm or less, the shoes should be replaced.

3   Inspect the drum portion of the disc for scoring, grooves or cracks due to heat. If any of these conditions exist or there is significant wear to the drum, the disc must be replaced.

4   Wash off the brake parts with brake sys-

tem cleaner. **Note:** *Work on one side at a time using the opposite side for reference as necessary.*

5   Turn the star wheel adjuster until it's at its shortest setting (the threads disappear into the adjuster) **(see illustration)**.

6   Remove the adjuster springs **(see illustration)**.

7   Remove the adjuster **(see illustration)**.

8   Remove the return spring **(see illustration)**.

9   Remove the rear shoe hold-down retainer, spring, pin and shoe.

10   Remove the forward shoe hold-down retainer, spring, pin and shoe.

11   Clean the backing plate thoroughly, then lubricate the brake shoe contact areas on the plate with high-temperature grease (some areas are hidden) **(see illustration)**.

12   Clean and lubricate the adjuster **(see**

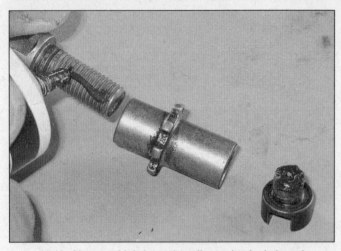

12.12 Clean and lubricate the adjuster (typical shown)

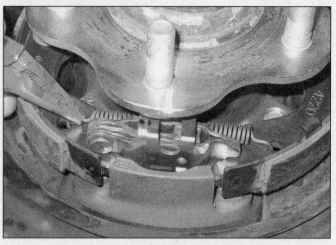

12.15 Install the return spring (typical shown)

illustration)

13   Place the forward shoe in position and install the hold-down retainer, spring and pin.

14   Place the rear shoe in position and install the hold-down retainer, spring and pin.

15   Install the return spring **(see illustration)**.

16   Install the adjuster **(see illustration)**.

17   Install the adjuster spring **(see illustration)**.

18   Install the brake disc and use two lug nuts with washers (for spacing) to hold the disc securely against the hub.

19   Adjust the rear parking brake shoes (see Section 13).

20   Install the caliper bracket (see Section 5) and brake caliper (see Section 4). Be sure to tighten the bolts to the torque values listed in this Chapter's Specifications.

21   Install the wheel and tighten the lug nuts to the torque specified in Chapter 1.

22   Check the parking brake for proper operation and adjust it again if necessary (see Section 13).

13   Parking brake - adjustment

Refer to illustrations 13.4 and 13.7

Note: *If the parking brake shoe clearance or cable adjusting nut require a significant amount of adjustment, it is advisable to inspect the brake shoe lining thickness (see Section 12).*

1   The parking brake should fully engage with a moderate number of clicks of the parking brake lever. If the lever can be moved quite a lot, the parking brake may not hold the vehicle adequately on an incline, allowing the car to roll. If the lever cannot be moved much (three clicks or less), there's a chance the parking brake may not be releasing completely and might be dragging. Generally, the parking brake is fully engaged after about 5 to 7 clicks.

2   Block the front wheels, raise the rear of the vehicle and support it securely on jackstands. Remove the rear wheels.

3   If you're working on a model with rear drum brakes, ensure the rear brake shoes are

properly adjusted (see Section 6).

4   If you're working on a model with rear disc brakes, use the access hole in the disc to adjust the parking brake shoe clearance **(see illustration)**. Turn the star-wheel adjuster until the disc cannot be rotated, then reverse the adjuster five notches. Rotate the discs to make sure the shoes aren't dragging. Adjust both wheels, then reinstall the rubber plugs that cover the access holes.

5   Set the parking brake fully and compare the number of clicks to those specified in Step 1. If more adjustment is necessary, move on to the next adjustment at the parking brake lever assembly.

6   Remove the center console (see Chapter 11).

7   Locate the adjusting nut near the base of the hand lever assembly and either tighten or loosen it to achieve a moderate number of clicks to set **(see illustration)**. Tightening the nut (turning it clockwise) decreases the number of clicks, while the opposite is achieved by loosening the nut (turning it counterclockwise).

12.16 Install the adjuster (typical shown)

12.17 Install the adjuster spring (typical shown)

13.4 Turn the star adjuster until the brake disc cannot be moved, then back off the adjustment 5 notches (typical shown)

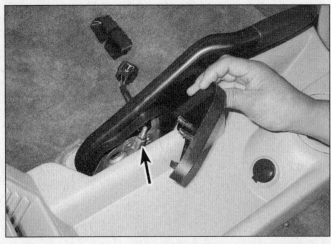

13.7 The adjuster nut location at the hand lever parking brake assembly

8   Confirm that the parking brake is fully engaged within about 5 to 7 clicks.

9   Release the parking brake and confirm that the brakes don't drag when the rear wheels are turned.

14   Brake light switch/brake pedal position sensor - replacement

Refer to illustration 14.2

Note: *On 2007 and later model vehicles, the brake light switch is replaced with a brake pedal position sensor. The manufacturer states that the sensor must be calibrated after service to it. Calibration requires the use of a scan tool by a dealership service department or a qualified repair location and is beyond the scope of the home mechanic.*

1   Remove the insulator panel beneath the instrument panel (see Chapter 11). The brake light switch is located at the top of the brake pedal bracket.

Brake light switch

2   Disconnect the wiring connector(s) from the switch **(see illustration)**.

3   Depress the brake pedal, then compress the switch's lock tabs. Rotate the switch counterclockwise while pulling and remove the switch.

4   To install the new switch, first make sure that the pedal is all the way at the top of its travel. Rotate the switch clockwise while inserting it into the mounting bracket, then connect the electrical connector(s). The switch should be fully seated against the pedal when the pedal is fully released.

5   Apply the brake pedal, release the pedal and verify that the brake lights go off when the pedal is released.

Brake pedal position sensor

Note: *The manufacturer states that a special*

scan tool is necessary for calibrating this component when it is installed. They also state that a new mounting bolt should be installed whenever it is removed.

6   Disconnect the brake pedal position sensor electrical connector.

7   Mark the relationship of the sensor to its mount, then remove the fastener and the sensor.

8   Installation is the reverse of removal. Align the sensor exactly with the marks made in Step 7.

Caution: *If the sensor requires calibration after it's installed, the circuit may set a diagnostic trouble code (DTC) and a warning lamp may illuminate on the dash. The vehicle will have to be taken to a dealership service department or other qualified repair shop to calibrate the sensor and clear any codes. For more information on diagnostic trouble codes, (see Chapter 6).*

15   Power brake booster pump - replacement

2.4L models

Note: *On 2.4L models the brake booster pump is located on the front side of the engine, just behind the radiator at the bottom.*

1   Raise the vehicle and support it securely on jackstands.

2   Depress the brake pedal several times until the brake pedal is firm to remove the vacuum from the brake booster.

3   Disconnect the brake booster pump quick-connect hose from the pump.

4   Disconnect the electrical connector to the pump, and then disconnect the electrical connector from the pump bracket.

5   Remove the pump lower mounting bolts then remove the upper mounting bolts and remove the pump.

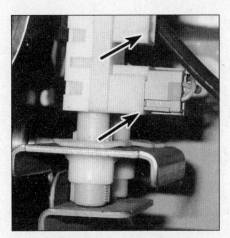

14.2 Disconnect the electrical connector(s) from the brake light switch, then turn it counterclockwise, press the lock tabs and pull the switch out

6   Installation is the reverse of removal.

3.0L models

Note: *On 3.0L models the brake booster pump is located at the back of the engine compartment just below the master cylinder.*

7   Depress the brake pedal several times until the brake pedal is firm to remove the vacuum from the brake booster.

8   Disconnect the brake booster quick-connect hose from the pump.

9   Disconnect the electrical connector to the pump, and then disconnect the electrical connector from the pump bracket.

10   Remove the pump mounting nuts and remove the pump from the bracket.

11   Installation is the reverse of removal.

2013 and later 3.6L models

Note: *On these models the brake booster pump is located at the back of the engine* compartment to the right of the master cylinder.

12   Remove the engine oil filler cap and engine cover fastener, then lift the engine cover up and off. Reinstall the engine oil filler cap.

13   Disconnect the brake booster vacuum hose quick-connect fitting and remove the hose from the mounting clips on the power brake booster pump.

14   Disconnect the power brake booster pump electrical connector.

15   Disconnect the electrical connector to the pump, and then disconnect the electrical connector from the pump bracket.

16   Remove the power brake booster pump nuts and pump.

17   Installation is the reverse of removal.

Chapter 10
Suspension and steering systems

Contents

Specifications

Torque specifications
Ft-lbs (unless otherwise indicated)

Note: *One foot-pound (ft-lb) of torque is equivalent to 12 inch-pounds (in-lbs) of torque. Torque values below approximately 15 foot-pounds are expressed in inch-pounds, because most foot-pound torque wrenches are not accurate at these smaller values.*

Front suspension
Balljoint
Balljoint-to-steering knuckle nut
2005 models
Balljoint with silver bottom 44
Balljoint with black bottom 30
2006 through 2008 models 30
2009 and later models
Step 1 .... 89 in-lbs
Step 2 .... Tighten an additional 165 degrees
Balljoint-to-control arm nuts and bolts 50
Hub/wheel bearing assembly mounting bolts*
2008 and earlier models 96
2009 and later models 74
Control arm-to-subframe bolts
Front bolts**
2005 models 148
2006 through 2009 models
Sport models 107
All other models 140
2010 and later models (through bolt/nut)
Step 1 74
Step 2 .... Loosen two full turns
Step 3 85
Step 4 .... Tighten an additional 100 degrees
Rear bolts/nuts**
2008 and earlier models 52
2009 models
Step 1 85
Step 2 .... Tighten an additional 90 degrees
2010 and later models (vertical through bolts/nuts) 52
Control arm rear bushing nut 111
Stabilizer bar
Stabilizer bar link nuts
2005 models 48
2006 through 2009 models 63

*Use high-strength thread-locking compound during installation
**New fasteners recommended by manufacturer

Torque specifications (continued)

Ft-lbs (unless otherwise indicated)

Note: *One foot-pound (ft-lb) of torque is equivalent to 12 inch-pounds (in-lbs) of torque. Torque values below approximately 15 foot-pounds are expressed in inch-pounds, because most foot-pound torque wrenches are not accurate at these smaller values.*

Front suspension (continued)

Stabilizer bar link nuts (continued)
    2010 and later models
        Upper link nut 63
        Lower link nut 55
  Stabilizer bushing/retainer bolts 37
Strut
  Strut upper mounting fasteners
    2008 and earlier models 18
    2009 models 22
    2010 and later models 18
  Strut-to-suspension support (damper shaft) nut
    2005 and 2006 models 55
    2007 and 2008 models 63
    2009 and later models 41
  Strut-to-steering knuckle bolts/nuts
    2005 and 2006 models 133
    2007 and 2008 models 148
    2009 and later models 137
Subframe bolts**
  Subframe-to-body bolts 114
  Subframe bracket-to-body bolts 43
  Subframe-to-rear transmission mount bolts 81

Rear suspension

Shock absorber mounting bolts 81
Stabilizer bar
  Stabilizer bar link-to-lower control arm nut .... 132 in-lbs
  Stabilizer bar link-to-stabilizer bar nut
    2006 and earlier 42
    2007 through 2009 models 37
    2010 and later models 70
  Stabilizer bar bushing/retainer bolts 52
Toe link pivot bolts 118
Trailing arm mounting bolts
  Trailing arm-to-knuckle bolts
    2009 and earlier models 81
    2010 and later models
        Step 1 44
        Step 2 .... Tighten an additional 50 degrees
  Trailing arm bracket-to-body bolts 81
  Trailing arm through-bolt nut 118
Control arm pivot bolts
  Upper control arm
    To knuckle 118
    To suspension support 121
  Lower control arm
    To knuckle 118
    To suspension support
        2006 and earlier 81
        2007 and later 92
Hub/wheel bearing mounting bolts*
  2006 and earlier 55
  2007 through 2009 models 62
  2010 and later models 52
Driveaxle/hub nut (AWD models) .... See Chapter 8

Steering system

Steering column mounting nut/bolts
     2007 and earlier models 18
     2010 and later models 16
Steering gear mounting bolts... 81
Steering shaft universal joint pinch bolt** 25
Steering wheel nut
     2007 and earlier models 31
     2010 and later models 37
Tie-rod end-to-steering knuckle nut**
     Step 1 18
     Step 2 .... Tighten an additional 90-degrees
Wheel lug nuts... See Chapter 1

*Use high-strength thread-locking compound during installation
**New fasteners recommended by manufacturer

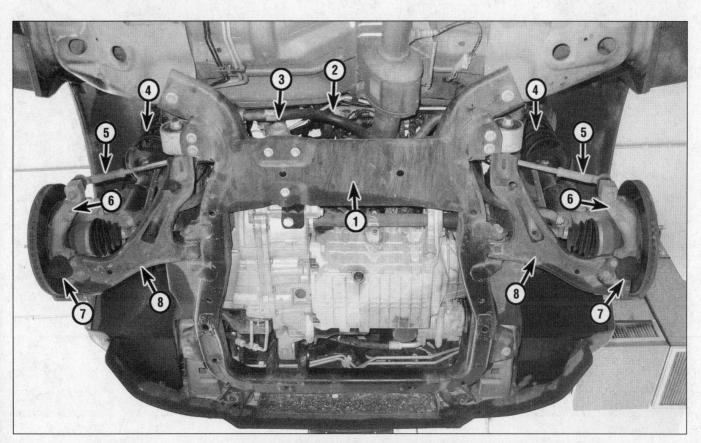

1.1 Front suspension and steering components:

1	Subframe (cradle)	4	Strut/coil spring assembly	7	Balljoint
2	Stabilizer bar	5	Tie-rod end	8	Control arm
3	Steering gear	6	Steering knuckle		

1 General information

Front suspension

Refer to illustration 1.1

The front suspension is a MacPherson strut design. The upper end of each strut is attached to the vehicle's body strut support. The lower ends of the struts are connected to the upper ends of the steering knuckles. The steering knuckles are attached to balljoints mounted on the outer ends of the suspension control arms **(see illustration)**.

Rear suspension

Refer to illustration 1.2

The rear suspension uses coil springs and shock absorbers **(see illustration)**. Knuckles are used to mount the hubs and brake assemblies; the knuckles are located by four suspension arms.

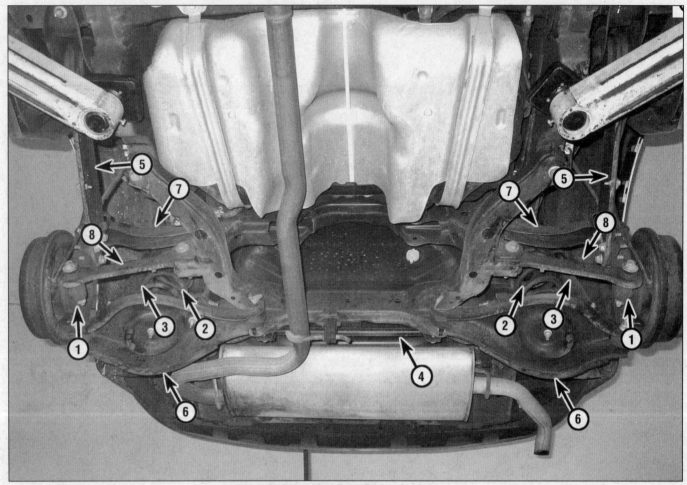

1.2 Rear suspension components:

1	Rear hub and knuckle	4	Stabilizer bar	7	Upper control arm
2	Coil spring	5	Trailing arm	8	Toe link
3	Shock absorber	6	Lower control arm		

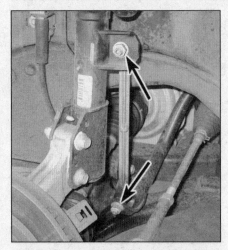

2.4 To detach the stabilizer bar link from the bar, remove the lower nut; if you're removing the strut (or replacing the link), remove the upper nut

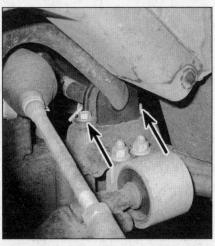

2.6 Stabilizer bar bushing retainer bolts

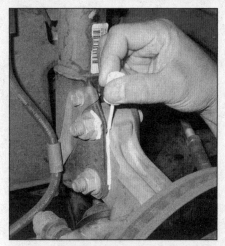

3.4 Mark the position of each strut on the steering knuckle before loosening the bolts

Precautions

Frequently, when working on the suspension or steering system components, you may come across fasteners that seem impossible to loosen. These fasteners on the underside of the vehicle are continually subjected to water, road grime, mud, etc., and can become rusted or frozen, making them extremely difficult to remove. In order to unscrew these stubborn fasteners without damaging them (or other components), be sure to use lots of penetrating oil and allow it to soak in for a while. Using a wire brush to clean exposed threads will also ease removal of the nut or bolt and prevent damage to the threads. Sometimes a sharp blow with a hammer and punch will break the bond between nut and bolt threads, but care must be taken to prevent the punch from slipping off the fastener and ruining the threads. Heating the stuck fastener and surrounding area with a torch sometimes helps too, but isn't recommended because of the obvious dangers associated with fire. Long breaker bars and extension, or cheater, pipes will increase leverage, but never use an extension pipe on a ratchet - the ratcheting mechanism could be damaged. Sometimes tightening the nut or bolt first will help to break it loose. Fasteners that require drastic measures to remove should always be replaced with new ones.

Since most of the procedures dealt with in this Chapter involve jacking up the vehicle and working underneath it, a good pair of jackstands will be needed. A hydraulic floor jack is the preferred type of jack to lift the vehicle, and it can also be used to support certain components during various operations. **Warning:** *Never, under any circumstances, rely on a jack to support the vehicle while working on it. Whenever any of the suspension or steering fasteners are loosened or removed they must be inspected and, if necessary, replaced* with new ones of the same part number or of original equipment quality and design. Torque specifications must be followed for proper reassembly and component retention. Never attempt to heat or straighten any suspension or steering components. Instead, replace any bent or damaged part with a new one.

2 Stabilizer bar and bushings (front) - removal and installation

Refer to illustrations 2.4 and 2.6

Note: *The steering gear must be detached from the subframe to remove the stabilizer bar.*

1   Loosen the front wheel lug nuts, then raise the front of the vehicle and support it securely on jackstands.

2   Remove the front wheels.

3   On 2007 through 2009 models, detach the catalytic converter from the exhaust manifold (see Chapter 4).

4   Disconnect the stabilizer bar links from the bar (**see illustration**). If the ballstud turns with the nut, hold the stud with a Torx or Allen wrench, as applicable.

5   Disconnect the left tie-rod end from the steering knuckle (see Section 18).

6   Detach both stabilizer bar bushing retainers from the subframe (**see illustration**).

7   On 2007 through 2009 models, detach the steering gear from the subframe (see Section 20).

8   Carefully remove the stabilizer bar out the left wheel well. **Caution:** *Be careful not to catch the shift cable during removal.*

9   While the stabilizer bar is detached, slide off the retainer bushings and inspect them. If they're cracked, worn or deteriorated, replace them. **Note:** *It is not necessary to remove the bar to replace the bushings.*

10   Clean the bushing area of the stabilizer bar with a stiff wire brush to remove any rust or dirt.

11   Lubricate the inside and outside of the new bushings with vegetable oil (used in cooking) to simplify reassembly. **Caution:** *Don't use petroleum or mineral-based lubricants or brake fluid - they will lead to deterioration of the bushings.*

12   Install the retainers and bolts, tightening the bolts to the torque listed in this Chapter's Specifications.

13   Install the links, tightening the link nuts to the torque listed in this Chapter's Specifications.

14   On 2007 through 2009 models, install the steering gear and catalytic converter.

15   Install the wheel and lug nuts, then lower the vehicle. Tighten the lug nuts to the torque listed in the Chapter 1 Specifications.

3 Strut assembly (front) - removal, inspection and installation

Removal

Refer to illustrations 3.4 and 3.5

1   Loosen the wheel lug nuts, raise the vehicle and support it securely on jackstands. Remove the wheel. Support the control arm with a floor jack.

2   Unbolt the brake hose bracket from the strut.

3   Disconnect the upper end of the stabilizer bar link from the strut (**see illustration 2.4**).

4   Mark the exact position of the strut to the steering knuckle (**see illustration**). Remove the strut-to-knuckle bolts and nuts.

5   Support the strut and spring assembly with one hand (or have an assistant hold it) and remove the three strut upper mounting

3.5 Strut upper mounting fasteners

4.3 Install the spring compressor in accordance with the tool manufacturer's instructions and compress the spring until all force is relieved from the upper spring seat

4.4 Remove the damper shaft nut

fasteners **(see illustration)**. Remove the strut assembly out from the fenderwell.

Inspection

6   Check the strut body for leaking fluid, dents, cracks and other obvious damage that would warrant repair or replacement.
7   Check the coil spring for chips or cracks in the spring coating (this can cause premature spring failure due to corrosion). Inspect the spring seat for cuts, hardness and general deterioration.
8   If any undesirable conditions exist, proceed to the strut disassembly procedure (see Section 4).

Installation

9   Guide the strut assembly into position and install the fasteners so the strut won't fall back through. This is most easily accom-

4.5 Lift the top mount assembly off the damper shaft; don't handle the plastic part - the bearing could fall apart (typical unit shown)

plished with the help of an assistant, as the strut is quite heavy and awkward.
10   Slide the steering knuckle into the strut flange and insert the two bolts. Install the nuts, match the position of the strut to the steering knuckle that was previously marked and tighten them to the torque listed in this Chapter's Specifications.
11   Reattach the brake hose bracket and the stabilizer bar link.
12   Install the wheel and lug nuts, then lower the vehicle and tighten the lug nuts to the torque listed in the Chapter 1 Specifications.
13   Tighten the upper mounting fasteners to the torque listed in this Chapter's Specifications.
14   It's a good idea to have the front wheel alignment checked and adjusted, if necessary, after this job has been performed.

4   Strut/coil spring assembly - replacement

1   If the struts or coil springs exhibit the telltale signs of wear (leaking fluid, loss of damping capability, chipped, sagging or cracked coil springs) explore all options before beginning any work. The strut/shock absorber assemblies are not serviceable and must be replaced if a problem develops. However, strut assemblies complete with springs may be available on an exchange basis, which eliminates much time and work. Whichever route you choose to take, check on the cost and availability of parts before disassembling your vehicle. **Warning:** *Disassembling a strut is potentially dangerous and utmost attention must be directed to the job, or serious injury may result. Use only a high-quality spring compressor and carefully follow the manufacturer's instructions furnished with the tool. After removing the coil spring from the strut assembly, set it aside in a safe, isolated area.*

Disassembly

Refer to illustrations 4.3, 4.4, 4.5, 4.6 and 4.7

2   Remove the strut and spring assembly (see Section 3). Mount the strut assembly in a vise. Line the vise jaws with wood or rags to prevent damage to the unit. Don't tighten the vise excessively.
3   Following the tool manufacturer's instructions, install the spring compressor (which can be obtained at most auto parts stores or equipment yards on a daily rental basis) on the spring and compress it sufficiently to relieve all force from the upper spring seat **(see illustration)**. This can be verified by wiggling the spring.
4   Loosen the damper shaft nut **(see illustration)**.
5   Remove the nut and suspension support **(see illustration)**. Inspect the bearing in the suspension support for smooth operation. If it doesn't turn smoothly, replace the suspension support. Check the rubber portion of the suspension support for cracking and general deterioration. If there is any separation of the rubber, replace it.
6   Remove the top mount assembly from the damper shaft **(see illustration)**. Remove the upper spring seat. Check the spring seat for cracking; replace it if necessary.
7   Carefully lift the compressed spring from the assembly **(see illustration)** and set it in a safe place. **Warning:** *Never place your head near the end of the spring!*
8   Remove the dust boot. Slide the rubber bumper off the damper shaft.
9   Check the lower insulator for wear, cracking and hardness and replace it if necessary.

Reassembly

Refer to illustration 4.11

10   If the lower insulator is being replaced, set it into position with the dropped portion seated in the lowest part of the seat. Replace

the dust boot. Extend the damper rod to its full length and install the rubber bumper. **Note:** *The spring end locating tab must be 180-degrees from the knuckle mounting bracket.*

11 Carefully place the coil spring onto the lower insulator, with the end of the spring resting in the lowest part of the insulator and against the rotation tab **(see illustration)**.

12 Install the upper spring seat onto the shaft. Align its flat with the knuckle mounting bracket.

13 Install the top mount with its flat 180-degrees from the flat on the upper spring seat. This will align it with the rotation tab of the upper spring seat.

14 Install the damper nut and tighten it to the torque listed in this Chapter's Specifications.

15 Remove the spring compressor tool.

4.6 Remove the spring seat from the damper shaft

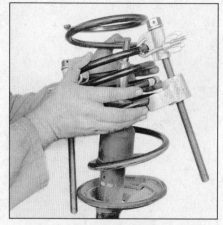

4.7 Remove the compressed spring assembly - keep the ends of the spring pointed away from your body

---

5 Control arm (front) - removal, inspection and installation

Removal

Refer to illustration 5.3

1 Loosen the wheel lug nuts on the side to be dismantled, raise the front of the vehicle, support it securely on jackstands and remove the wheel.

2 Remove the cotter pin, loosen the balljoint nut and separate the balljoint from the steering knuckle before removing the nut (see Section 6). If possible, use a screw-type balljoint separator rather than a wedge or pickle fork tool to avoid damaging the rubber boot.

3 Remove the bolts that attach the rear of the control arm to the engine cradle **(see illustration)**.

4 Remove the pivot bolt and nut that attach the front of the control arm to the engine cradle.

5 Remove the control arm.

Inspection and bushing replacement

6 Make sure the control arm is straight. If it's bent, replace it. Do not attempt to straighten a bent control arm.

7 Inspect the bushings. If they're cracked, torn or worn, replace them.

8 To replace the rear bushing, remove the nut and slide the bushing off its post. When installing the new bushing, tighten the nut to the torque listed in this Chapter's Specifications.

9 If the front pivot bushing is in need of replacement, take the control arm to an automotive machine shop to have the old bushing pressed out and the new bushing pressed in.

Installation

10 Installation is the reverse of removal. Be sure to tighten all fasteners to the torque listed in this Chapter's Specifications. **Note 1:** *The manufacturer recommends that the control arm mounting bolts be replaced with new ones when they are removed.* **Note 2:** *Before tightening the control arm front pivot bolt, raise*

the outer end of the control arm with a floor jack to simulate normal ride height. **Note 3:** *Be sure to use a new cotter pin on the balljoint stud nut. If necessary, tighten the balljoint nut a little more to align the hole in the ballstud with the slots in the nut - don't loosen the nut to achieve this alignment.*

11 Install the wheel and lug nuts, lower the vehicle and tighten the lug nuts to the torque listed in the Chapter 1 Specifications.

12 It's a good idea to have the front wheel alignment checked, and if necessary, adjusted after this job has been performed.

---

6 Balljoints - replacement

1 Loosen the wheel lug nuts, raise the vehicle and support it securely on jackstands. Remove the wheel. **Note:** *If you're going to remove the balljoint using a puller (as described in Steps 4 through 6), loosen the driveaxle/hub nut before raising the vehicle (see Chapter 8).*

4.11 When installing the spring, make sure the end fits against the locating tab

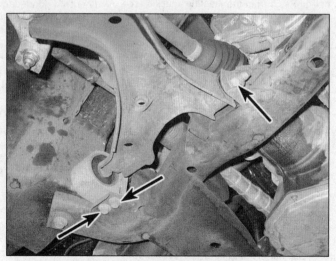

5.3 Control arm mounting fasteners

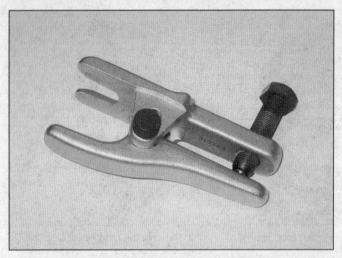

6.5a A balljoint separator tool like this one is available at most auto parts stores and will not damage the balljoint boot when used correctly

6.5b Strike the knuckle sharply with a hammer at this spot to try to pop the balljoint loose; there is little clearance above the balljoint for a typical puller - special pullers that avoid having to remove the driveaxle are available at auto parts stores

Picklefork/wedge method

Caution: *The following procedure is the quickest way to detach a balljoint from the steering knuckle, but it will very likely damage the balljoint boot. If you want to save the boot, proceed to Step 4.*

2   Remove the cotter pin from the balljoint stud and loosen the nut a few turns (but don't remove it yet).

3   Separate the balljoint from the steering knuckle by hammering in a picklefork-type balljoint wedge. Remove the balljoint stud nut.

Puller method

Refer to illustrations 6.5a and 6.5b

4   Remove the cotter pin from the balljoint stud and loosen the nut a few turns (but don't remove it yet).

5   Install a special puller and pop the balljoint stud from the steering knuckle **(see illustration)**. Because of driveaxle interference, a special puller is needed that works from below the balljoint. If the special puller isn't available, try striking the steering knuckle near the side of the balljoint with a hammer while pulling the balljoint out **(see illustration)**. The impact may break the balljoint loose. If these methods don't work, remove the driveaxle first.

6   Remove the nut and pull the balljoint stud out of the steering knuckle.

All models

Refer to illustration 6.8

7   Refer to Section 5 and remove the control arm. **Note:** *The control arm can be left on the vehicle, but it's more difficult to drill out the rivets in that position.*

8   Drill out the balljoint rivets **(see illustration)**. Begin with a 5/16-inch (8 mm) starter hole and complete each with a 31/64-inch (12

mm) drill bit. Remove all drilling debris.

9   Install the new balljoint using the supplied bolts. Make sure to put the bolt heads on top and the nuts below. Tighten the bolts to the torque listed in this Chapter's Specifications.

10   The remainder of the installation is the reverse of removal. Tighten the balljoint-to-knuckle nuts and the control arm bolts to the torques listed in this Chapter's Specifications. **Note:** *Before tightening the control arm front pivot bolt, raise the outer end of the control arm with a floor jack to simulate normal ride height. Be sure to use a new cotter pin. If necessary, tighten the balljoint nut a little more to align the hole in the ballstud with the slots in the nut - don't loosen the nut to achieve this alignment.*

11   Install the wheel and lug nuts, lower the vehicle and tighten the lug nuts to the torque listed in the Chapter 1 Specifications.

12   It's a good idea to have the front wheel alignment checked, and if necessary, adjusted after this job has been performed.

---

7   Hub and bearing assembly (front) - removal and installation

---

Refer to illustration 7.7

Warning: *Dust created by the brake system is harmful to your health. Never blow it out with compressed air and don't inhale any of it. Do not, under any circumstances, use petroleum-based solvents to clean brake parts. Use brake system cleaner only.*

1   Loosen the driveaxle/hub nut (see Chapter 8).

2   Loosen the wheel lug nuts, raise the vehicle and support it securely on jackstands. Remove the wheel.

3   Refer to Chapter 9 and remove the brake disc.

4   Disconnect the wheel speed sensor electrical connector and unclip the harness from the bracket.

5   Remove the driveaxle/hub nut.

6   Use wire to support the driveaxle (don't let it hang by the inner CV joint after the hub and bearing has been removed).

7   Remove the hub mounting bolts **(see illustration)**. **Note:** *On 2007 and later models, be sure to clean the hub mounting bolt threads and holes and apply new high-strength thread-locking compound during installation **(see illustrations 3.5g and 3.5h** in Chapter 9).*

8   Remove the hub and bearing unit. If it's stuck to the driveaxle, tap the driveaxle end with a hammer and a block of wood to avoid damaging the threads.

9   Installation is the reverse of removal. Tighten the bolts to the torque listed in this Chapter's Specifications.

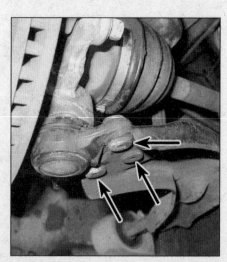

6.8 These large rivets must be drilled out to remove the old balljoint

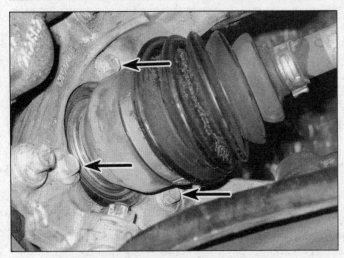

7.7 Front hub-to-knuckle mounting bolts

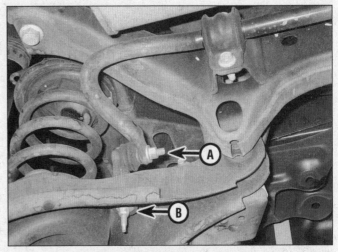

9.2 The rear stabilizer bar attachment to the link (A) and link attachment to the control arm (B)

8 Steering knuckle - removal and installation

Removal

1   Loosen the driveaxle/hub nut (see Chapter 8). Loosen the wheel lug nuts, raise the vehicle and support it securely on jackstands. Remove the wheel.
2   Mark the position of the strut to the steering knuckle **(see illustration 3.4)**.
3   Loosen, but do not remove, the strut-to-steering knuckle bolts.
4   Separate the tie-rod end from the steering knuckle arm (see Section 18).
5   If the hub and bearing assembly is to be removed from the knuckle, remove it now (see Section 7).
6   Disconnect the balljoint from the steering knuckle (see Section 6).
7   Push the driveaxle from the hub. Support the end of the driveaxle with a piece of wire.
8   Separate the steering knuckle from the strut.

Installation

9   Guide the knuckle and hub assembly into position, inserting the driveaxle into the hub.
10   Push the knuckle into the strut flange and install the bolts and nuts, but don't tighten them yet.
11   Connect the balljoint to the steering knuckle and tighten the nut to the torque listed in this Chapter's Specifications. Install a new cotter pin. If necessary, tighten the balljoint nut a little more to align the hole in the ball-stud with the slots in the nut - don't loosen the nut to achieve this alignment.
12   Attach the tie-rod end to the steering knuckle arm (see Section 18). Tighten the strut bolt nuts, the balljoint-to-control arm bolt and nuts and the tie-rod nut to the torque values listed in this Chapter's Specifications.

Replace all cotter pins with new ones.
13   Place the brake disc on the hub and install the caliper and wheel speed sensor as outlined in Chapter 9.
14   Install a **new** driveaxle/hub nut and tighten it securely (final tightening will be carried out when the vehicle is lowered).
15   Install the wheel and lug nuts, lower the vehicle and tighten the lug nuts to the torque listed in the Chapter 1 Specifications.
16   Tighten the driveaxle/hub nut to the torque listed in the Chapter 8 Specifications.

9 Stabilizer bar and bushings (rear) - removal and installation

Refer to illustrations 9.2 and 9.3

1   Raise the rear of the vehicle and support it securely on jackstands.
2   Disconnect the stabilizer bar links from the bar **(see illustration)**. If the ballstud turns with the nut, use a Torx or Allen wrench, as applicable, to hold the stud.
3   Unbolt the stabilizer bar bushing retainers **(see illustration)**. Remove the two retainer brackets from the body, if necessary.
4   On 2007 and later models, lower the rear subframe enough to remove the stabilizer bar. **Note:** *It is not necessary to remove the bar to replace the bushings.*
5   The stabilizer bar can now be removed from the vehicle. Remove the bushings from the stabilizer bar, noting their positions and orientation.
6   Check the bushings for wear, hardness, distortion, cracking and other signs of deterioration, replacing them if necessary. Check the stabilizer bar links as described in Section 2.
7   Using a wire brush, clean the areas of the bar where the bushings ride. Lubricate the inside and outside of the new bushings with vegetable oil (used in cooking). **Caution:** *Don't use petroleum or mineral-based lubri-*

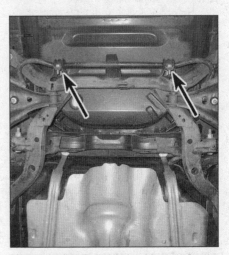

9.3 Rear stabilizer bar bushing retainers

cants or brake fluid - they will lead to deterioration of the bushings.
8   Installation is the reverse of removal. Tighten the links and the bushing retainer fasteners to the torque values listed in this Chapter's Specifications.

10 Shock absorber (rear) - replacement

Refer to illustrations 10.4a and 10.4b
Note: *When installing any rear suspension components, move the suspension to its normal ride-height angle and position, then fully tighten the bolts.*
1   Loosen the rear wheel lug nuts, raise the rear of the vehicle and support it securely on jackstands. Remove the wheel.
2   Remove the inner fender splash shields (see Chapter 11, Section 11).
3   Slightly raise the outer end of the lower

control arm with a floor jack so it can't move downward when the shock absorber mounting bolts are removed. **Warning:** *The jack must support the lower control arm until the shock absorber is reinstalled.*

4   Remove the upper and lower shock absorber mounting bolts and remove the shock absorber **(see illustrations)**.

5   Installation is the reverse of removal. Tighten the mounting fasteners to the torque listed in this Chapter's Specifications.

11   Coil spring (rear) - removal and installation

Note: *When installing any rear suspension components, move the suspension to its normal ride-height angle and position, then fully tighten the bolts.*

1   Loosen the rear wheel lug nuts, raise the rear of the vehicle and support it securely on jackstands. Remove the wheel.

2   Refer to Section 9 and disconnect the stabilizer bar link from the lower control arm.

3   On 2006 and earlier models, unbolt the bracket at the front of the trailing arm from the body **(see illustration 12.9)**.

4   Slightly raise the outer end of the lower control arm with a floor jack so it can't move downward when the shock absorber mounting bolt is removed.

5   Remove the lower shock absorber mounting bolt **(see illustration 10.4a)**.

6   Loosen the inner through-bolt at the lower control arm pivot.

7   Remove the outer through-bolt from the lower control arm.

8   Slowly and carefully lower the jack until the coil spring extends fully.

9   Installation is the reverse of removal. Make sure that the spring insulators are properly installed on the spring before putting the spring into place.

10   Tighten all fasteners to the torque values listed in this Chapter's Specifications.

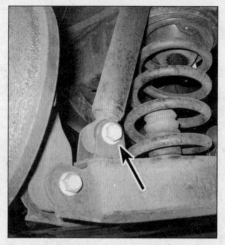

10.4a Shock absorber lower mounting bolt

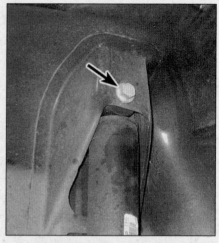

10.4b Shock absorber upper mounting bolt

12   Suspension arms (rear) - removal and installation

Note: *When installing any rear suspension components, move the suspension to its normal ride-height angle and position, then fully tighten the bolts.*

1   Loosen the rear wheel lug nuts, raise the rear of the vehicle and support it securely on jackstands. Remove the rear wheel.

Toe link

Refer to illustration 12.3

2   Mark the positions of the inner bolt heads or cam adjusters to the subframe prior to loosening any bolts. **Note:** *The inner mounting holes are slotted. Nuts with adjuster cams (for alignment purposes) are available at dealer parts departments.*

3   Remove the inner and outer pivot bolts **(see illustration)**. Remove the toe link.

4   Installation is the reverse of removal. Install the bolts with their heads facing forward.

5   Align the bolt heads or adjuster cams with the marks made in Step 2. Make sure that all the adjusters, if present, are in the same positions as when you marked them.

6   Before tightening the mounting bolts, raise the outer end of the lower control arm with a floor jack to simulate normal ride height.

7   Tighten the bolts to the torque values listed in this Chapter's Specifications.

8   It's a good idea to have the wheel alignment checked and, if necessary, adjusted.

Trailing arm

Refer to illustrations 12.9 and 12.11

9   Remove the three mounting bolts from the trailing arm-to-body bracket **(see illustration)**.

10   Unbolt the parking brake cable clip from the trailing arm.

11   Remove the three bolts securing the arm to the knuckle and lift off the arm **(see illustration)**.

12   If necessary, remove the through-bolt and separate the bracket from the trailing arm.

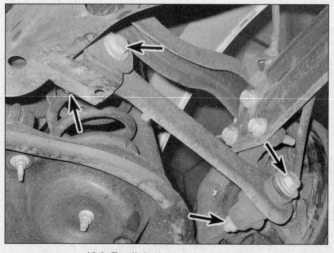

12.3 Toe link pivot nuts and bolts

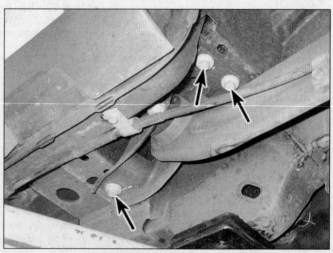

12.9 Trailing arm bracket-to-body bolts

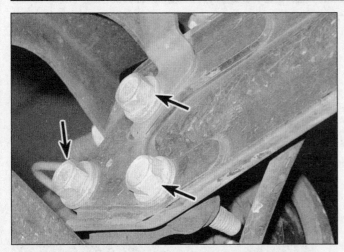

12.11 Trailing arm-to-knuckle bolts

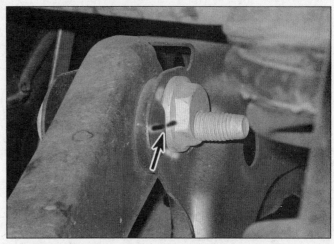

12.15 Mark the position of the cam adjuster on the inner pivot bolt of the upper control arm

13 Installation is the reverse of removal. Install all of the mounting bolts finger tight before tightening any of them. Tighten all of the fasteners to the torques listed in this Chapter's Specifications. It's a good idea to have the wheel alignment checked and, if necessary, adjusted.

Upper control arm

Refer to illustrations 12.15 and 12.17

14 Unbolt the trailing arm bracket from the body **(see illustration 12.9)**.

15 Mark the position of the cam adjuster on the inner end of the upper control arm **(see illustration)**.

16 Disconnect the ABS wiring from the upper control arm.

17 Support the lower control arm with a floor jack. Remove the through-bolt from each end of the upper control arm pivot **(see illustration)**. Lift out the arm.

18 To install the arm, first loosely install the bolt at each end.

19 After the adjuster marks are aligned, raise the lower control arm to simulate normal ride height, then tighten the fasteners to the torque listed in this Chapter's Specifications.

20 The remainder of installation is the reverse of removal. It's a good idea to have the wheel alignment checked and, if necessary, adjusted.

Lower control arm

Refer to illustration 12.26

21 Refer to Section 9 and disconnect the stabilizer bar link from the lower control arm.

22 Unbolt the trailing arm bracket from the body **(see illustration 12.9)**.

23 Position a floor jack under the outer end of the control arm and raise it slightly.

24 Remove the shock absorber lower mounting bolt.

25 Remove the bumper mounting nut at the center of the bottom of the spring.

26 Loosen the lower control arm inner pivot bolt **(see illustration)**, remove the outer bolt and slowly lower the arm, allowing the coil spring to extend.

27 Remove the spring and its lower bumper.

28 Remove the arm inner bolt and lift out the arm.

29 Installation is the reverse of removal. Raise the lower control arm to simulate normal ride height, then tighten the fasteners to the torque listed in this Chapter's Specifications.

30 Have the wheel alignment checked by a dealer service department or an alignment shop.

---

13 Hub and bearing assembly (rear) - removal and installation

---

Refer to illustration 13.6

Warning: *Dust created by the brake system is harmful to your health. Never blow it out with compressed air and don't inhale any of it. Do not, under any circumstances, use petroleum-based solvents to clean brake parts. Use brake system cleaner only.*

12.17 Upper control arm pivot bolts

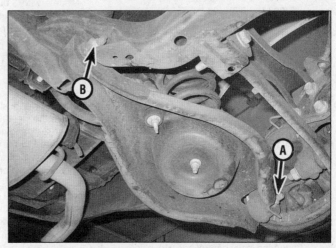

12.26 Lower control arm-to-rear knuckle bolt/nut (A) and inner pivot bolt/nut (B)

13.6 The rear hub is attached to the knuckle by four bolts

16.3 Insert blunt tools into the holes in the back of the steering wheel to release the airbag module

1   On AWD models, loosen the rear drive-axle/hub nut.
2   Loosen the wheel lug nuts, raise the vehicle and support it securely on jackstands. Remove the wheel.
3   Refer to Chapter 9 and remove the brake drum or disc.
4   On AWD models, remove the driveaxle/hub nut. Use wire to support the end of the driveaxle.
5   Disconnect the electrical connector for the wheel speed sensor.
6   Remove the hub mounting bolts **(see illustration)**, then guide the hub and bearing out from the brake assembly.
7   Installation is the reverse of removal. After all bolts have been installed, tighten them to the torque listed in this Chapter's Specifications. **Note:** *On 2007 and later models, be sure to clean the hub mounting bolt threads and holes and apply new high-strength thread-locking compound during installation* **(see illustrations 3.5g and 3.5h in Chapter 9)**.
8   On AWD models, install a new driveaxle/hub nut and tighten it securely (final tightening will be carried out after the vehicle has been lowered).
9   Install the brake drum or disc and the wheel. Lower the vehicle and tighten the lug nuts to the torque listed in the Chapter 1 Specifications. On AWD models, tighten the driveaxle/hub nut to the torque listed in the Chapter 9 Specifications.

14   Suspension knuckle (rear) - removal and installation

1   On models with rear disc brakes, detach the parking brake cable from the bracket and actuator lever at the rear of the knuckle.
2   Refer to Section 13 and remove the hub and bearing assembly.
3   Support the lower control arm with a floor jack.
4   Unbolt the knuckle from the upper and lower control arms.
5   Remove the knuckle bolts from the trailing arm and the toe link.
6   Separate the knuckle from the other components and lift it out.
7   Installation is the reverse of removal. Install all of the mounting bolts finger tight for now.
8   When the components are all connected, raise the lower control arm to simulate normal ride height, then tighten the fasteners to the torque values listed in this Chapter's Specifications in the following order: lower control arm, upper control arm, toe link and the trailing arm.

15   Steering system - general information

All models are equipped with rack-and-pinion steering. The steering gear is bolted to the subframe and operates the steering knuckles via tie-rods. The inner ends of the tie-rods are protected by rubber boots that should be inspected periodically for secure attachment, tears and leaking lubricant.
Models with 3.0L and 3.6L V6 engines are equipped with hydraulic power steering. A belt-driven pump on the engine supplies hydraulic pressure to the steering gear.
Models with the 2.4L four-cylinder and 3.4L V6 engines are equipped with electric power steering. The systems consist of a torque sensor which is in the steering column, an electric power assist motor that has been mounted to the steering column on 3.4L models, and on 2.4L engine models the motor has been mounted to the steering gear, and the Power Steering Control Module (PSCM). The PSCM gets inputs from the torque sensor as well as other sources and instructs the electric motor to apply the proper amount of assist to the steering column motor or steering gear motor. If the PSCM detects a problem in the system, it will turn on a warning light on the instrument panel. System troubles must be diagnosed and repaired by a dealer service department or other qualified repair shop.

The steering wheel operates the steering shaft, which actuates the steering gear through universal joints. Looseness in the steering can be caused by wear in the steering shaft universal joints, the steering gear, the tie-rod ends or loose retaining bolts.

16   Steering wheel - removal and installation

Warning: *The models covered by this manual are equipped with Supplemental Restraint Systems (SRS), more commonly known as airbags. Always disable the airbag system before working in the vicinity of any airbag system component to avoid the possibility of accidental deployment of the airbag(s), which could cause personal injury (see Chapter 12).*

Removal

Refer to illustrations 16.3, 16.4a, 16.4b and 16.6

1   Turn the steering wheel so that the wheels are pointing straight ahead. Turn the ignition key to Off, then disconnect the cable from the negative terminal of the battery (see Chapter 5). Wait at least two minutes for the airbag system to lose power.
2   Locate the holes on the back of the steering wheel.
3   Insert small blunt tools into the holes to release the airbag module on that side, then repeat the procedure for the other side while carefully turning the steering wheel as necessary **(see illustration)**.
4   Lift the airbag module off the steering wheel and disconnect the electrical connectors for the module **(see illustrations)**. **Warning:** *Carry the airbag module with the trim side facing away from you and set it down in an isolated area with the trim side facing up.*
5   Return the steering wheel to the centered position.
6   Remove the steering wheel retaining nut, then mark the relationship of the steering shaft to the hub (if marks don't already exist or don't line up) to simplify installation and

16.4a Remove the airbag module carefully and disconnect the electrical connectors

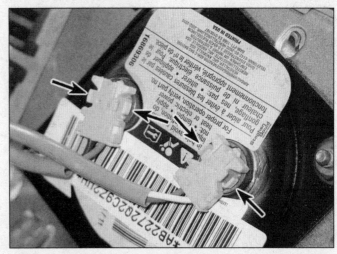

16.4b Squeeze the small tabs then pull the connectors directly out to release them from the airbag module

ensure steering wheel alignment **(see illustration)**.

7   Wiggle the steering wheel while pulling on it with moderate force to remove it from the steering shaft. **Caution:** *Don't hammer on the shaft or the steering wheel in an attempt to remove the wheel. If it will not come of with moderate effort, use a steering wheel puller to remove it.*

8   If it's necessary to remove the airbag clockspring, remove the steering column covers (see Chapter 11). Disengage the clockspring retaining fingers while detaching it from the steering column, then follow the wiring harnesses and disconnect the electrical connectors.

Installation

Refer to illustration 16.11

Note: *Steps 10 and 11 need not be done if you're sure the airbag clockspring hasn't been rotated after steering wheel removal.*

9   Make sure that the front wheels are facing straight ahead.

10   Turn the hub of the clockspring counterclockwise by hand until it becomes harder to turn the cable (don't apply too much force). Rotate it clockwise and count the turns until the same resistance is met.

11   Rotate the clockspring back to half that number of turns to center it **(see illustration)**. Install the clockspring, making sure its hub doesn't rotate in relation to the housing.

12   To install the steering wheel, align the mark on the hub with the mark on the shaft and slip the wheel onto the shaft. Install the nut and tighten it to the torque listed in this Chapter's Specifications. **Note:** *If a new clockspring is installed, remove the yellow alignment tab through the opening in the steering wheel.*

13   Plug in the electrical connectors for the airbag module and any other connectors. Make sure the connector locks are pushed back into position for the module.

14   Install the airbag module and make sure all of the retainers are snapped securely.

15   Connect the negative battery cable (see Chapter 5).

17   Steering column - removal and installation

Warning: *The models covered by this manual are equipped with Supplemental Restraint Systems (SRS), more commonly known as airbags. Always disable the airbag system before working in the vicinity of any airbag system component to avoid the possibility of accidental deployment of the airbag(s), which could cause personal injury (see Chapter 12).*
Note: *Vehicles equipped with electric power steering require calibration after service to the steering column. Take the vehicle to a dealership service department or other qualified repair location for calibration with the use of a scan tool.*

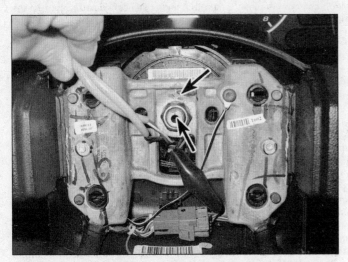

16.6 If there are no marks present, mark the steering wheel so it can be reinstalled in the same position on the shaft

16.11 Line up the index marks on the clockspring after rotating it to center it

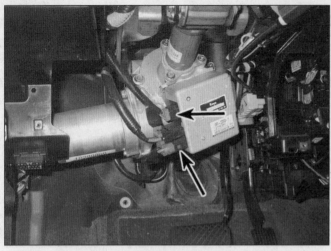

17.11 Disconnect the electrical connectors from the electric power steering unit

17.12 Steering column lower through-bolt (electric power assist type shown - hydraulic type similar)

17.13 Upper steering shaft pinch bolt (electric power assist type shown - hydraulic type similar)

17.14 Steering column upper mounting bolts

Removal

Refer to illustrations 17.11, 17.12, 17.13 and 17.14

1   Park the vehicle with the wheels pointing straight ahead. Disconnect the cable from the negative terminal of the battery (see Chapter 5).

2   Remove the steering wheel (see Section 16) and the steering column covers (see Chapter 11).

3   Unsnap the airbag clockspring electrical connectors at their connections below the steering column. Disconnect the airbag electrical connectors from the instrument panel harness.

4   Use a flat-blade screwdriver to release the clips from the airbag clockspring and slide it from the steering column.

5   Carefully pry the bezel from the lock cylinder.

6   Remove the multi-function switches (see Chapter 12).

7   Disconnect the electrical connectors from the ignition switch assembly.

8   Release the lock tabs on the park/lock cable end and remove it from the ignition lock cylinder housing.

9   Remove the two bolts from the lock cylinder housing. Remove the housing and slide the multi-function switch bracket from the column.

10   Refer to Chapter 11 and remove the driver's knee bolster trim panel and the lower insulator panel.

11   Disconnect the electrical connectors from the power steering unit, if equipped **(see illustration)**.

12   Remove the through-bolt from the lower part of the steering column **(see illustration)**.

13   Mark the relationship of the steering shaft U-joint to the intermediate shaft, then remove the pinch bolt **(see illustration)**.

14   Remove the steering column upper mounting fasteners **(see illustration)**, lower the column and pull it to the rear, making

sure nothing is still connected. Separate the intermediate shaft from the steering shaft and remove the column.

Installation

15   Guide the steering column into position, align the marks and connect the intermediate shaft, then install the three mounting bolts, but don't tighten them yet. **Note:** *The manufacturer recommends that a new pinch bolt be installed whenever it is removed.* Tighten the pinch bolt to the torque values listed in this Chapter's Specifications.

16   Slide the multi-function switch housing back onto the column, making sure that the lock tab is fully seated.

17   Install the ignition switch assembly and hand tighten the bolts so that there is a 1/8-inch gap at each end. Gradually and evenly tighten both bolts until you reach the torque listed in this Chapter's Specifications.

18   Tighten the left upper column mount-

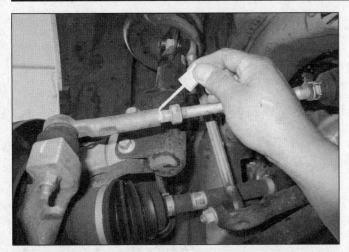

18.2 Back off the jam nut and mark the exposed threads to ensure that the new tie-rod end is threaded on the same number of turns

18.4 Install a small puller to separate the tie-rod end from the steering knuckle

ing bolt to the torque listed in this Chapter's Specifications.

19 Tighten the lower column mounting through-bolt to the torque listed in this Chapter's Specifications.

20 Actuate the tilt lever and move the column up and down several times, finally setting it at its top position.

21 Tighten the top right column mounting bolt to the same torque as the other top bolt.

22 The remainder of installation is the reverse of removal.

18 Tie-rod ends - removal and installation

Removal

Refer to illustrations 18.2 and 18.4

1 Loosen the wheel lug nuts. Raise the front of the vehicle, support it securely on jackstands, block the rear wheels and set the parking brake. Remove the front wheel.

2 Hold the tie-rod with a pair of locking pliers or wrench and loosen the jam nut enough to mark the position of the tie-rod end in relation to the threads **(see illustration)**.

3 Loosen the nut from the tie-rod end ball-stud about 1/8-inch.

4 Disconnect the tie-rod from the steering knuckle arm with a puller **(see illustration)**. Remove the nut and separate the tie-rod. Discard the used nut.

5 Unscrew the tie-rod end from the tie-rod.

Installation

6 Thread the tie-rod end on to the marked position and insert the tie-rod stud into the steering knuckle arm. Tighten the jam nut securely.

7 Install a new nut on the stud and tighten it to the torque listed in this Chapter's Specifications. **Note:** *The manufacturer recommends that these nuts be replaced with new ones whenever they are removed.*

8 Install the wheel and lug nuts. Lower the vehicle and tighten the lug nuts to the torque listed in the Chapter 1 Specifications.

9 Have the wheel alignment checked and, if necessary, adjusted.

19 Steering gear boots - replacement

Refer to illustration 19.3

Note: *The steering gear can be left in place to perform this job, however it may be easier to remove it and replace the boots on a workbench.*

1 Loosen the lug nuts, raise the vehicle and support it securely on jackstands. Remove the wheel.

2 Remove the tie-rod end and jam nut (see Section 18).

3 Remove the steering gear boot clamps **(see illustration)** and slide off the boot.

4 Before installing the new boot, wrap the threads and serrations on the end of the steering rod with a layer of tape so the small end of the new boot isn't damaged. Apply grease from the supplied grease packet to the tie-rod

in the area where the boot seats, and to the inner end of the inside of the boot.

5 Slide the new boot into position on the steering gear until it seats in the groove on the housing and the groove in the steering rod. Install the clamps.

6 Remove the tape and install the tie-rod end (see Section 18).

7 Install the wheel and lug nuts. Lower the vehicle and tighten the lug nuts to the torque listed in the Chapter 1 Specifications.

8 Have the wheel alignment checked and, if necessary, adjusted.

20 Steering gear - removal and installation

Warning: *Make sure the steering shaft is not turned while the steering gear is removed or you could damage the clockspring for the airbag system. To prevent the shaft from turning, place the ignition key in the lock position or thread the seat belt through the steering wheel and clip it into place. Don't turn the front wheels.*

19.3 Remove the outer clamp from the steering gear boot with a pair of pliers; the inner clamp must be cut or pried off

20.3 Steering intermediate shaft pinch bolt

20.13 Left-side steering gear mounting bolt

Removal

Refer to illustrations 20.3 and 20.13

Warning: *On models equipped with electronic power assisted steering, electrostatic discharge (ESD) can damage many of the electrical components. Some components are labeled with an ESD symbol and some are not. Always disconnect the battery and make sure you handle all electrical components very carefully.*

Note: *Some models are equipped with electronic power assist steering. On some versions, a small electric motor has been mounted to the steering column; on other models an electric motor has been mounted to the steering gear to help reduce steering effort.*

1   Disconnect the cable from the negative terminal of the battery (see Chapter 5).

2   Park the vehicle with the front wheels pointing straight ahead. Loosen the front wheel lug nuts, raise the front of the vehicle and support it securely on jackstands. Apply the parking brake and remove the wheels.

3   Mark the relationship of the intermediate shaft to the steering shaft U-joint and remove the pinch bolt **(see illustration)**.

4   Separate the tie-rod ends from the steering knuckles (see Section 18).

5   Detach the links from the ends of the stabilizer bar.

6   Remove the heat shield fasteners and heat shield.

7   On 2.4L models, remove the catalytic converter and front exhaust pipe. On 3.0L and 3.6L models remove the front "Y" exhaust pipe (see Chapter 4).

8   Remove the rear transmission mount (see Chapter 7A).

9   On AWD models, disconnect the driveshaft (see Chapter 8).

10   On 2.4L models, disconnect the electrical connectors to the electronic power assist motor.

11   On hydraulically assisted power steering gears, detach the pressure and return lines from the steering gear and the bracket that secures them. **Note:** *Place a drain pan under the steering gear when detaching fluid hoses or lines.*

12   On hydraulically assisted power steering gears, remove the transmission mount mounting fasteners (located just above the steering gear) and move the mount aside (see Chapter 7).

13   Remove the steering gear mounting bolts **(see illustration)**.

14   On later models, partially loosen and lower the subframe to allow for removal clearance of electronic power assist motor.

15   Separate the intermediate shaft U-joint from the steering gear input shaft, then remove the steering gear assembly out the right side of the vehicle (models with electric power steering), or the left side (models with hydraulic power steering).

Installation

16   Raise the steering gear into position and connect the intermediate shaft U-joint to the steering input shaft, aligning the marks. Tighten the pinch bolt to the torque listed in this Chapter's Specifications.

17   Install the steering gear mounting bolts and tighten them to the torque listed in this Chapter's Specifications.

18   Connect the tie-rod ends to the steering knuckle arms (see Section 18).

19   The remainder of installation is the reverse of removal. Be sure to tighten all of the fasteners to the torques listed in this Chapter's Specifications.

20   Have the wheel alignment checked and, if necessary, adjusted.

---

21   Power steering pump - removal and installation

Note: *The following procedure applies to 2009 and earlier models with the 3.6L V6 engine only.*

Removal

1   Remove the drivebelt (see Chapter 1).

2   Remove as much fluid as possible from the power steering fluid reservoir.

3   Position a drain pan under the power steering pump. Disconnect the feed hose and the pressure line from the rear of the pump. Plug or cap all openings to prevent contaminants from entering any part of the system.

4   Remove the pump mounting fasteners and guide the pump out the right fenderwell while taking care not to spill fluid on painted surfaces. **Note:** *If necessary, transfer the pulley from the old pump to the new one. Pulley removal and installation tools can be found at most auto parts stores.*

Installation

5   Position the pump and install the mounting fasteners; tighten them to the torque listed in this Chapter's Specifications.

6   Connect the pressure line and feed hose to the pump making sure that all connections are secure.

7   Install the drivebelt.

8   Fill the power steering reservoir with the recommended fluid (see Chapter 1) and bleed the system following the procedure described in Section 22.

---

22   Power steering system - bleeding

Note: *The following procedure applies to hydraulically assisted power steering systems.*

1   Following any operation in which the power steering fluid lines have been disconnected, the power steering system must be bled to remove all air and obtain proper steering performance.

2   With the front wheels in the straight ahead position, check the power steering fluid level and, if low, add fluid until it reaches the COLD fill mark (see Chapter 1).

3   Raise the front of the vehicle just enough for the wheels to clear the ground and support

23.13 Subframe/cradle mounting bolts

it securely on jackstands. Apply the parking brake.

4   Turn the key to the ON position with the engine off, then turn the steering wheel fully in each direction (stop-to-stop) 12 times.

5   Check the fluid level and add more if necessary to reach the COLD fill mark.

6   Start the engine and allow it to warm up.

7   Turn the steering wheel from side-to-side and confirm that there is no noise coming from the system due to aeration of the fluid. **Caution:** *DO NOT hold the steering wheel to the stops (in the extreme right or left position) because this could damage the power steering pump.*

8   Road test the vehicle to be sure the steering system is functioning normally and noise free.

9   Recheck the fluid level to be sure it's up to the HOT fill mark while the engine is at normal operating temperature. Add fluid if necessary.

23 Subframe/cradle - removal and installation

Refer to illustration 23.13

1   Disconnect the cable from the negative battery terminal (see Chapter 5).

2   Loosen the front wheel lug nuts, raise the front of the vehicle and support it securely on jackstands. Remove both front wheels. **Note:** *The jackstands must be behind the front suspension subframe, not supporting the vehicle by the subframe.*

3   Use wire to tie the upper part of the radiator/condenser assembly to the front upper body structure.

4   Remove the plastic side splash shields.

5   Detach the front air dam from the subframe.

6   Unbolt the front and rear transaxle mounts from the subframe (see Chapter 7).

7   Remove the stabilizer bar bracket mounting bolts.

8   Disconnect the balljoints from the knuckles (see Section 6).

9   Remove the steering gear mounting bolts and use wire to tie the gear to the exhaust

system.

10   Inspect the subframe for any hose, line or harness brackets that may be attached and detach them.

11   Support the engine from above using an engine hoist or support fixture (see Chapter 2C). **Warning:** *DO NOT place any part of your body under the engine when it's supported only by a hoist or other support device.*

12   Using two floor jacks, support the subframe. Position one jack on each side of the subframe, midway between the front and rear mounting points.

13   With the jacks sufficiently supporting the subframe, remove the fasteners securing the subframe brackets and the subframe **(see illustration)**. **Note:** *The manufacturer states that the subframe bolts should be replaced with new ones whenever they are removed.*

14   With the use of an assistant to steady the subframe, carefully lower the jacks until the subframe is sufficiently resting on the ground.

15   Remove the control arms if necessary.

16   Installation is the reverse of removal. Install *new* subframe mounting bolts and tighten them to the torque listed in this Chapter's Specifications.

24 Wheels and tires - general information

Refer to illustration 24.1

1   All vehicles covered by this manual are equipped with metric-sized fiberglass or steel belted radial tires **(see illustration)**. Use of other size or type of tires may affect the ride and handling of the vehicle. Don't mix different types of tires, such as radials and bias belted, on the same vehicle as handling may be seriously affected. It's recommended that tires be replaced in pairs on the same axle, but if only one tire is being replaced, be sure it's the same size, structure and tread design as the other.

2   Because tire pressure has a substantial effect on handling and wear, the pressure

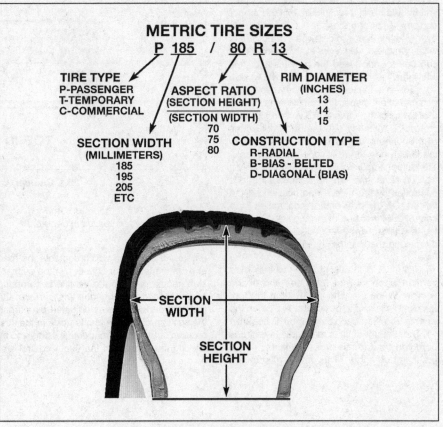

24.1 Metric tire size code

on all tires should be checked at least once a month or before any extended trips (see Chapter 1).

3   Wheels must be replaced if they are bent, dented, leak air, have elongated bolt holes, are heavily rusted, out of vertical symmetry or if the lug nuts won't stay tight. Wheel repairs that use welding or peening are not recommended.

4   Tire and wheel balance is important in the overall handling, braking and performance of the vehicle. Unbalanced wheels can adversely affect handling and ride characteristics as well as tire life. Whenever a tire is installed on a wheel, the tire and wheel should be balanced by a shop with the proper equipment.

25   Wheel alignment - general information

Refer to illustration 25.1

A wheel alignment refers to the adjustments made to the wheels so they are in proper angular relationship to the suspension and the ground. Wheels that are out of proper alignment not only affect vehicle control, but also increase tire wear. The alignment angles normally measured are camber, caster and toe-in **(see illustration).**

Getting the proper wheel alignment is a very exacting process, one in which complicated and expensive machines are necessary to perform the job properly. Because of this, you should have a technician with the proper equipment perform these tasks. We will, however, use this space to give you a basic idea of what is involved with a wheel alignment so you can better understand the process and deal intelligently with the shop that does the work.

Toe-in is the turning in of the wheels. The purpose of a toe specification is to ensure parallel rolling of the wheels. In a vehicle with zero toe-in, the distance between the front edges of the wheels will be the same as the distance between the rear edges of the wheels. The actual amount of toe-in is normally only a fraction of an inch. On the front end, toe-in is controlled by the tie-rod end position on the tie-rod. On the rear end, it's controlled by a cam at the inner end of the toe link. Incorrect toe-in will cause the tires to wear improperly by making them scrub against the road surface.

Camber is the tilting of the wheels from vertical when viewed from one end of the vehicle. When the wheels tilt out at the top, the camber is said to be positive (+). When the wheels tilt in at the top the camber is negative (-). The amount of tilt is measured in degrees from vertical and this measurement is called the camber angle. This angle affects the

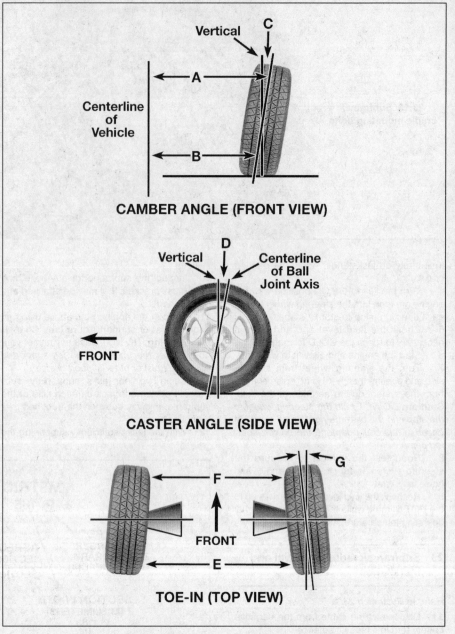

25.1 Camber, caster and toe-in angles

A minus B = C (degrees camber)
D = caster (expressed in degrees)

E minus F = toe-in (measured in inches)
G = toe-in (expressed in degrees)

amount of tire tread which contacts the road and compensates for changes in the suspension geometry when the vehicle is cornering or traveling over an undulating surface. On the front end, camber is adjusted by altering the position of the steering knuckle in the suspension strut. On the rear end, it's adjusted by altering the position of the upper control arm on the subframe.

Caster is the tilting of the front steering axis from the vertical. A tilt toward the rear is positive caster and a tilt toward the front is negative caster. Too little caster will make the front end wander, while too much caster can make the steering effort higher. Caster is not adjustable on these vehicles; if it isn't within specification, check for bent suspension components.

Notes

Notes

Chapter 11   Body

Contents

1   General information

These models feature a unibody layout, using a floor pan with integral side frame rails which support the body components, front and rear suspension systems and other mechanical components.

Certain components are particularly vulnerable to accident damage and can be unbolted and repaired or replaced. Among these parts are the body moldings, bumpers, front fenders, the hood and trunk lid, doors and all glass.

Only general body maintenance practices and body panel repair procedures within the scope of the do-it-yourselfer are included in this Chapter.

Fastener and trim removal

Refer to illustrations 1.6a through 1.6h and 1.7

There are a variety of plastic fasteners used to hold trim panels, splash shields and other parts in place in addition to typical screws, nuts and bolts. Once you are familiar with them, they can usually be removed without too much difficulty.

The proper tools and approach can prevent added time and expense to a project by minimizing the number of broken fasteners and/or parts.

The following illustrations show various types of fasteners that are typically used on most vehicles and how to remove and install them **(see illustrations)**. Replacement fasteners are commonly found at most auto parts stores, if necessary.

Trim panels are typically made of plastic and their flexibility can help during removal. The key to their removal is to use a tool to pry the panel near its retainers (or fasteners) to release it without damaging surrounding areas or breaking-off any retainers. The retainers or fasteners will usually snap out of their designated slot or hole after force is applied to them. Stiff plastic tools designed for prying on trim panels can be found at specialty automotive or tool supply stores and are ideal

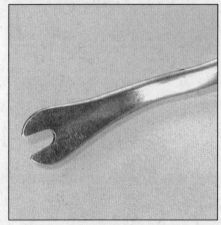

1.6a This tool is designed to remove special fasteners. A small pry tool used for removing nails will also work well in place of this tool

1.6b A Phillips head screwdriver can be used to release the center portion of this fastener, but light pressure must be used because the plastic is easily damaged. Once the center is up, the fastener can easily be pried from its hole

1.6c Here is a view with the center portion fully released. Install the fastener as shown, then press the center in to set it

1.6d This type of fastener is commonly used for interior panels. Press in the small pin at the center in to release it . . .

1.6e . . . the pin will stay with the fastener in the released position

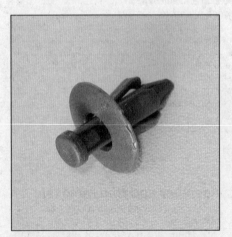

1.6f Reset the fastener for installation by moving the pin out. Install the fastener, then press the pin flush with the fastener to set it

1.6g This fastener is used for exterior panels and shields. The center portion must be pried up to release the fastener. Install the fastener with the center up, then press the center in to set it

1.6h This fastener is used for exterior and interior panels. It has no moving parts. Simply pry the fastener from its hole like the claw of a hammer removes a nail

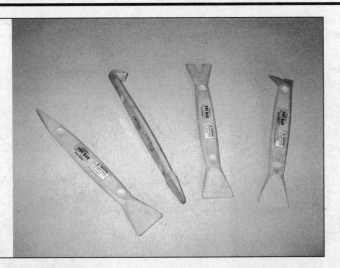

1.7 These small plastic pry tools are ideal for prying off trim panels and removing fasteners

for ongoing projects **(see illustration)**. Tools that are tapered and wrapped in protective tape, such as a screwdriver or small pry tool, are also very effective when used with care.

2   Body - maintenance

1   The condition of your vehicle's body is very important, because the resale value depends a great deal on it. It's much more difficult to repair a neglected or damaged body than it is to repair mechanical components. The hidden areas of the body, such as the wheel wells, the frame and the engine compartment, are equally important, although they don't require as frequent attention as the rest of the body.
2   Once a year, or every 12,000 miles, it's a good idea to have the underside of the body steam-cleaned. All traces of dirt and oil will be removed and the area can then be inspected carefully for rust, damaged brake lines, frayed electrical wires, damaged cables and other problems.
3   At the same time, clean the engine and the engine compartment with a steam cleaner or water-soluble degreaser.
4   The wheel wells should be given close attention, since undercoating can peel away and stones and dirt thrown up by the tires can cause the paint to chip and flake, allowing rust to set in. If rust is found, clean down to the bare metal and apply an anti-rust paint.
5   The body should be washed about once a week. Wet the vehicle thoroughly to soften the dirt, then wash it down with a soft sponge and plenty of clean soapy water. If the surplus dirt is not washed off very carefully, it can wear down the paint.
6   Spots of tar or asphalt thrown up from the road should be removed with a cloth soaked in kerosene. Scented lamp oil is available in most hardware stores and the smell is easier to work with than straight kerosene.
7   Once every six months, wax the body and chrome trim. If a chrome cleaner is used to remove rust from any of the vehicle's plated parts, remember that the cleaner also removes part of the chrome, so use it sparingly. On any plated parts where chrome cleaner is used, use a good paste wax over the plating for extra protection.

3   Vinyl trim - maintenance

Don't clean vinyl trim with detergents, caustic soap or petroleum-based cleaners. Plain soap and water works just fine, with a soft brush to clean dirt that may be ingrained. Wash the vinyl as frequently as the rest of the vehicle.

After cleaning, application of a high quality rubber and vinyl protectant will help prevent oxidation and cracks. The protectant can also be applied to weather stripping, vacuum lines and rubber hoses, which often fail as a result of chemical degradation, and to the tires.

4   Upholstery and carpets - maintenance

1   Every three months remove the floormats and clean the interior of the vehicle (more frequently if necessary). Use a stiff whisk broom to brush the carpeting and loosen dirt and dust, then vacuum the upholstery and carpets thoroughly, especially along seams and crevices.
2   Dirt and stains can be removed from carpeting with basic household or automotive carpet shampoos available in spray cans. Follow the directions and vacuum again, then use a stiff brush to bring back the nap of the carpet.
3   Most interiors have cloth or vinyl upholstery, either of which can be cleaned and maintained with a number of material-specific cleaners or shampoos available in auto supply stores. Follow the directions on the product for usage, and always spot-test any upholstery cleaner on an inconspicuous area (bottom edge of a backseat cushion) to ensure that it doesn't cause a color shift in the material.

4   After cleaning, vinyl upholstery should be treated with a protectant. **Note:** *Make sure the protectant container indicates the product can be used on seats - some products may make a seat too slippery.* **Caution:** *Do not use protectant on steering wheels.*
5   Leather upholstery requires special care. It should be cleaned regularly with saddle-soap or leather cleaner. Never use alcohol, gasoline, nail polish remover or thinner to clean leather upholstery.
6   After cleaning, regularly treat leather upholstery with a leather conditioner, rubbed in with a soft cotton cloth. Never use car wax on leather upholstery.
7   In areas where the interior of the vehicle is subject to bright sunlight, cover leather seating areas of the seats with a sheet if the vehicle is to be left out for any length of time.

5   Body repair - minor damage

Flexible plastic body panels

The following repair procedures are for minor scratches and gouges. Repair of more serious damage should be left to a dealer service department or qualified auto body shop. Below is a list of the equipment and materials necessary to perform the following repair procedures on plastic body panels. Although a specific brand of material may be mentioned, it should be noted that equivalent products from other manufacturers may be used instead.

> *Wax, grease and silicone removing*
> *solvent*
> *Cloth-backed body tape*
> *Sanding discs*
> *Drill motor with three-inch disc holder*
> *Hand sanding block*
> *Rubber squeegees*
> *Sandpaper*
> *Non-porous mixing palette*
> *Wood paddle or putty knife*
> *Curved-tooth body file*
> *Flexible parts repair material*

1   Remove the damaged panel, if necessary or desirable. In most cases, repairs can be carried out with the panel installed.
2   Clean the area(s) to be repaired with a wax, grease and silicone removing solvent applied with a water-dampened cloth.
3   If the damage is structural, that is, if it extends through the panel, clean the backside of the panel area to be repaired as well. Wipe dry.
4   Sand the rear surface about 1-1/2 inches beyond the break.
5   Cut two pieces of fiberglass cloth large enough to overlap the break by about 1-1/2 inches. Cut only to the required length.
6   Mix the adhesive from the repair kit according to the instructions included with the kit, and apply a layer of the mixture approximately 1/8-inch thick on the backside of the panel. Overlap the break by at least 1-1/2 inches.

These photos illustrate a method of repairing simple dents. They are intended to supplement *Body repair - minor damage* in this Chapter and should not be used as the sole instructions for body repair on these vehicles.

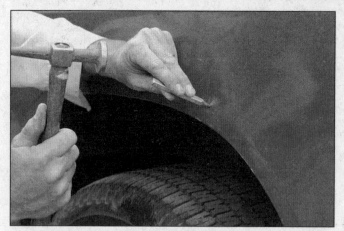

1 If you can't access the backside of the body panel to hammer out the dent, pull it out with a slide-hammer-type dent puller. In the deepest portion of the dent or along the crease line, drill or punch hole(s) at least one inch apart . . .

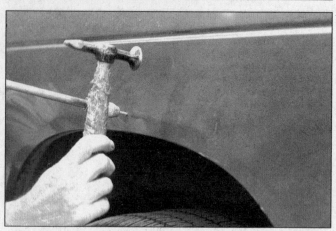

2 . . . then screw the slide-hammer into the hole and operate it. Tap with a hammer near the edge of the dent to help 'pop' the metal back to its original shape. When you're finished, the dent area should be close to its original contour and about 1/8-inch below the surface of the surrounding metal

3 Using coarse-grit sandpaper, remove the paint down to the bare metal. Hand sanding works fine, but the disc sander shown here makes the job faster. Use finer (about 320-grit) sandpaper to feather-edge the paint at least one inch around the dent area

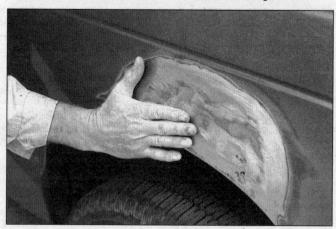

4 When the paint is removed, touch will probably be more helpful than sight for telling if the metal is straight. Hammer down the high spots or raise the low spots as necessary. Clean the repair area with wax/silicone remover

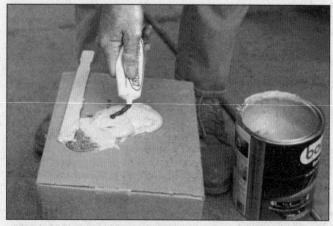

5 Following label instructions, mix up a batch of plastic filler and hardener. The ratio of filler to hardener is critical, and, if you mix it incorrectly, it will either not cure properly or cure too quickly (you won't have time to file and sand it into shape)

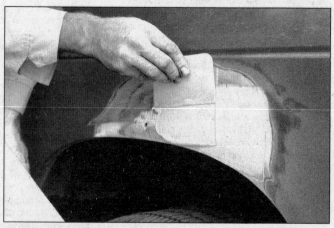

6 Working quickly so the filler doesn't harden, use a plastic applicator to press the body filler firmly into the metal, assuring it bonds completely. Work the filler until it matches the original contour and is slightly above the surrounding metal

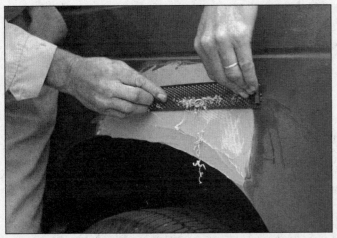

7 Let the filler harden until you can just dent it with your fingernail. Use a body file or Surform tool (shown here) to rough-shape the filler

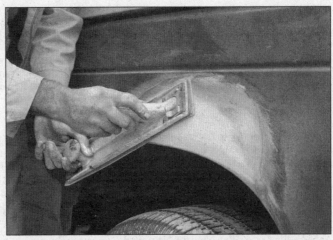

8 Use coarse-grit sandpaper and a sanding board or block to work the filler down until it's smooth and even. Work down to finer grits of sandpaper - always using a board or block - ending up with 360 or 400 grit

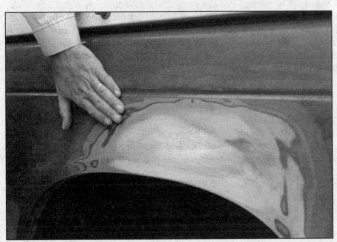

9 You shouldn't be able to feel any ridge at the transition from the filler to the bare metal or from the bare metal to the old paint. As soon as the repair is flat and uniform, remove the dust and mask off the adjacent panels or trim pieces

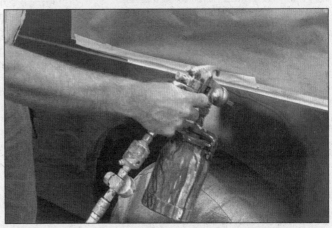

10 Apply several layers of primer to the area. Don't spray the primer on too heavy, so it sags or runs, and make sure each coat is dry before you spray on the next one. A professional-type spray gun is being used here, but aerosol spray primer is available inexpensively from auto parts stores

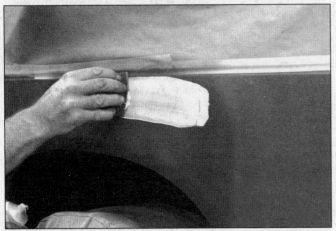

11 The primer will help reveal imperfections or scratches. Fill these with glazing compound. Follow the label instructions and sand it with 360 or 400-grit sandpaper until it's smooth. Repeat the glazing, sanding and respraying until the primer reveals a perfectly smooth surface

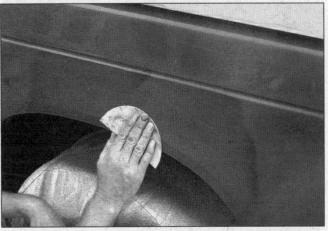

12 Finish sand the primer with very fine sandpaper (400 or 600-grit) to remove the primer overspray. Clean the area with water and allow it to dry. Use a tack rag to remove any dust, then apply the finish coat. Don't attempt to rub out or wax the repair area until the paint has dried completely (at least two weeks)

7   Apply one piece of fiberglass cloth to the adhesive and cover the cloth with additional adhesive. Apply a second piece of fiberglass cloth to the adhesive and immediately cover the cloth with additional adhesive in sufficient quantity to fill the weave.

8   Allow the repair to cure for 20 to 30 minutes at 60-degrees to 80-degrees F.

9   If necessary, trim the excess repair material at the edge.

10   Remove all of the paint film over and around the area(s) to be repaired. The repair material should not overlap the painted surface.

11   With a drill motor and a sanding disc (or a rotary file), cut a "V" along the break line approximately 1/2-inch wide. Remove all dust and loose particles from the repair area.

12   Mix and apply the repair material. Apply a light coat first over the damaged area; then continue applying material until it reaches a level slightly higher than the surrounding finish.

13   Cure the mixture for 20 to 30 minutes at 60-degrees to 80-degrees F.

14   Roughly establish the contour of the area being repaired with a body file. If low areas or pits remain, mix and apply additional adhesive.

15   Block sand the damaged area with sandpaper to establish the actual contour of the surrounding surface.

16   If desired, the repaired area can be temporarily protected with several light coats of primer. Because of the special paints and techniques required for flexible body panels, it is recommended that the vehicle be taken to a paint shop for completion of the body repair.

Steel body panels

See photo sequence on previous pages

Repair of minor scratches

17   If the scratch is superficial and does not penetrate to the metal of the body, repair is very simple. Lightly rub the scratched area with a fine rubbing compound to remove loose paint and built up wax. Rinse the area with clean water.

18   Apply touch-up paint to the scratch, using a small brush. Continue to apply thin layers of paint until the surface of the paint in the scratch is level with the surrounding paint. Allow the new paint at least two weeks to harden, then blend it into the surrounding paint by rubbing with a very fine rubbing compound. Finally, apply a coat of wax to the scratch area.

19   If the scratch has penetrated the paint and exposed the metal of the body, causing the metal to rust, a different repair technique is required. Remove all loose rust from the bottom of the scratch with a pocket knife, then apply rust inhibiting paint to prevent the formation of rust in the future. Using a rubber or nylon applicator, coat the scratched area with glaze-type filler. If required, the filler can be mixed with thinner to provide a very

thin paste, which is ideal for filling narrow scratches. Before the glaze filler in the scratch hardens, wrap a piece of smooth cotton cloth around the tip of a finger. Dip the cloth in thinner and then quickly wipe it along the surface of the scratch. This will ensure that the surface of the filler is slightly hollow. The scratch can now be painted over as described earlier in this Section.

Repair of dents

20   When repairing dents, the first job is to pull the dent out until the affected area is as close as possible to its original shape. There is no point in trying to restore the original shape completely as the metal in the damaged area will have stretched on impact and cannot be restored to its original contours. It is better to bring the level of the dent up to a point which is about 1/8-inch below the level of the surrounding metal. In cases where the dent is very shallow, it is not worth trying to pull it out at all.

21   If the back side of the dent is accessible, it can be hammered out gently from behind using a soft-face hammer. While doing this, hold a block of wood firmly against the opposite side of the metal to absorb the hammer blows and prevent the metal from being stretched.

22   If the dent is in a section of the body which has double layers, or some other factor makes it inaccessible from behind, a different technique is required. Drill several small holes through the metal inside the damaged area, particularly in the deeper sections. Screw long, self tapping screws into the holes just enough for them to get a good grip in the metal. Now the dent can be pulled out by pulling on the protruding heads of the screws with locking pliers.

23   The next stage of repair is the removal of paint from the damaged area and from an inch or so of the surrounding metal. This is easily done with a wire brush or sanding disk in a drill motor, although it can be done just as effectively by hand with sandpaper. To complete the preparation for filling, score the surface of the bare metal with a screwdriver or the tang of a file or drill small holes in the affected area. This will provide a good grip for the filler material. To complete the repair, see the Section on filling and painting.

Repair of rust holes or gashes

24   Remove all paint from the affected area and from an inch or so of the surrounding metal using a sanding disk or wire brush mounted in a drill motor. If these are not available, a few sheets of sandpaper will do the job just as effectively.

25   With the paint removed, you will be able to determine the severity of the corrosion and decide whether to replace the whole panel, if possible, or repair the affected area. New body panels are not as expensive as most people think and it is often quicker to install a new panel than to repair large areas of rust.

26   Remove all trim pieces from the affected area except those which will act as a guide to the original shape of the damaged body, such as headlight shells, etc. Using metal snips or a hacksaw blade, remove all loose metal and any other metal that is badly affected by rust. Hammer the edges of the hole in to create a slight depression for the filler material.

27   Wire brush the affected area to remove the powdery rust from the surface of the metal. If the back of the rusted area is accessible, treat it with rust inhibiting paint.

28   Before filling is done, block the hole in some way. This can be done with sheet metal riveted or screwed into place, or by stuffing the hole with wire mesh.

29   Once the hole is blocked off, the affected area can be filled and painted. See the following subsection on filling and painting.

Filling and painting

30   Many types of body fillers are available, but generally speaking, body repair kits which contain filler paste and a tube of resin hardener are best for this type of repair work. A wide, flexible plastic or nylon applicator will be necessary for imparting a smooth and contoured finish to the surface of the filler material. Mix up a small amount of filler on a clean piece of wood or cardboard (use the hardener sparingly). Follow the manufacturer's instructions on the package, otherwise the filler will set incorrectly.

31   Using the applicator, apply the filler paste to the prepared area. Draw the applicator across the surface of the filler to achieve the desired contour and to level the filler surface. As soon as a contour that approximates the original one is achieved, stop working the paste. If you continue, the paste will begin to stick to the applicator. Continue to add thin layers of paste at 20-minute intervals until the level of the filler is just above the surrounding metal.

32   Once the filler has hardened, the excess can be removed with a body file. From then on, progressively finer grades of sandpaper should be used, starting with a 180-grit paper and finishing with 600-grit wet-or-dry paper. Always wrap the sandpaper around a flat rubber or wooden block, otherwise the surface of the filler will not be completely flat. During the sanding of the filler surface, the wet-or-dry paper should be periodically rinsed in water. This will ensure that a very smooth finish is produced in the final stage.

33   At this point, the repair area should be surrounded by a ring of bare metal, which in turn should be encircled by the finely feathered edge of good paint. Rinse the repair area with clean water until all of the dust produced by the sanding operation is gone.

34   Spray the entire area with a light coat of primer. This will reveal any imperfections in the surface of the filler. Repair the imperfections with fresh filler paste or glaze filler and once more smooth the surface with sandpaper. Repeat this spray-and-repair procedure

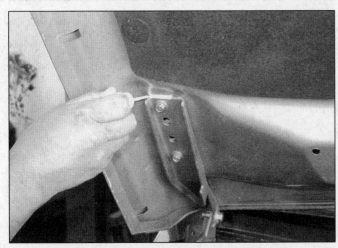

9.2 Draw alignment marks around the hood hinges to ensure proper alignment of the hood when it's reinstalled

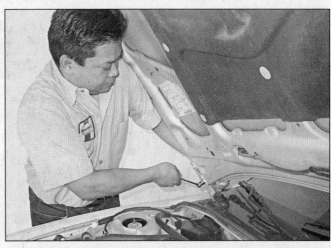

9.4 Support the hood with your shoulder while removing the hood bolts (typical shown)

until you are satisfied that the surface of the filler and the feathered edge of the paint are perfect. Rinse the area with clean water and allow it to dry completely.

35   The repair area is now ready for painting. Spray painting must be carried out in a warm, dry, windless and dust free atmosphere. These conditions can be created if you have access to a large indoor work area, but if you are forced to work in the open, you will have to pick the day very carefully. If you are working indoors, dousing the floor in the work area with water will help settle the dust which would otherwise be in the air. If the repair area is confined to one body panel, mask off the surrounding panels. This will help minimize the effects of a slight mismatch in paint color. Trim pieces such as chrome strips, door handles, etc., will also need to be masked off or removed. Use masking tape and several thickness of newspaper for the masking operations.

36   Before spraying, shake the paint can thoroughly, then spray a test area until the spray painting technique is mastered. Cover the repair area with a thick coat of primer. The thickness should be built up using several thin layers of primer rather than one thick one. Using 600-grit wet-or-dry sandpaper, rub down the surface of the primer until it is very smooth. While doing this, the work area should be thoroughly rinsed with water and the wet-or-dry sandpaper periodically rinsed as well. Allow the primer to dry before spraying additional coats.

37   Spray on the top coat, again building up the thickness by using several thin layers of paint. Begin spraying in the center of the repair area and then, using a circular motion, work out until the whole repair area and about two inches of the surrounding original paint is covered. Remove all masking material 10 to 15 minutes after spraying on the final coat of paint. Allow the new paint at least two weeks to harden, then use a very fine rubbing compound to blend the edges of the new paint into the existing paint. Finally, apply a coat of wax.

6   Body repair - major damage

1   Major damage must be repaired by an auto body shop specifically equipped to perform unibody repairs. These shops have the specialized equipment required to do the job properly.

2   If the damage is extensive, the body must be checked for proper alignment or the vehicle's handling characteristics may be adversely affected and other components may wear at an accelerated rate.

3   Due to the fact that some of the major body components (hood, fenders, doors, etc.) are separate and replaceable units, any seriously damaged components should be replaced rather than repaired. Sometimes the components can be found in a wrecking yard that specializes in used vehicle components, often at considerable savings over the cost of new parts.

7   Hinges and locks - maintenance

Once every 3000 miles, or every three months, the hinges and latch assemblies on the doors, hood and trunk (or liftgate) should be given a few drops of light oil or lock lubricant. The door latch strikers should also be lubricated with a thin coat of grease to reduce wear and ensure free movement. Lubricate the door and trunk (or liftgate) locks with spray-on graphite lubricant.

8   Windshield and fixed glass - replacement

Replacement of the windshield and fixed glass requires the use of special fast-setting adhesive/caulk materials and some specialized tools and techniques. These operations should be left to a dealer service department or a shop specializing in glass work.

9   Hood - removal, installation and adjustment

Note: *The hood is awkward to remove and install; at least two people should perform this procedure.*

Removal and installation
Refer to illustrations 9.2 and 9.4

1   Open the hood, then place blankets or pads over the fenders and cowl area of the body. This will protect the body and paint as the hood is lifted off.

2   Make marks around the hood hinge to ensure proper alignment during installation **(see illustration)**.

3   On Torrent and all 2010 and later models, remove the hood strut **(see illustration 21.6)**.

4   Have an assistant support one side of the hood. Take turns removing the hinge-to-hood bolts and lift off the hood **(see illustration)**.

5   Installation is the reverse of removal. Align the hinge bolts with the marks made in Step 2.

Adjustment
Refer to illustration 9.9

6   Fore-and-aft and side-to-side adjustment of the hood is done by moving the hinges after loosening the hinge-to-body bolts.

7   Loosen the bolts and move the hood into correct alignment. Move it only a little at a time. Tighten the hinge bolts and carefully lower the hood to check the position.

8   If necessary after installation, the entire hood latch assembly can be adjusted from side-to-side on the radiator support so the hood closes properly **(see illustration 10.1)**. Scribe a line or mark around the hood latch mounting bolts to provide a reference point, then loosen them and reposition the latch assembly, as necessary. Following adjustment, retighten the mounting bolts. **Note:** *The hood latch on Torrent models is mounted vertically, but the same procedure is followed for adjustment.*

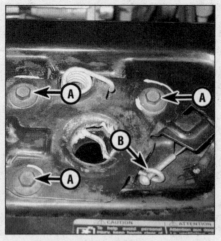

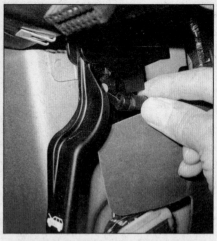

9.9   To adjust the vertical height of the leading edge of the hood so that it's flush with the fenders, turn each edge cushion clockwise (to lower the hood) or counterclockwise (to raise the hood)

10.3   Hood latch bolts (A) and cable attachment (B) (early Equinox model shown)

10.6   Pull the cable out of the retainer, then slide the end out of the handle

9   Finally, adjust the hood bumpers on the radiator support so the hood, when closed, is flush with the fenders **(see illustration)**.

10   The hood latch assembly, as well as the hinges, should be periodically lubricated with white, lithium-base grease to prevent binding and wear.

---

10   Hood latch and release cable - removal and installation

Latch

Refer to illustration 10.3

1   Open the hood and scribe a line around the latch and bolt heads to aid alignment when installing.

2   On 2010 and later models, insert long-handled trim tool or screwdriver into the hood latch opening (under the plastic trim panel) and disconnect the electrical connector fastener, then reach behind and under the radiator support and disconnect the electrical connector from the latch.

3   Remove the mounting bolts securing the hood latch to the radiator support **(see illustration)**. Remove the latch.

4   Disconnect the hood release cable by disengaging the cable from the latch.

5   Installation is the reverse of removal.

Cable

Refer to illustration 10.6

6   Working in the passenger compartment, release the cable from the retainer and slide the end out of the handle **(see illustration)**.

7   Working in the engine compartment, disconnect the hood release cable from the latch as described in Steps 1 through 4. Unclip all the cable retaining clips on the radiator support and inner fenderwell.

Note: *On 2010 and later models, the front bumper cover must be removed to access the cable retaining clips (see Section 11).*

8   Installation is the reverse of removal.

11   Bumper covers - removal and installation

Note: *Refer to Section 1 for fastener and trim removal.*

Front bumper cover

Refer to illustrations 11.2, 11.3, 11.5, 11.7a and 11.7b

1   Apply the parking brake, raise the vehicle and support it securely on jackstands.

2   Remove the front air dam mounting fasteners, then remove it (if equipped) **(see illustration)**.

3   Remove the front inner fender splash shields from the bumper cover **(see illustration)**. Move the shields aside or remove them completely.

4   Disconnect the fog lights (if so equipped).

5   Remove the screws joining the bumper cover and the lower front part of each fender **(see illustration)**.

6   On 2013 and later models, working in the engine compartment, remove the four plastic push-pin retainers and covers above

11.2   Remove the air dam fasteners

11.3   Inner fender splash shield fasteners (front)

11.5 The bumper cover-to-fender screw location

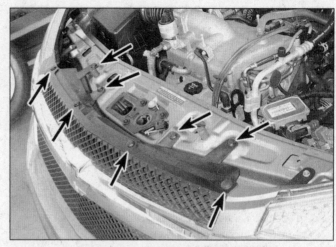

11.7a Bumper cover top mounting fasteners

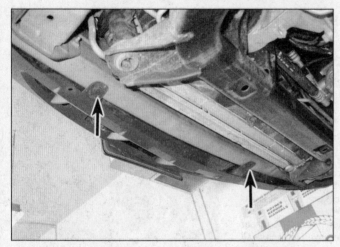

11.7b Bumper cover bottom mounting fasteners

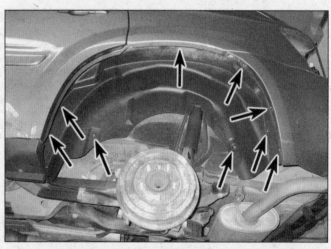

11.10 Inner fender splash shield fasteners (rear)

the headlight housings.

7   Remove the fasteners from the top and the bottom of the bumper cover **(see illustrations)**.

8   Carefully pull the bumper cover away from the front and sides of the vehicle and remove it completely.

9   Installation is the reverse of removal. An assistant is helpful at this point.

Rear bumper cover

Refer to illustrations 11.10, 11.11, 11.12 and 11.13

10   Remove the inner fender splash shield

fasteners that attach to the bumper cover **(see illustration)**.

11   Open the liftgate, pry off the bolt covers from the top of the bumper cover, then remove the bolts **(see illustration)**.

12   Remove the screws joining the bumper cover and the lower part of each fender **(see**

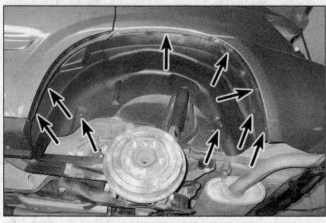

11.11 The rear bumper cover upper mounting bolts are located under these small covers

11.12 The bumper cover-to-fender screw location

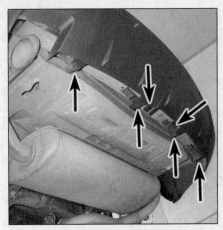

11.13 Rear bumper cover lower mounting fasteners

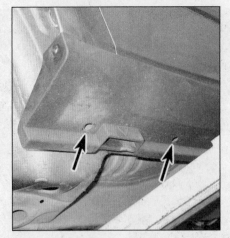

12.4 Remove the front fasteners for the rocker molding

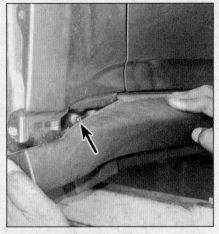

12.5 Pull the molding away and remove the fender lower mounting bolt

illustration).

13   Detach the fasteners securing the bottom of the bumper cover **(see illustration)**.

14   Disconnect any electrical connectors, if necessary.

15   Pull the bumper cover out and away from the vehicle while disengaging the existing tabs.

16   Installation is the reverse of removal.

12   Front fender - removal and installation

Refer to illustrations 12.4, 12.5, 12.6, 12.7 and 12.8

Note: *Refer to Section 1 for fastener and trim removal.*

1   Loosen the front wheel lug nuts. Raise the vehicle, support it securely on jackstands and remove the front wheel.

2   Remove the bezel from the base of the antenna, if equipped (see Chapter 12).

3   Remove the inner fender splash shield **(see illustration 11.3)**.

4   Remove the front fasteners for the rocker molding so that it can be pulled away slightly **(see illustration)**.

5   Pull the rocker molding away enough to remove the fender lower mounting bolt **(see illustration)**.

6   Remove the upper fender mounting bolt inside the doorjamb **(see illustration)**.

7   Remove the bumper cover-to-fender screw at the end of the bumper cover **(see illustration 11.5)**, then pull the bumper cover away enough to remove the bumper cover bracket-to-fender mounting bolt **(see illustration)**.

Note: *On 2010 and later models, the front bumper cover must be removed to access all the fender mounting bolts (see Section 11).*

8   Remove the fasteners along the top edge of the fender, then lift off the fender **(see illustration)**. It's a good idea to have an assistant support the fender while it's being moved away from the vehicle to prevent damage to the surrounding body panels.

12.6 Remove the upper fender mounting bolt in the doorjamb area

9   Installation is the reverse of removal. Check the alignment of the fender to the hood and front edge of the door before final tightening of the fender fasteners.

12.7 Remove the bumper cover bracket-to-fender mounting fastener (bumper cover removed for clarity)

12.8 Remove the fender upper mounting bolts and the bolt for the headlamp housing

14.2 Remove the fasteners for the small cowl cover on the right side

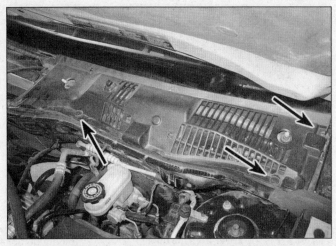

14.3 Remove the fasteners for the large cowl cover

13 Radiator grille - removal and installation

Note: *Refer to Section 1 for fastener and trim removal details.*

1   Refer to Section 11 and remove the front bumper cover.

Equinox and Terrain models

2   Remove the long outer trim by pressing the retainers through the grille to release it.

3   On the back of the grille, remove the screw at the lower center of the grille.

4   Release the retaining clips from around the grille to separate it from the bumper cover.

5   On SS models, remove the lower grille by carefully removing the energy absorber from the back of the bumper cover and releasing the tabs for the small outer trim on each end from the back of the bumper cover. Remove the grille mounting nuts and the grille.

6   Installation is the reverse of removal.

Torrent models

7   To remove an upper (small) grille(s), remove the grille bezel by releasing its tabs on the rear of the bumper cover.

8   Push the grille out at the bottom from the rear of the bumper cover.

9   To remove the lower grille, remove the grille bezel by removing the plastic fasteners on the bottom, then remove the grille.

10   On GXP models, remove the lower grille by carefully removing the energy absorber from inside the bumper cover, then release the retaining clips from around the grille to release it.

11   Installation is the reverse of removal.

14 Cowl cover - removal and installation

Refer to illustrations 14.2 and 14.3
Note: *Refer to Section 1 for fastener and trim removal.*

15.2 Remove the small trim and disconnect the electrical connector, if equipped (front door shown - rear door similar)

1   Remove the wiper arms (see Chapter 12). Make sure the wipers are in the parked position and note the locations of the blades on the windshield.

2   Using a trim tool, remove the weatherstrip along the front edge of the cowl. Remove the fasteners securing the small cowl cover on the right (passenger's) side, then remove it **(see illustration)**.

Note: *On 2013 and later models, the cowl has been changed to one long cover with two small extensions on each end of the cowl panel.*

3   Remove the remaining fasteners from the larger cowl cover **(see illustration)**.

4   On 2013 and later models, working in the engine compartment, remove the four plastic push-pin retainers and covers above the headlight housings.

5   Disconnect the windshield washer hose, then remove the cowl.

6   Installation is the reverse of removal.

15.3 Remove the screw for the release handle trim

15 Door trim panels - removal and installation

Caution: *Wear gloves when working inside the door openings to protect against sharp metal edges.*
Note: *Refer to Section 1 for fastener and trim removal.*

Front and rear doors

Refer to illustrations 15.2, 15.3, 15.4, 15.5, 15.6 and 15.7

1   Disconnect the cable from the negative battery terminal (see Chapter 5).

2   Pull or pry the small triangular panel from the top to release its retainer(s), then pull it out of the door trim **(see illustration)**. On front doors, disconnect the electrical connector for the speaker wiring, if equipped.

3   Remove the small trim cap (if equipped) and the screw for the release handle trim, then remove the trim **(see illustration)**.

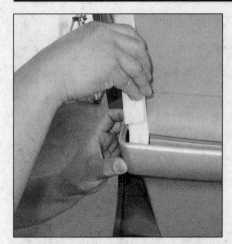

15.4 Pry the cover from the door pull handle

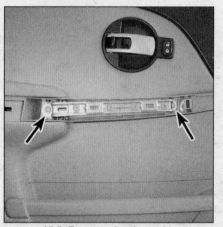

15.5 Remove the door trim mounting bolts (2009 and earlier models shown)

15.6 Pry the door trim away from the door around the edges

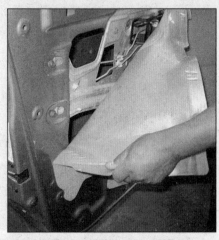

15.7 Carefully peel back the plastic watershield

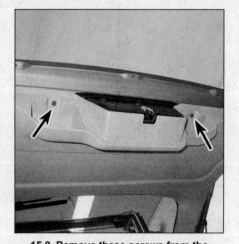

15.9 Remove these screws from the liftgate trim panel

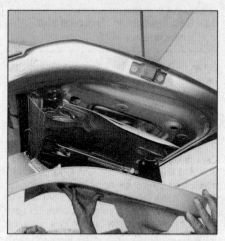

15.10 Pull or pry the lower panel away from the liftgate

4   On 2009 and earlier models, pry off the small trim that covers the pull handle, beginning with the rear first **(see illustration)**. On 2010 and later models, remove the small trim covers from under the pull handle to locate the mounting bolts.

5   Remove the mounting bolts behind the covers **(see illustration)**.

6   Detach the door trim fasteners by prying the door trim away from the door near the outside edge **(see illustration)**.

7   For access to the door outside handle or the door window regulator inside the door, raise the window fully, then carefully peel back the plastic watershield **(see illustration)**. On 2010 and later models, once the watershield is removed, use a trim tool to remove the plastic push-pins attaching the trim panel around the top of the door frame, then remove the trim panel.

8   Installation is the reverse of removal. Align the door trim fasteners with the holes in the door then push the door trim into place.

Liftgate

Refer to illustrations 15.9, 15.10, 15.11a and 15.11b

9   Open the liftgate and remove the trim mounting screws near the bottom **(see illustration)**.

10   Pull or pry the trim lower panel away from the liftgate to disengage the trim fasteners **(see illustration)**.

11   Remove the smaller trim panels along the top and sides by pulling or prying them directly away from the liftgate **(see illustrations)**.

12   Carefully peel back the plastic watershield as necessary to access other components.

13   Installation is the reverse of removal.

---

16   Door - removal, installation and adjustment

Warning: *The models covered by this manual are equipped with Supplemental Restraint Systems (SRS), more commonly known as airbags. Always disable the airbag system before working in the vicinity of any airbag system component to avoid the possibility of accidental deployment of the airbag, which could cause personal injury (see Chapter 12).*

Note: *The door is heavy and somewhat awkward to remove - at least two people should perform this procedure.*

Removal and installation

Refer to illustrations 16.6a, 16.6b and 16.6c

1   Raise the window completely in the door.

2   Disconnect the cable from the negative battery terminal (see Chapter 5).

3   Open the door all the way and support it with a jack or blocks covered with rags to prevent damaging the outer surface.

4   Remove the fender (see Section 12) if you're working on a front door and must replace or adjust the hinges.

5   Remove the inner door trim panel and watershield as described in Section 15.

6   Pull the rubber conduit off of the electrical connector for the door's wiring harness, then disconnect the connector **(see illustrations)**.

7   Unbolt the door stop strut.

8   Mark around the door hinges with a pen or a scribe to facilitate realignment during reassembly.

15.11a Pull or pry the upper panel away from the liftgate

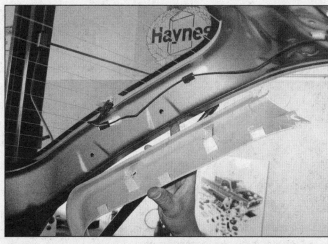

15.11b Pull or pry the side panels away from the liftgate after the upper and lower trim panels are removed first

9   With an assistant holding the door, remove the hinge-to-door bolts and lift off the door.
10   Installation is the reverse of removal.

Adjustment

Refer to illustration 16.14
11   Having proper door-to-body alignment is a critical part of a well-functioning door assembly. First check the door hinge pins for excessive play. Fully open the door and lift up and down on the door without lifting the body. If a door has 1/16-inch or more excessive play, the hinges should be replaced.
12   Door-to-body alignment adjustments are made by loosening the hinge-to-body bolts or hinge-to-door bolts and moving the door. Proper body alignment is achieved when the top of the doors are parallel with the roof section, the front door is flush with the fender, the rear door is flush with the rear quarter panel and the bottom of the doors are aligned with the lower rocker panel. If these goals can't be reached by adjusting the hinge-to-body or hinge-to-door bolts, body alignment shims may have to be purchased and inserted

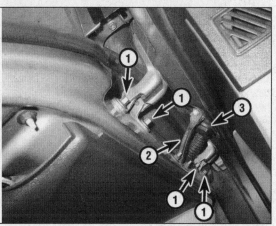

16.6a Door mounting details

1   *Hinge-to-door mounting bolts*
2   *Door stop strut*
3   *Wiring harness rubber conduit*

behind the hinges to achieve correct alignment. **Note:** *It is necessary to remove the front fender to access the hinge-to-body bolts (see Section 12).*
13   To adjust the door-closed position, scribe a line or mark around the striker plate to provide a reference point, then check that the door latch is contacting the center of the latch

striker. If not, adjust the up and down position first.
14   Finally adjust the latch striker sideways position, so that the door panel is flush with the center pillar or rear quarter panel and provides positive engagement with the latch mechanism **(see illustration)**.

16.6b After exposing the connector, press the small tabs on the edge to release it from the body

16.6c Raise the locking tab on the connector and disconnect it

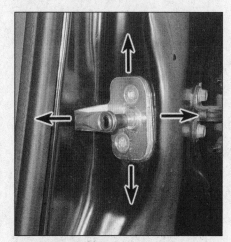

16.14 Adjust the door lock striker by loosening the mounting screws and gently tapping the striker in the desired direction

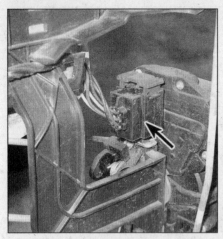

17.2 Door latch electrical connector

17.5 Remove the bolt cover

17.6 Pull the handle out and remove the small cover, then remove the handle

17 Door latch, lock cylinder and handles - removal and installation

Note: *Based on the modular design of the window regulator, some of the following components are removed as an assembly along with the window regulator module. Refer to Section 19 for component removal as listed below.*

Door latch

Refer to illustration 17.2

1   Remove the window regulator module (see Section 19).
2   Disconnect the electrical connector from the latch **(see illustration)**.
3   Detach the rods from the latch and remove it from the window regulator module.
4   Installation is the reverse of removal.

Outside handle

Refer to illustrations 17.5 and 17.6

5   Remove the bolt cover, then remove the bolt beneath it **(see illustration)**.

6   On 2009 and earlier models, pull the door handle out and remove the small cover, then remove the handle **(see illustration)**. Be careful not to lose or damage the small seal beneath the cover.

2010 and later models

7   Unclip the lock trim cover then remove the cover.
8   Pull the door handle part way open and locate the bellcrank lever. The bellcrank lever is spring-loaded to keep the handle closed.
9   Using a hooked tool, insert the tool through the small opening between the handle and hook the bellcrank lever, holding it in the partially opened position.
Caution: *Use care not to break the bellcrank when removing it from the handle housing.*
10   While holding the bellcrank in the partially open position, push the handle closed to relieve the spring pressure, push the handle rearward (towards the front) to clear the bellcrank lever, then pull the rear of the handle outwards and remove the handle.
11   Installation is the reverse of removal.

Door lock cylinder

Refer to illustration 17.13

12   Remove the door trim panel (see Section 15) and, on 2009 and earlier models, remove the window regulator module (see Section 19).
13   Release the retaining clip from the rear of the door lock cylinder **(see illustration)**.
14   Push the lock out of the frame for the outside door handle, detach the rod, then remove the lock.
15   Installation is the reverse of removal.

Inside door handle

Refer to illustration 17.17

Note: *On 2010 and later models, the handle is mounted to the door trim panel, not the door assembly.*

16   Remove the door trim panel and the plastic watershield (see Section 15).
17   Remove the mounting screws and detach the handle assembly from the door **(see illustration)**.
18   Disconnect the rods and the electrical connector from the assembly, then remove it.

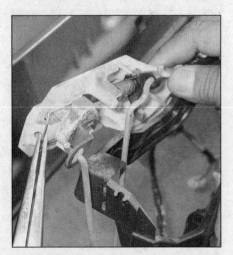

17.13 Remove the retaining clip, then push the lock out of the handle frame

17.17 Inside door handle mounting screws

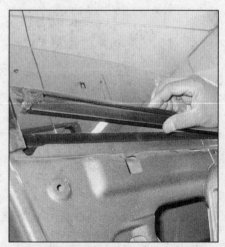

18.2 Pull the seal off of the pinch weld at the window opening

18.3 Use a screwdriver or equivalent on the release tab to separate the window glass from the regulator guide plate

19.3a Exterior trim mounting screws (A) and the lower window trim (B)

19.3b Pull or pry the exterior trim to release the retainers

19   Installation is the reverse of removal.

18   Door window glass - removal and installation

Caution: *Wear gloves when working inside the door openings to protect against cuts from sharp metal edges.*
Note: *Refer to Section 1 for fastener and trim removal.*

Door glass
Refer to illustrations 18.2 and 18.3
1   Remove the door trim panel and the plastic watershield (see Section 15). On 2010 and later models, use a trim tool to remove the plastic push-pins attaching the trim panel around the door window frame, then remove the trim panel.
2   Pull the window seal off of the pinch weld at the window opening **(see illustration)**.
3   On 2009 and earlier models, lower the window until the release tab at the bottom of the glass is accessible **(see illustration)**. On 2010 and later models, lower the window until the pinch bolt is accessible.
4   On 2009 and earlier models, press the release tab and on 2010 and later models loosen the pinch bolt, pull the window out of the regulator guide plate, then slide the glass down and out of the guide channels.
5   Guide the window up and tilt it outward to remove it through the window opening.
6   Installation is the reverse of removal.

Liftgate, windshield and quarter window glass
7   Replacement of these glass pieces requires the use of special fast-setting urethane adhesive materials, some specialized tools and techniques. These operations should be left to a dealer service department or a shop specializing in glass work.

19   Door window glass regulator module - removal and installation

2009 and earlier models
Refer to illustrations 19.3a, 19.3b, 19.4a, 19.4b, 19.5, 19.6, 19.7 and 19.9
Caution: *Wear gloves when working inside the door openings to protect against cuts from sharp metal edges.*
Note: *Refer to Section 1 for fastener and trim removal.*
1   Remove the window glass (see Section 18).
2   On front doors, disconnect the electrical connector for the outside mirror (see Section 20).
3   Remove the exterior trim on the side(s) and the bottom of the window opening, as applicable **(see illustrations)**.
4   Remove the rear guide channel fasteners and pull the channel out of the window opening and secure it aside. Be careful not to damage the rubber portion of the channel

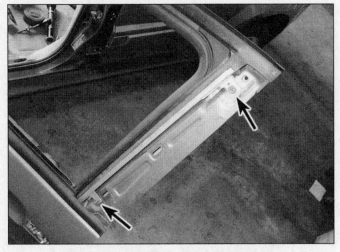

19.4a The upper run channel (rear) fasteners

19.4b The lower run channel (rear) fastener

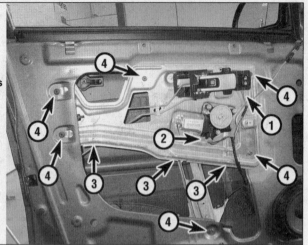

19.5 Window regulator module and related details

1 Electrical connector for the door locks
2 Electrical connector for the window motor
3 Door wiring harness fasteners attached to the module
4 Module mounting screws

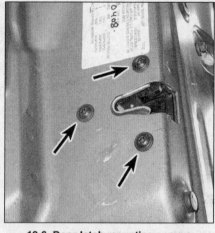

19.6 Door latch mounting screws

near the top of the window opening **(see illustrations)**. **Note:** *Remove the front and rear guide channels and the door speakers (see Chapter 12), if necessary.*

5 Disconnect the electrical connectors for the outside mirror, window regulator motor and door lock switch, then move the wiring harness away from the regulator module **(see illustration 20.2 and the accompanying illustration)**.

6 Remove the latch mounting screws **(see illustration)**.

7 Loosen the screw for the outside handle frame **(see illustration)**.

8 Detach the wiring harness fasteners from the module.

9 Loosen the two rear regulator module mounting fasteners, then remove the others. Carefully guide the module out from the door along with the door latch and outside handle frame **(see illustration)**. Detach the wiring harness and any other electrical connectors as the module is being removed.

10 Installation is the reverse of removal. Lubricate the rollers and wear points on the regulator with white grease before installation.

2010 and later models

11 Remove the door trim panel (see Section 15).

12 Remove the water shield and remove the window glass (see Section 18).

13 Lower the window regulator mechanism until it is three quarters of the way down.

14 Remove the regulator assembly mounting bolts, then remove the regulator from inside the door through the large opening and disconnect the electrical connectors.

15 Installation is the reverse of removal.

20 Mirrors - removal and installation

Outside mirrors

Refer to illustration 20.2

1 Refer to Step 2 in Section 15 to remove the small interior trim panel that covers the mirror's mounting nuts.

Note: *On 2010 and later Equinox models, the door panel and door window frame panel must be removed to access the electrical con-*

nectors for the mirrors (see Section 15).

2 Remove the small insulator, then disconnect the electrical connector from the back of the mirror **(see illustration)**.

3 Remove the mirror mounting nuts, then detach it from the vehicle **(see illustration 20.2)**.

4 Installation is the reverse of removal.

Inside mirror

Refer to illustration 20.7

5 Disconnect the electrical connector from the mirror, if equipped.

6 Remove the setscrew located at the base of the mirror stalk.

7 Slide the mirror upwards to remove it from its mount **(see illustration)**.

8 If the mount plate itself has come off the windshield, adhesive kits are available at auto parts stores to resecure it. Follow the instructions included with the kit. Be sure to position the flanged part of the mount away from the glass and the narrower part pointing up.

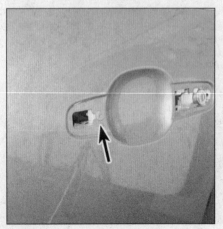

19.7 Outside door handle frame screw

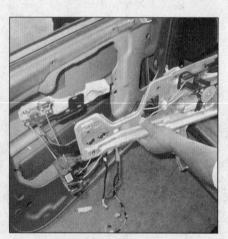

19.9 Carefully remove the module and the attached components from the door

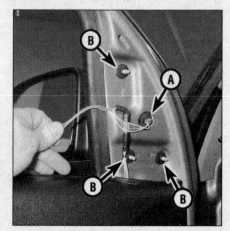

20.2 The electrical connector (A) and the outside mirror fasteners (B)

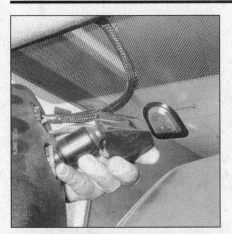

20.7 Remove the mirror set screw, then carefully lift the mirror upwards towards the top of the glass

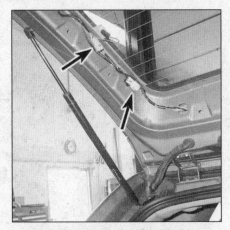

21.4 Disconnect the liftgate wiring harness connectors

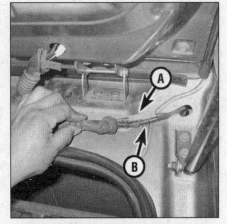

21.5 Disconnect the liftgate wiring harness connector (A) and washer hose (B)

21 Liftgate - removal, installation and adjustment

Note: *The liftgate is heavy and awkward to remove - at least two people should perform this procedure.*

Removal and installation

Refer to illustrations 21.4, 21.5, 21.6 and 21.7

1   Disconnect the cable from the negative battery terminal (see Chapter 5).
2   Open the liftgate and support it securely.
3   Remove the liftgate trim panel (see Section 15).
4   On the driver's side of the liftgate, disconnect the two electrical connectors, detach the rubber conduit from the liftgate and guide the wiring harnesses out of the hole in the liftgate **(see illustration)**.
5   On the passenger's side, detach the rubber conduit from the body and pull the harness and washer hose out. Disconnect the washer hose and electrical connector. Tape the hose and harness to keep them from falling into the hole in the body **(see illustration)**.
6   With an assistant holding the liftgate, disconnect the support struts by prying out the small clips, then pulling the struts from their ballstuds **(see illustration)**.
7   With an assistant holding the liftgate, remove the hinge-to-liftgate bolts and lift the liftgate off **(see illustration)**. **Note:** *Draw a reference line around the hinges on the liftgate before removing the bolts.*
8   Installation is the reverse of removal.

Adjustment

9   Having proper liftgate-to-body alignment is a critical part of a well-functioning liftgate assembly. First check the liftgate hinge pins for excessive play. Fully open the liftgate and lift up and down on the liftgate without lifting the body. If a liftgate has 1/8-inch or more excessive play, the hinges should be replaced.
10   Liftgate-to-body alignment adjustments are made by loosening the hinge-to-body bolts or hinge-to-liftgate bolts and moving the liftgate. Proper body alignment is achieved when the top of the liftgate is parallel with the roof section and the sides of the liftgate are flush with the rear quarter panels and the bottom of the liftgate is aligned with the lower liftgate sill. If these goals can't be reached by adjusting the hinge-to-body or hinge-to-liftgate bolts, body alignment shims may have to be purchased and inserted behind the hinges to achieve correct alignment.
11   To adjust the liftgate-closed position, scribe a line or mark around the striker plate to provide a reference point, then check that the liftgate latch is contacting the center of the latch striker. If not, adjust the up and down position first.
12   Finally adjust the latch striker sideways position, so that the liftgate panel is flush with the rear quarter panel and provides positive engagement with the latch mechanism.

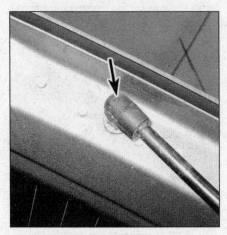

21.6 Use a small screwdriver to release the clip, then pull the liftgate support strut off its ballstud

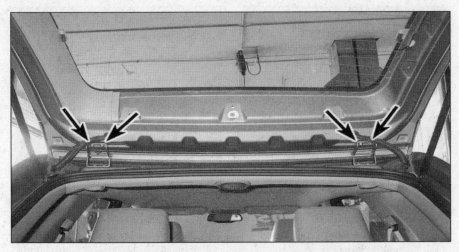

21.7 Have an assistant support the liftgate as you detach the liftgate hinges

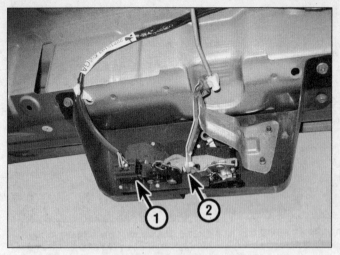

22.3 Liftgate latch details

22.4 Liftgate latch bolts

1   Liftgate latch electrical connector
2   Liftgate actuating rod

22   Liftgate latch, lock cylinder and handle - removal and installation

Manually operated models

1   Disconnect the cable from the negative battery terminal (see Chapter 5).
2   Open the liftgate and remove the door trim panel and watershield as described in Section 15.

Liftgate latch

Refer to illustrations 22.3 and 22.4
3   Disconnect the electrical connector from the latch **(see illustration)**.
4   Unbolt and remove the latch while detaching it from the actuating rod **(see illustration)**.
5   Installation is the reverse of removal.

Liftgate outside handle

Refer to illustration 22.6

6   Disconnect the outside handle rod from the handle **(see illustration)**.
7   Remove the handle mounting nuts, remove the license plate lamps from the handle, then detach it from the liftgate.
8   Installation is the reverse of removal.

Electrically operated models
Liftgate latch

9   If the liftgate cannot be opened electrically, a manual over-ride is provided. Remove the plug in the liftgate trim and insert a blunt tool in the hole to release the latch and open the liftgate.
10   Remove the door trim panel as described in Section 15.
11   Disconnect the electrical connector from the latch.
12   Remove the mounting bolts, then remove the latch.
13   Installation is the reverse of removal.

23   Center cupholder console - removal and installation (2009 and earlier)

Refer to illustrations 23.1 and 23.2
1   Remove the screw covers and screws from the lower sides of the console **(see illustration)**.
2   Lift up the console and disconnect the electrical connector for the wiring harness **(see illustration)**.
3   Remove the console from the floor.
4   Installation is the reverse of removal.

24   Dashboard trim panels - removal and installation

Warning: *Models covered by this manual are equipped with a Supplemental Restraint System (SRS), more commonly known as*

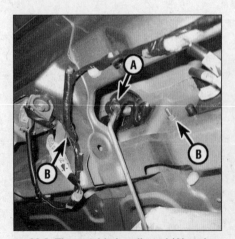

22.6 The outside handle rod (A) and mounting nuts (B)

23.1 Remove the lower screws and lift the console (seats removed for clarity) (right side shown - left side similar)

23.2 Disconnect the electrical connector(s) before removing the console entirely

24.3 Pull the small shifter bezel down to expose the mounting screw

24.4 Carefully pry the bezel up . . .

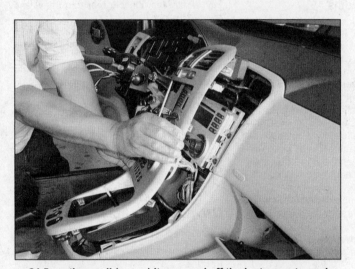

24.5 . . . then pull (or pry) it rearward off the instrument panel

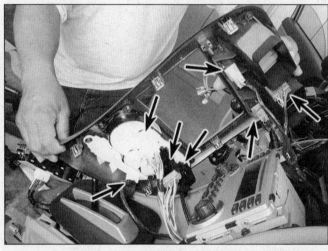

24.6 Disconnect all of the electrical connectors and the cable for the air conditioning control unit

airbags. *Always disable the airbag system before working in the vicinity of any airbag system component to avoid the possibility of accidental deployment of the airbag, which could cause personal injury (see Chapter 12).*
Note: *Refer to Section 1 for fastener and trim removal.*

1   These panels provide access to various instrument panel mounting screws. If you're going to remove the entire instrument panel, remove all of the covers.

2   Turn the ignition key to the ON position and turn the temperature control knob to the extreme cold position, then remove the key. Disconnect the cable from the negative terminal of the battery (see Chapter 5). Disable the airbag system (see Chapter 12).

Instrument panel center bezel
2009 and earlier models
Refer to illustrations 24.3, 24.4, 24.5 and 24.6

3   Remove the shifter knob:

a)   *Place the shifter into Neutral.*

b)   *Pull or pry the small shifter knob bezel down approximately 1/2 inch* **(see illustration).**

c)   *Loosen the small shifter knob screw at the base of the knob.*

d)   *Pull the shifter knob off of the shifter.*

4   Pry the bottom of the bezel up **(see illustration).**

5   Pry the upper portion of the bezel rearward **(see illustration).**

6   Lift the bezel and disconnect all wiring and the cable for the air conditioning control unit, then remove the bezel **(see illustration).**

7   Installation is the reverse of removal.

2010 and later models

8   Using a trim tool, carefully pry the instrument panel accessory trim plate up from the instrument panel, then disconnect the electrical connector to the sun/headlamp automatic control light sensor.

9   On some models, the trim plate has been replaced with a small storage compartment. If your vehicle has a storage compart-

ment, open the lid and remove the two mounting screws, then remove the storage unit from the top of the instrument panel.

10   Remove the instrument cluster bezel (see Steps 60 through 62).

11   Remove the instrument panel end covers (see Steps 41 and 42).

12   Remove the glove box (see Step 39).

13   Using a trim tool, carefully release the retainer clips securing the AM/FM stereo and clock and audio disc player radio control assembly to the instrument panel and pull the control assembly away from the instrument panel.

14   Disconnect the electrical connectors and remove the control assembly.

Note: *If equipped with navigation display, remove the display unit (see Chapter 12).*

15   With the radio and navigation display removed, remove the center compartment bracket fasteners and bracket from the instrument panel upper trim panel.

16   Disconnect the electrical connectors to the passenger's side airbag module (see

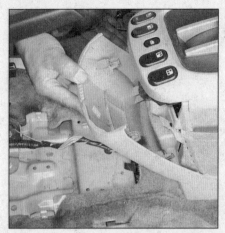

24.19 Pull the cover from the rear of the shifter console

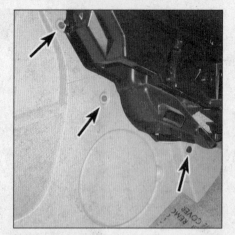

24.22a Remove the screws from the sides of the side trim panels . . .

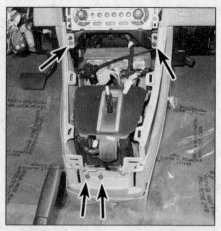

24.22b . . . and the screws that were hidden by the other panels

Chapter 12).

17   Remove the upper trim panel mounting fasteners and remove the panel taking note of how the main wiring harness is routed.

18   Installation is the reverse of removal.

Shifter console and trim panels

2009 and earlier models

Refer to illustrations 24.19, 24.22a and 24.22b

19   Pull off the cover from the rear of the shifter console **(see illustration)**.

20   Remove the instrument panel center bezel (see Steps 3 through 6).

21   Remove the glove box (see Steps 38 and 39) and the driver's knee bolster (see Steps 35 and 36).

22   Remove all fasteners from the side trim panels, then remove the panels **(see illustrations)**.

23   Installation is the reverse of removal.

2010 and later models

24   Using a trim tool, carefully pry the cup holder trim out from the top edge of the holder.

25   Using a trim tool, carefully pry the shifter console extensions from the upper trim panel and shifter console to remove the panels.

26   Move the seat all the way to the rear then remove the plastic push-pin from the sides of the lower side trim panels. Carefully pull the panel outwards to release the mounting clips along the edge of the panel from the floor console and remove the panel. Repeat the same procedure for the opposite panel.

27   Remove the lower extension panel-to-instrument panel fasteners the remove the panel.

28   Use a small screwdriver to depress the tabs around the shifter trim ring and remove the ring.

29   Unclip the console rear trim panel and remove the panel.

30   Remove the console arm rest bolts from the rear of the arm rest and remove the arm-rest.

31   Remove the center console mounting screws then use a trim tool to carefully pry up the top of the center console.

32   Disconnect the range selector lever cable from the transmission control assembly (see Chapter 7A).

33   Remove the fasteners along the side of the console and remove the console from the floor.

34   Installation is the reverse of removal.

Driver's knee bolster

Refer to illustrations 24.35 and 24.36

35   Remove the mounting screws **(see illustration)**.

24.35 The knee bolster mounting screw locations

24.36 Pull or pry the knee bolster away from the instrument panel to release its clips

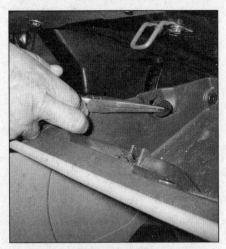

24.38 Twist the glove box stops to remove them (2009 and earlier models)

36   Pull the panel directly rearward to detach its clips **(see illustration)**.

37   Installation is the reverse of removal.

Glove box

Refer to illustrations 24.38 and 24.39

38   On 2009 and earlier models, twist the door stop pins to remove them **(see illustration)**.

39   Open the glove box and, on 2009 and earlier models, pry out the pivot pins **(see illustration)**. On 2010 and later models, push the door stops upwards and lower the glove box. Remove the glove box.

40   Installation is the reverse of removal.

Instrument panel end covers

Refer to illustrations 24.41 and 24.42

41   Remove the left end cover by pulling or prying it to disengage the clips **(see illustration)**.

42   Open the glove box, remove the fasteners, then pull or pry the cover to remove it **(see illustration)**.

43   Installation is the reverse of removal.

Front trim panels

44   Remove the upper panel cover (see Steps 64 through 66).

45   Remove the instrument cluster bezel (see Step 58).

46   Remove the instrument panel center bezel (see Steps 3 through 6).

Left panel

Refer to illustrations 24.49 and 24.50

47   Remove the left instrument panel end cover (see Step 41).

48   Remove the knee bolster (see Steps 35 and 36).

49   Pry the small air outlet panel from the trim panel and disconnect the electrical connector **(see illustration)**.

50   Remove the trim mounting screws, then remove the trim **(see illustration)**.

51   Installation is the reverse of removal.

24.39 Pry the glove box pivots inward (2009 and earlier models)

24.41 Pull the instrument panel end cover away to release its clips

24.42 Push the centers of the fasteners in with a pointed tool, then remove the fasteners

24.49 Pry out the small air outlet then disconnect the electrical connectors

24.50 Remove the left trim mounting screws

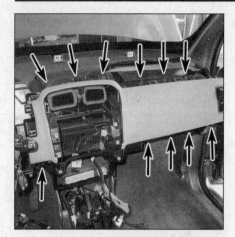

24.53 Remove the right trim mounting screws

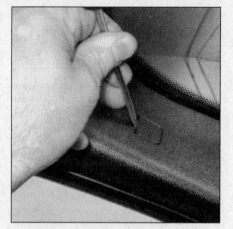

24.55 The windshield pillar cover retaining screw is under this small plastic cover

24.58 Pry the cluster bezel rearward to release its clips

Right panel

Refer to illustration 24.53

52   Remove the left instrument panel end cover (see Step 41).
53   Remove the trim mounting screws, then remove the trim **(see illustration)**.
54   Installation is the reverse of removal.

Windshield pillar covers

Refer to illustration 24.55

55   Pry off the screw covers and remove the screws **(see illustration)**.
56   Gently but firmly pull the covers from the pillars.
57   Installation is the reverse of removal.

Instrument cluster bezel
2009 and earlier models

Refer to illustration 24.58

58   Pull or pry the bezel rearward to unclip it **(see illustration)**, then remove it from the instrument panel.
59   Installation is the reverse of removal.

2010 and later models

60   Using a trim tool, pry the instrument cluster trim pad up, then disconnect the plastic tether and remove the pad from the top of the instrument panel.
61   Remove the instrument cluster bezel fasteners.
62   Using a flat blade screwdriver, release the retaining tabs to the steering column seal and the instrument panel, then remove the instrument cluster bezel.
63   Installation is the reverse of removal.

Upper panel cover
2009 and earlier models

Refer to illustrations 24.66, 24.67a, 24.67b and 24.67c

64   Remove the windshield pillar covers (see Steps 55 and 56).
65   Remove the instrument cluster bezel (see Step 58).
66   Pry the small plastic covers from the instrument panel on each end, then remove the mounting bolts underneath **(see illustration)**.
67   Pull or pry the front of the panel upward

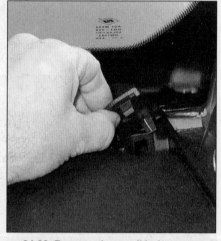

24.66 Remove the small bolt covers

at its edges to release its clips **(see illustrations)**, then remove the light sensor from underneath the panel **(see illustration)**.
68   Installation is the reverse of removal.

24.67a Pry the panel upward to release its clips . . .

24. 67b . . . then lift it up slightly

24.67c Remove the light sensor from underneath the cover

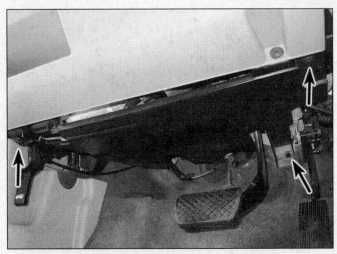

24.80 Left-side under-dash insulator mounting fasteners

24.82 Right-side under-dash insulator mounting fasteners

2010 and later models

69   Using a trim tool, carefully pry the instrument panel accessory trim plate up from the instrument panel, then disconnect the electrical connector to the sun/headlamp automatic control light sensor.

70   Remove the instrument cluster bezel (see Steps 60 through 62).

71   Remove the instrument panel end covers (see Steps 41 and 42).

72   Remove the glove box (see Steps 38 and 39).

73   On models equipped with a storage compartment, open the storage compartment lid and remove the two mounting screws, then remove the storage unit from the top of the instrument panel. On models equipped with an upper center trim panel, use a trim panel tool and carefully pry the trim panel from the top of instrument panel.

74   Remove the radio front speaker (see Chapter 12).

75   Remove the radio/heater air conditioning control assembly and radio (see Chapter 12).
Note: *If equipped with navigation display, remove the display unit (see Chapter 12).*

76   With the radio and navigation display removed, remove the center compartment bracket fasteners and bracket from the instrument panel upper trim panel.

77   Disconnect the electrical connectors to the passenger's side airbag module (see Chapter 12).

78   Remove the upper trim panel mounting fasteners and remove the panel, taking note on how the main wiring harness is routed.

79   Installation is the reverse of removal.

Insulator panels

Left panel

Refer to illustration 24.80

80   Remove the trim mounting screws and fasteners, then remove the trim **(see illustration)**.

81   Installation is the reverse of removal.

Right panel

Refer to illustration 24.82

82   Remove the trim mounting screws, then

remove the trim **(see illustration)**.

83   Installation is the reverse of removal.

25   Steering column covers - removal and installation

Refer to illustrations 25.2, 25.3 and 25.4

Warning: *Models covered by this manual are equipped with a Supplemental Restraint System (SRS), more commonly known as airbags. Always disable the airbag system before working in the vicinity of any airbag system component to avoid the possibility of accidental deployment of the airbag, which could cause personal injury (see Chapter 12).*

1   Disconnect the cable from the negative terminal of the battery (see Chapter 5).

2   Remove the screws from the bottom of the lower cover **(see illustration)**.

3   Pry the upper cover up until it unclips from the lower cover(s) **(see illustration)**.

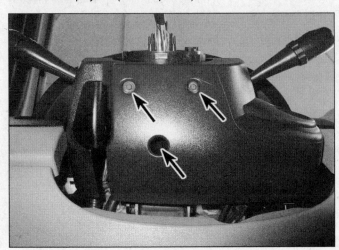

25.2 The steering column cover screws are accessed from the bottom

25.3 Separate the steering column covers

25.4 Pull or pry the small trim cover from the ignition lock (2009 and earlier models)

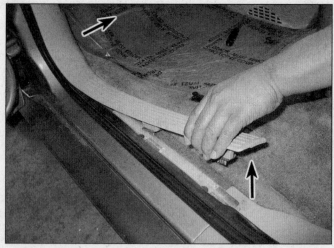

26.10 Pull or pry up the door threshold portion, then remove the kick panel portion by pulling it towards the middle of the vehicle. Remove the panels on both sides (driver's side shown - passenger's side similar)

4   On 2009 and earlier models, remove the small trim from around the ignition lock (**see illustration**). On 2010 and later models, telescope the steering column to is most extended position, then remove the forward facing mounting screws (turn steering wheel to gain access to the screws.

5   Remove the covers.

6   Installation is the reverse of the removal procedure.

26   Instrument panel - removal and installation

Refer to illustrations 26.10, 26.13a, 26.13b, 26.13c, 26.13d, 26.13e and 26.13f

Warning: *Models covered by this manual are equipped with a Supplemental Restraint System (SRS), more commonly known as airbags. Always disable the airbag system before working in the vicinity of any airbag system component to avoid the possibility of accidental deployment of the airbag, which could cause personal injury (see Chapter 12).*

Note 1: *This is a difficult procedure for the home mechanic. There are many hidden fasteners, difficult angles to work in and many*

electrical connectors to tag and disconnect/ connect. We recommend that this procedure be done only by an experienced do-it-yourselfer.

Note 2: *During removal of the instrument panel, make careful notes of how each piece comes off, where it fits in relation to other pieces and what holds it in place. If you note how each part is installed before removing it, getting the instrument panel back together again will be much easier.*

Note 3: *It is not mandatory, but it is suggested to remove both front seats to allow additional working space and lessen the chance of damage to the seats during this procedure.*

1   Disconnect the cable from the negative battery terminal (see Chapter 5).

2   Remove the instrument panel upper cover and the glove box (see Section 24)

3   Remove the instrument cluster and the radio (see Chapter 12).

4   Remove the center storage console (see Section 24).

5   Remove the shifter console side and rear covers (see Section 24).

6   Remove the air outlets that are at the ends of the instrument panel.

7   Unclip the instrument cluster wiring from the instrument panel (**see illustration 26.13a**).

8   Working at the right end of the panel, disconnect the large wiring harnesses from the body control module.

9   Remove the diagnostic link connector by releasing its retaining tabs and pushing it through the instrument panel.

10   Remove the kick panels/door thresholds (**see illustration**).

11   Remove the passenger airbag module (see Chapter 12).

12   Several electrical connectors and ground wires must be disconnected in order to remove the instrument panel. Most are designed so that they will only fit on the matching connector, but if there is any doubt, mark the connectors with masking tape and a marking pen before disconnecting them.

13   Remove all of the fasteners holding the instrument panel to the body (**see illustrations**). Once all are removed, lift the panel, then pull it away from the windshield and take it out through a door opening (**see illustration**). **Note:** *This is a two-person job.*

14   Installation is the reverse of removal. Make sure that there is an insulator at each matching point where the instrument panel attaches to the support. Install the mounting screws and tighten them starting from the middle and working outward.

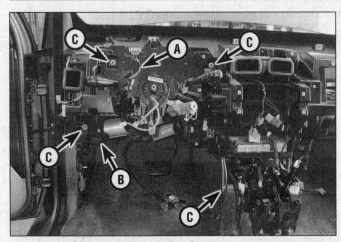

26.13a Release the wiring harness (A), detach the Data Link
Connector (B), then remove the mounting fasteners from the left
side of the instrument panel (C)

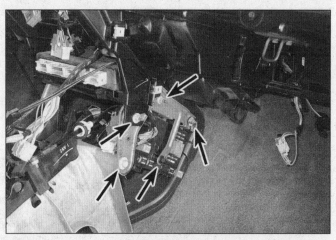

26.13b Remove these fasteners from the lower middle area of the
instrument panel

26.13c Remove the mounting fasteners from the right side of the
instrument panel

26.13d There is a nut on the other side of the stud shown here
that must be removed from below

26.13e Remove the ground wire on the right end of the
instrument panel

26.13f Have an assistant help you guide the panel out the
door opening

27   Seats - removal and installation

Front seat

Refer to illustrations 27.1a, 27.1b, 27.1c and 27.2

1   On the passenger's side seat, remove the lower trim to access the shoulder-belt release **(see illustrations)**. On the driver's side, simply pull the trim away enough to access the shoulder-belt release **(see illustration)**.

2   Remove the mounting bolts at the rear of the seats **(see illustration)**.

3   Disconnect all electrical connectors for the seat wiring harness.

4   Disengage the front seat mount hooks from the floor and lift out the seat.

5   Installation is the reverse of removal.

Rear seat

Refer to illustration 27.8

6   Remove the cargo area mat as equipped.

7   Remove the seat mounting bolts that are under the cover behind the seat at the floor.

8   Remove the two bolts that are under the seat cushion at the front **(see illustration)**.

9   Lift out the seat. Installation is the reverse of removal.

28   Rear spoiler - removal and installation

1   Remove the liftgate rear trim panel (see Section 15).

2   Remove the spoiler mounting bolts from the inside of the liftgate.

3   Lift the spoiler up and disconnect the rear wiper washer hose and the electrical connector to the center high-mounted stop light.

4   Remove the spoiler assembly.

5   Installation is the reverse of removal.

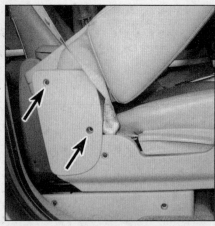

27.1a Remove the small trim on the side of the seat

27.1b Pull the trim cover out slightly for access to the belt release

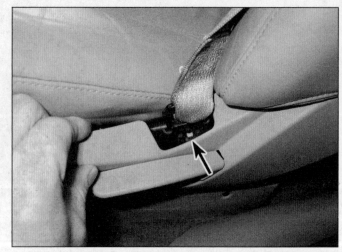

27.1c Push in the quick-release tab to remove the seat belt

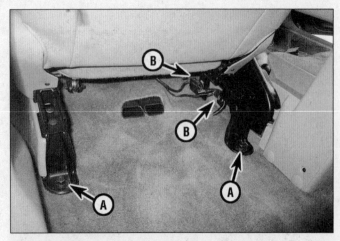

27.2 The front seats are retained by bolts (A) at the rear - pull the seat rearward to disengage the front clips after disconnecting any electrical connectors (B) to the seat's wiring harness

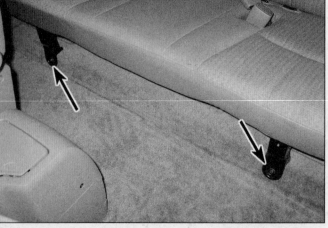

27.8 Remove the seat cushion mounting fasteners

Chapter 12
Chassis electrical system

Contents

1   General information

1   The electrical system is a 12-volt, negative ground type. Power for the lights and all electrical accessories is supplied by a lead/acid-type battery that is charged by the alternator.

2   This Chapter covers repair and service procedures for the various electrical components not associated with the engine. Information on the battery, alternator, ignition system and starter motor can be found in Chapter 5.

3   It should be noted that when portions of the electrical system are serviced, the negative cable should be disconnected from the battery to prevent electrical shorts and/or fires.

2   Electrical troubleshooting - general information

Refer to illustrations 2.5a, 2.5b, 2.6 and 2.9

1   A typical electrical circuit consists of an electrical component, any switches, relays, motors, fuses, fusible links or circuit breakers related to that component and the wiring and connectors that link the component to both the battery and the chassis. To help you pinpoint an electrical circuit problem, wiring diagrams are included at the end of this Chapter.

2   Before tackling any troublesome electrical circuit, first study the appropriate wiring diagrams to get a complete understanding of what makes up that individual circuit. Trouble spots, for instance, can often be narrowed down by noting if other components related to the circuit are operating properly. If several components or circuits fail at one time, chances are the problem is in a fuse or ground connection, because several circuits are often routed through the same fuse and ground connections.

3   Electrical problems usually stem from simple causes, such as loose or corroded connections, a blown fuse, a melted fusible link or a failed relay. Visually inspect the condition of all fuses, wires and connections in a problem circuit before troubleshooting the circuit.

4   If test equipment and instruments are going to be utilized, use the diagrams to plan ahead of time where you will make the necessary connections in order to accurately pinpoint the trouble spot.

5   The basic tools needed for electrical troubleshooting include a circuit tester or voltmeter (a 12-volt bulb with a set of test leads can also be used), a continuity tester, which includes a bulb, battery and set of test leads, and a jumper wire, preferably with a circuit breaker incorporated, which can be used to bypass electrical components **(see illustrations)**. Before attempting to locate a problem with test instruments, use the wiring diagram(s) to decide where to make the connections.

Voltage checks

6   Voltage checks should be performed if a circuit is not functioning properly. Connect one lead of a circuit tester to either the negative battery terminal or a known good ground. Connect the other lead to a connector in the circuit being tested, preferably nearest to the battery or fuse **(see illustration)**. If the bulb of the tester

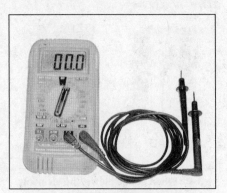

2.5a The most useful tool for electrical troubleshooting is a digital multimeter that can check volts, amps, and test continuity

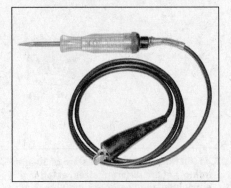

2.5b A test light is a very handy tool for checking voltage

2.6 In use, a basic test light's lead is clipped to a known good ground, then the pointed probe can test connectors, wires or electrical sockets - if the bulb lights, the part being tested has battery voltage

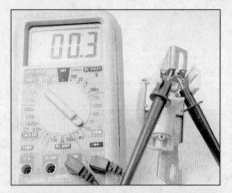

2.9 With a multimeter set to the ohms scale, resistance can be checked across two terminals - when checking for continuity, a low reading indicates continuity, a high reading indicates lack of continuity

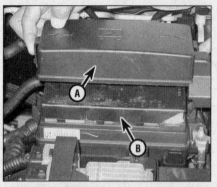

3.1a The main fuse/relay panel is in the engine compartment; disengage the locking tabs and remove the outer cover (A) then the inner cover (B) for access to the fuses and relays

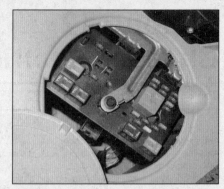

3.1b The interior fuse/relay panel is located under a cover on the right side of the center console

lights, voltage is present, which means that the part of the circuit between the connector and the battery is problem free. Continue checking the rest of the circuit in the same fashion. When you reach a point at which no voltage is present, the problem lies between that point and the last test point with voltage. Most of the time the problem can be traced to a loose connection. **Note:** *Keep in mind that some circuits receive voltage only when the ignition key is in the Accessory or Run position.*

Finding a short

7   One method of finding shorts in a circuit is to remove the fuse and connect a test light or voltmeter in place of the fuse terminals. There should be no voltage present in the circuit. Move the wiring harness from side-to-side while watching the test light. If the bulb goes on, there is a short to ground somewhere in that area, probably where the insulation has rubbed through. The same test can be performed on each component in the circuit, even a switch.

Ground check

8   Perform a ground test to check whether a component is properly grounded. Disconnect the battery and connect one lead of a continuity tester or multimeter (set to the ohms scale), to a known good ground. Connect the other lead to the wire or ground connection being tested. If the resistance is low (less than 5 ohms), the ground is good. If the bulb on a self-powered test light does not go on, the ground is not good.

Continuity check

9   A continuity check is done to determine if there are any breaks in a circuit - if it is passing electricity properly. With the circuit off (no power in the circuit), a self-powered continuity tester or multimeter can be used to check the circuit. Connect the test leads to both ends of the circuit (or to the power end and a good ground), and if the test light comes on the circuit is passing current properly **(see illustration)**. If the resistance is low (less than 5 ohms), there is continuity; if the reading is 10,000 ohms or higher, there is a break somewhere in the circuit. The same procedure can be used to test a switch, by connect-

ing the continuity tester to the switch terminals. With the switch turned On, the test light should come on (or low resistance should be indicated on a meter).

Finding an open circuit

10   When diagnosing for possible open circuits, it is often difficult to locate them by sight because the connectors hide oxidation or terminal misalignment. Merely wiggling a connector on a sensor or in the wiring harness may correct the open circuit condition. Remember this when an open circuit is indicated when troubleshooting a circuit. Intermittent problems may also be caused by oxidized or loose connections.

11   Electrical troubleshooting is simple if you keep in mind that all electrical circuits are basically electricity running from the battery, through the wires, switches, relays, fuses and fusible links to each electrical component (light bulb, motor, etc.) and to ground, from which it is passed back to the battery. Any electrical problem is an interruption in the flow of electricity to and from the battery.

Connectors

12   Most electrical connections on these vehicles are made with multiwire plastic con-

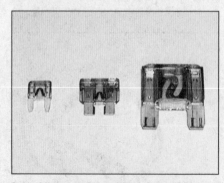

3.4a All three of these fuses are of 30-amp rating, yet are different sizes, at left is a small fuse, the center is a medium, and at right is a large - make sure you get the right amperage and size when purchasing replacement fuses

nectors. The mating halves of many connectors are secured with locking clips molded into the plastic connector shells. The mating halves of large connectors, such as some of those under the instrument panel, are held together by a bolt through the center of the connector.

13   To separate a connector with locking clips, use a small screwdriver to pry the clips apart carefully, then separate the connector halves. Pull only on the shell, never pull on the wiring harness as you may damage the individual wires and terminals inside the connectors. Look at the connector closely before trying to separate the halves. Often the locking clips are engaged in a way that is not immediately clear. Additionally, many connectors have more than one set of clips.

14   Each pair of connector terminals has a male half and a female half. When you look at the end view of a connector in a diagram, be sure to understand whether the view shows the harness side or the component side of the connector. Connector halves are mirror images of each other, and a terminal shown on the right side end-view of one half will be on the left side end-view of the other half.

15   It is often necessary to take circuit voltage measurements with a connector connected. Whenever possible, carefully insert a small straight pin (not your meter probe)

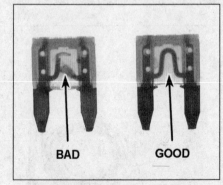

3.4b When a fuse blows, the element between the terminals melts

into the rear of the connector shell to contact the terminal inside, then clip your meter lead to the pin. This kind of connection is called "backprobing." When inserting a test probe into a terminal, be careful not to distort the terminal opening. Doing so can lead to a poor connection and corrosion at that terminal later. Using the small straight pin instead of a meter probe results in less chance of deforming the terminal connector.

---

3 Fuses and fusible links - general information

Fuses

Refer to illustrations 3.1a, 3.1b, 3.4a and 3.4b

1   The electrical circuits of the vehicle are protected by a combination of fuses, circuit breakers and fusible links. The main fuse/relay panel is in the engine compartment **(see illustration)**, while the interior fuse/relay panel is located in the right center console, floor panel, to the left of the glove box **(see illustration)**. The covers have a legend on the underside to identify the fuses and relays.

2   Each of the fuses is designed to protect a specific circuit, and the various circuits are identified on the fuse panel itself.

3   Several sizes of fuses are employed in the fuse blocks. There are small, medium and large sizes of the same design, all with the same blade terminal design. The medium and large fuses can be removed with your fingers, but the small fuses require the use of pliers or the small plastic fuse-puller tool found in most fuse boxes.

4   If an electrical component fails, always check the fuse first. The best way to check the fuses is with a test light. Check for power at the exposed terminal tips of each fuse **(see illustration)**. If power is present at one side of the fuse but not the other, the fuse is blown. A blown fuse can also be identified by visually inspecting it **(see illustration)**.

5   Be sure to replace blown fuses with the correct type. Fuses (of the same physical size) of different ratings may be physically interchangeable, but only fuses of the proper rating should be used. Replacing a fuse with one of a higher or lower value than specified is not recommended. Each electrical circuit needs a specific amount of protection. The amperage value of each fuse is molded into the top of the fuse body.

6   If the replacement fuse immediately

Electrical connectors

Most electrical connectors have a single release tab that you depress to release the connector

Some electrical connectors have a retaining tab which must be pried up to free the connector

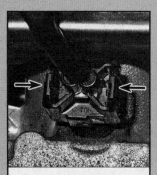

Some connectors have two release tabs that you must squeeze to release the connector

Some connectors use wire retainers that you squeeze to release the connector

Critical connectors often employ a sliding lock (1) that you must pull out before you can depress the release tab (2)

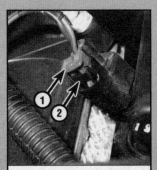

Here's another sliding-lock style connector, with the lock (1) and the release tab (2) on the side of the connector

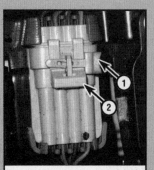

On some connectors the lock (1) must be pulled out to the side and removed before you can lift the release tab (2)

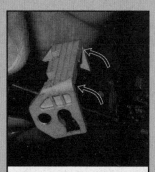

Some critical connectors, like the multi-pin connectors at the Powertrain Control Module employ pivoting locks that must be flipped open

3.8 The fusible link is located in the wiring from the starter to the alternator B+ terminal

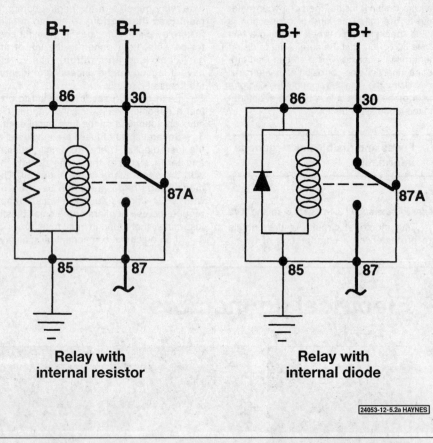

Relay with internal resistor   **Relay with internal diode**

24053-12-5.2a HAYNES

5.2a Typical ISO relay designs, terminal numbering and circuit connections

fails, don't replace it again until the cause of the problem is isolated and corrected. In most cases, this will be a short circuit in the wiring caused by a broken or deteriorated wire.

Fusible links

Refer to illustration 3.8

1   Some circuits are protected by fusible links. The links are used in circuits that are not ordinarily fused, such as the alternator circuit.

2   The fusible link for the alternator circuit is located in the wire from the starter to the alternator, and is easily identified. The link is a short length of heavy wire that is marked "fusible link" on the outer cover **(see illustration)**.

3   Although the fusible links appear to be a heavier gauge than the wires they're protecting, the appearance is due to the thick insulation. All fusible links are several wire gauges smaller than the wire they're designed to protect. Fusible links can't be repaired, but a new link of the same size wire can be installed. The procedure is as follows:

a) *Disconnect the cable from the negative terminal of the battery (see Chapter 5).*

b) *Cut the damaged fusible link out of the wire just behind the connector.*

c) *Strip the insulation back approximately 1-inch.*

d) *Spread the strands of the exposed wire apart, push them together and twist them in place.*

e) *Use rosin core solder and solder the wires together to obtain a good connection.*

f) *Use plenty of electrical tape around the soldered joint. No wires should be exposed.*

g) *Connect the negative battery cable. Test the circuit for proper operation.*

4   Circuit breakers - general information and check

1   Circuit breakers protect certain circuits, such as the power windows and power seats. Depending on the vehicle's accessories, there may be two 25-amp circuit breakers for the door locks and one 30-amp circuit breaker for the power seats, located in the interior fuse/relay box under the left rear seat.

2   Because the circuit breakers reset automatically, an electrical overload in a circuit-breaker-protected system will cause the circuit to fail momentarily, then come back on. If the circuit does not come back on, check it immediately.

3   For a basic check, pull the circuit breaker up out of its socket on the fuse panel, but just far enough to probe with a voltmeter. The breaker should still contact the sockets.

4   With the voltmeter negative lead on a good chassis ground, touch each end prong of the circuit breaker with the positive meter probe. There should be battery voltage at each end. If there is battery voltage only at one end, the circuit breaker must be replaced.

5   Relays - general information and testing

General information

1   Several electrical accessories in the vehicle, such as the fuel injection system, horns, starter, and fog lamps use relays to transmit the electrical signal to the component. Relays use a low-current circuit (the control circuit) to open and close a high-current circuit (the power circuit). If the relay is defective, that component will not operate properly. Most relays are mounted in the engine compartment and interior fuse/relay boxes **(see illustrations 3.1a and 3.1c)**. If a faulty relay is suspected, it can be removed and tested using the procedure below or by a dealer service department or a repair shop. Defective relays must be replaced as a unit.

Testing

Refer to illustrations 5.2a and 5.2b

2   Most of the relays used in these vehicles are of a type often called "ISO" relays, which refers to the International Standards Organization. The terminals of ISO relays are numbered to indicate their usual circuit connections and functions. There are two basic layouts of terminals on the relays used in these

5.2b Most relays are marked on the outside to easily identify the control circuit and power circuit - this one is of the four-terminal type

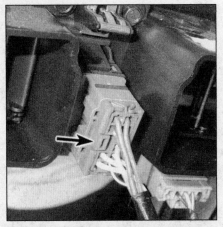

6.5 Depress the tab and disconnect the electrical connector, then release the tabs and remove the switch through the front of the trim bezel

vehicles **(see illustrations)**.

3   Refer to the wiring diagram for the circuit to determine the proper connections for the relay you're testing. If you can't determine the correct connection from the wiring diagrams, however, you may be able to determine the test connections from the information that follows.

4   Two of the terminals are the relay control circuit and connect to the relay coil. The other relay terminals are the power circuit. When the relay is energized, the coil creates a magnetic field that closes the larger contacts of the power circuit to provide power to the circuit loads.

5   Terminals 85 and 86 are normally the control circuit. If the relay contains a diode, terminal 86 must be connected to battery positive (B+) voltage and terminal 85 to ground. If the relay contains a resistor, terminals 85 and 86 can be connected in either direction with respect to B+ and ground.

6   Terminal 30 is normally connected to the battery voltage (B+) source for the circuit loads. Terminal 87 is connected to the ground side of the circuit, either directly or through a load. If the relay has several alternate terminals for load or ground connections, they usually are numbered 87A, 87B, 87C, and so on

7   Use an ohmmeter to check continuity through the relay control coil.

a) *Connect the meter according to the polarity shown in the illustration for one check; then reverse the ohmmeter leads and check continuity in the other direction.*

b) *If the relay contains a resistor, resistance should be indicated on the meter, and should be the same value with the ohmmeter in either direction.*

c) *If the relay contains a diode, resistance should be higher with the ohmmeter in the forward polarity direction than with the meter leads reversed.*

d) *If the ohmmeter shows infinite resistance in both directions, replace the relay.*

8   Remove the relay from the vehicle and use the ohmmeter to check for continuity

between the relay power circuit terminals. There should be no continuity between terminal 30 and 87 with the relay de-energized.

9   Connect a fused jumper wire to terminal 86 and the positive battery terminal. Connect another jumper wire between terminal 85 and ground. When the connections are made, the relay should click.

10   With the jumper wires connected, check for continuity between the power circuit terminals. Now, there should be continuity between terminals 30 and 87.

11   If the relay fails any of the above tests, replace it.

---

6   Turn signal and hazard flashers - general information

2005 and 2006 models

Refer to illustration 6.5

Warning: *The models covered by this manual are equipped with Supplemental Restraint Systems (SRS), more commonly known as airbags. Always disable the airbag system before working in the vicinity of any airbag system components to avoid the possibility of accidental deployment of the airbags, which could cause personal injury (see Section 28).*

1   The combination flasher - integrated with the hazard flasher switch - flashes the turn signals when the turn signal switch is operated. It operates all four signals when the hazard flasher switch - located between the instrument panel center vents - is operated.

2   When the flasher unit is functioning properly, an audible click can be heard during its operation. If the turn signal indicator on one side of the vehicle flashes much more rapidly than normal, a faulty turn signal bulb is indicated.

3   If both turn signals fail to blink, the problem may be due to a blown fuse, a faulty flasher unit, a broken switch or a loose or open connec-

tion. If a quick check of the fuse box indicates that the turn signal fuse has blown, check the wiring for a short before installing a new fuse.

4   The type of combination flasher unit used on these models have complex internal circuitry, and can't be tested using standard electrical test equipment. Refer to the wiring diagrams at the end of this Chapter and test the circuitry before replacing the flasher with a known-good unit.

5   To remove the flasher/emergency flasher switch, remove the center instrument panel trim bezel (see Chapter 11). Disconnect the electrical connector **(see illustration),** then disengage the tabs or remove the fasteners, as applicable, and pull the flasher/switch out through the front of the trim bezel.

6   Installation is the reverse of removal.

2007 and later models

7   There is no turn signal and hazard flasher relay or module on these models. This function is handled by the Body Control Module (BCM).

8   If a bulb on one side of the vehicle flashes much faster than normal but the bulb at the other end of the vehicle (on the same side) doesn't light at all, the bulb that doesn't flash is probably faulty. Replace the bulb (see Section 17).

9   If both the left and right front or rear turn signal bulbs are not flashing, the turn signal and hazard flasher relay function in the BCM is probably defective. Have the BCM replaced by a dealer service department or other repair facility equipped with the proper tool. This is not a job that you can do at home because the BCM must be programmed with a factory scan tool when it's replaced.

---

7   Steering column multi-function switch - replacement

Refer to illustration 7.5

Warning: *The models covered by this manual are equipped with Supplemental Restraint Systems (SRS), more commonly known as airbags. Always disable the airbag system before working in the vicinity of any airbag system components to avoid the possibility of accidental deployment of the airbags, which could cause personal injury (see Section 28).*

1   The multi-function switches are located on the left and right sides of the steering column. Each incorporates multiple functions; the turn signal, headlight, low-to-high beam and flash-to-pass are on the left. The windshield wiper/washer is on the right.

2   Release the tilt wheel lever and place the steering wheel in the center position.

3   Remove the steering column trim covers (see Chapter 11).

4   Depress the retaining tab and disconnect the multi-function switch electrical connector.

5   Depress the multi-function switch retaining tabs, then detach the switch from the

7.5 To remove either multi-function switch, depress the retaining tabs and pull it out of the housing

8.2 Detach the park lock cable adjuster from the pin on the park lock lever by sliding it off the pin towards the rear of the vehicle . . .

8.3 . . . then squeeze the cable retainer together to detach the park lock cable from the shifter assembly

steering column housing **(see illustrations)**.
6   Installation is the reverse of removal. **Note:** *After properly aligning, then installing a multi-function switch assembly, ensure the locking tabs are fully engaged.*

8   Park lock cable - removal, installation and adjustment

Warning: *The models covered by this manual are equipped with Supplemental Restraint Systems (SRS), more commonly known as airbags. Always disable the airbag system before working in the vicinity of any airbag system components to avoid the possibility of accidental deployment of the airbags, which could cause personal injury (see Section 28).*

Replacement

Refer to illustrations 8.2, 8.3, 8.5 and 8.6

1   Disconnect the cable from the negative terminal of the battery (see Chapter 5), then remove the center console (see Chapter 11).
2   Disconnect the shift lock cable adjuster from the park lock lever on the shifter assembly **(see illustration)**.
3   Remove the shift lock cable from its mounting location on the shifter assembly **(see illustration)**.
4   Follow the park lock cable up to the key

lock cylinder housing; note the routing of the cable.
5   Disengage the retainers and remove the cable retaining fitting from the key lock cylinder housing **(see illustration)**.
6   To remove the park lock cable end fitting, depress the release tab and separate the fitting from the cable **(see illustration)**.

Installation and adjustment

Refer to illustration 8.10

7   Properly align and install the park lock cable fitting onto the key lock cylinder housing; ensure that the retainers are properly engaged.
8   Route the cable into place, then connect the park lock cable to the park lock lever on the shifter assembly; ensure that the cable retainer clamp is properly engaged.
9   To avoid damage to the park lock adjuster retainer, carefully align and slide the cable adjuster onto the flanged pin on the end of the park lock lever.
10   Place the park lock cable adjuster in the unlocked position **(see illustration)**.

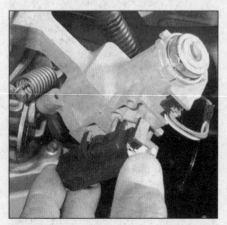

8.5 Using a small screwdriver, disengage the retainers

8.6 Depress the tab and pull the cable from the end fitting

8.10 Pull out the cable adjuster lock; the cable will then automatically adjust. Then push the lock back into place

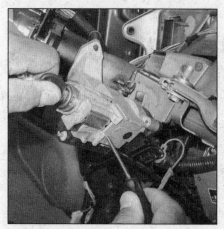

9.5 Push the retaining pin in to release the lock cylinder, then withdraw the key and cylinder

9.11 Pull out the connector lock, then unplug the electrical connector from the ignition switch

13   Slide the entire assembly off of the steering column.
14   Remove the ignition switch mounting screws and remove the ignition switch from the housing.

Installation

15   Turn the ignition switch to ACC, then place it into the housing. Tighten the ignition switch screws securely.
16   Put the lock cylinder housing onto the steering column and mate it to the multi-function lever bracket, then install the bolts. Make sure the lock tab on the lock cylinder housing engages with its slot in the steering column, then tighten the bolts securely.
17   Tighten the lower mounting bolt, then the upper mounting bolt, but only until they are seated firmly.
18   Finally, tighten the lower bolt, followed by the upper bolt, to the torque listed in this Chapter's Specifications.
19   The remainder of the installation procedure is the reverse of removal.

---

11   Set the parking brake.
12   Verify that the ignition key is removed from the key cylinder and the shift lever is set to the Park position.
13   Place the park lock cable adjuster into the lock position. **Note:** *The park lock cable is properly adjusted when the following conditions apply: (a) unable to shift out of the park position with the ignition key removed (b) ignition key will not return to the lock position when the shift lever is out of the Park position (c) the ignition key, when cycled to the run position and back to the lock position can be removed with ease with the shift lever set in the Park position. Repeat Steps 10 through 13 if one or more of the conditions listed above did not apply.*

---

9   Ignition switch and key lock cylinder - replacement

Warning: *The models covered by this manual are equipped with Supplemental Restraint Systems (SRS), more commonly known as airbags. Always disable the airbag system before working in the vicinity of any airbag system components to avoid the possibility of accidental deployment of the airbags, which could cause personal injury (see Section 28).*
1   Disconnect the cable from the negative terminal of the battery (see Chapter 5).
2   Remove the upper and lower steering column covers (see Chapter 11).
3   Disconnect the park lock cable from the lock cylinder housing (see Section 8).

Key lock cylinder
Refer to illustration 9.5

4   Insert the ignition key into the lock cylinder and turn it to the ACC position.
5   Insert a pick or other small tool into the hole in the back of the lock cylinder housing and push it in to depress the release button **(see illustration)**. Pull the lock cylinder out of the housing.

6   To install the lock cylinder, make sure the ignition switch is still in the ACC position. If it isn't, rotate it to the ACC position with a screwdriver or needle-nose pliers. The lock cylinder won't fit all the way into the bore if the switch is in any other position.
7   Insert the lock cylinder into the housing, making sure it engages with the switch and clicks into place.
8   Verify that the ignition switch operates correctly in the Off, ACC, Run and Start positions.

Housing and ignition switch
Removal
Refer to illustration 9.11

9   Refer to Chapter 10 and remove the steering wheel.
10   Refer to Section 8 and remove both multi-function switches. The wiring doesn't have to be disconnected.
11   Detach the electrical connectors from the ignition switch **(see illustration)**.
12   Remove the two ignition switch/key lock housing screws. Insert the key and turn it to the ACC position.

10   Instrument panel switches - replacement

Warning: *The models covered by this manual are equipped with Supplemental Restraint Systems (SRS), more commonly known as airbags. Always disable the airbag system before working in the vicinity of any airbag system components to avoid the possibility of accidental deployment of the airbags, which could cause personal injury (see Section 28).*

Center instrument panel switches
Refer to illustrations 10.1, 10.2a, 10.2b and 10.3

1   Refer to Chapter 11 and remove the center instrument panel bezel, pulling it out far enough to access the switches from the rear **(see illustration)**.

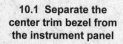

10.1 Separate the center trim bezel from the instrument panel

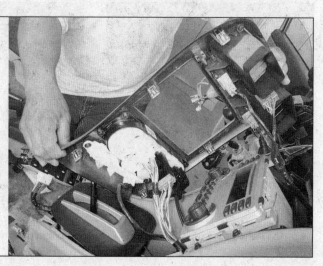

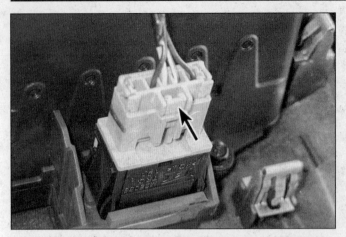

10.2a Depress the electrical connector release tab and unplug the connectors

10.2b Remove these switches by depressing the retaining tangs and pushing the switch out of the panel

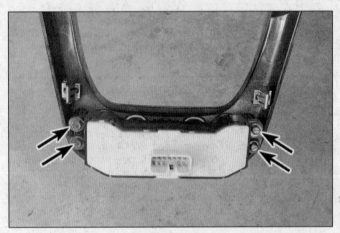

10.3 To remove the power window switch, remove the retaining screws

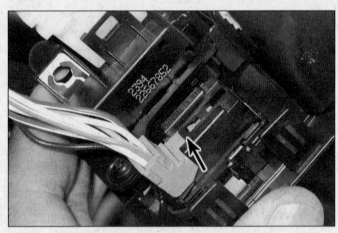

10.5a Lift the retaining clip and slide the connector from the outside power mirror switch

2   Disconnect the electrical connector from the switch and remove it from the bezel **(see illustrations)**.

3   Depress the retaining tangs and remove the switch **(see illustration).**

Left vent trim panel switches

Refer to illustrations 10.5a, 10.5b, 10.5c and 10.5d

4   Remove the vent trim bezel (see Chapter 11)

5   Disconnect the electrical connector, remove the screws, then separate the switch from the panel **(see illustrations)**.

6   Installation is the reverse of the removal procedure.

10.5b Depress the retaining tabs and pull the switch from the vent bezel

10.5c Lift the retaining clip and pull the connector from the dimmer switch

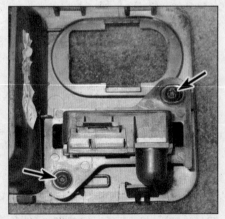

10.5d Remove the switch retaining screws and remove the switch from the bezel

11.4 Remove the four instrument cluster mounting screws

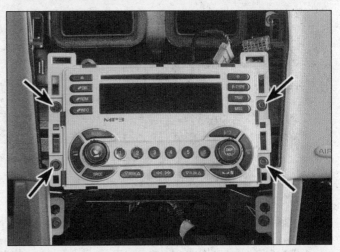

12.3 Remove the mounting screws and pull the radio from the instrument panel

11 Instrument cluster - removal and installation

Refer to illustration 11.4

Warning: *The models covered by this manual are equipped with Supplemental Restraint Systems (SRS), more commonly known as airbags. Always disable the airbag system before working in the vicinity of any airbag system components to avoid the possibility of accidental deployment of the airbags, which could cause personal injury (see Section 28).*

1   Disconnect the cable from the negative battery terminal (see Chapter 5).
2   Release the tilt wheel lever and lower the steering wheel to its lowest position.
3   Remove the instrument cluster bezel (see Chapter 11).
4   Remove the instrument cluster mounting screws **(see illustration).**
5   Carefully remove the instrument cluster from the instrument panel.
6   Installation is the reverse of removal.

12 Radio and speakers - removal and installation

Warning: *The models covered by this manual are equipped with Supplemental Restraint Systems (SRS), more commonly known as airbags. Always disable the airbag system before working in the vicinity of any airbag system components to avoid the possibility of accidental deployment of the airbags, which could cause personal injury (see Section 28).*

Radio

Refer to illustrations 12.3 and 12.4

1   Disconnect the cable from the negative battery terminal (see Chapter 5).
2   Remove the center trim panel (see Chapter 11). **Note:** *On 2010 and later models, the radio and air conditioning/heating control assembly is mounted to the center trim panel (see Chapter 11). Once the trim panel is removed the control assembly can be removed from the trim panel. Once the control unit is removed the radio unit may be removed as described in this Section.*

3   Remove the radio mounting screws and pull the radio from the instrument panel **(see illustration).**
4   Disconnect the electrical connector and the antenna cable from the back of the radio **(see illustration).**
5   Installation is the reverse of removal.

Speakers

Lower door speakers

Refer to illustration 12.7

6   Remove the door trim panel (see Chapter 11).
7   Remove the speaker mounting screws **(see illustration)**. Remove the speaker from the door, then disconnect the electrical connector.
8   Installation is the reverse of removal.

Upper front door speakers

Refer to illustrations 12.9 and 12.12

9   Pull off the upper speaker panel, starting at the top **(see illustration)**.
10 Lift it to slide the lower tabs out of the door panel.
11 Disconnect the electrical connector.

12.4 Disconnect the antenna cable and electrical connectors from the back of the radio

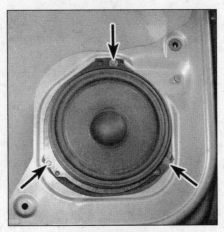

12.7 Remove the speaker mounting screws

12.9 Pull the trim panel from the door, then disconnect the tweeter electrical connector

12.12 To remove the tweeter from the trim panel, rotate it counterclockwise

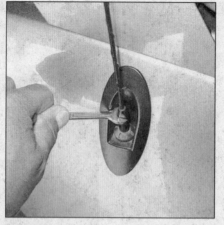

13.1 Use a small wrench to remove the antenna mast

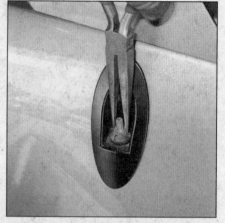

13.4 To remove the antenna base and cable, remove the base mounting nut

12   Twist the speaker to remove it from the panel **(see illustration)**.
13   Installation is the reverse of removal.

Rear speaker
14   Remove the carpet retaining trim piece.
15   Remove the pillar trim.
16   Remove the rear upper trim molding.
17   Remove the rear floor storage bin.
18   Remove the cargo hooks.
19   Pull the quarter trim panel out, disengaging its retaining clips.
20   Remove the speaker mounting screws, then remove the speaker and disconnect the electrical connector.

Instrument panel front center speaker
21   On models equipped with a storage compartment, open the storage compartment lid and remove the two mounting screws, then remove the storage unit from the top of the instrument panel. On models equipped with an upper center trim panel, use a trim panel tool and carefully pry the trim panel from the top of instrument panel.
22   Remove the speaker mounting screws, then remove the speaker and disconnect the electrical connector.
23   Installation is the reverse of removal.

---

13   Antenna - removal and installation

2009 and earlier models

Refer to illustrations 13.1 and 13.4

1   Use a small open-end wrench to unscrew the antenna mast **(see illustration)**. Be very careful - the tool could slip and scratch the body. It's a good idea to surround the base of the antenna with masking tape to prevent scratching.
2   If the antenna base/cable assembly must be replaced, remove the glove compartment and detach the lower rear portion of the inner fender liner (see Chapter 11).
3   Disconnect the antenna mast cable from

the radio cable where they join in a connector under the far right end of the instrument panel. Attach a fish wire to the end of the cable. Working inside the inner fender liner, pass the cable and grommet through the hole in the cowl into the fender opening.
4   Remove the antenna mast base mounting nut **(see illustration)** and pull out the base and its cable.
5   Attach the new cable to the fish wire and pull the wire back slowly and carefully into the body, routing it as the original cable had been.
6   If the extension cable (between the antenna base cable and the radio) must be replaced, remove the glove compartment (see Chapter 11) and the passenger's airbag module (see Section 28). Release the cable from the clips along the top of the instrument panel and disconnect it from the radio and antenna base cable.
7   Installation is the reverse of the removal procedure.

2010 and later models

Antenna base
Note: *The antenna base and mast are located on the roof just in front of the liftgate opening.*

14.2 Turn the bulb holder counterclockwise and remove it from the housing

8   Disconnect the cable from the negative terminal of the battery (see Chapter 5).
9   Remove the liftgate opening trim panel (see Chapter 11), then carefully pull the edge of the headliner down enough to access the base mounting bolt and remove the bolt.
10   Pull the antenna base up from the roof and disconnect the electrical connector to the base.
11   Installation is the reverse of removal.

Antenna mast
12   Rotate the mast counterclockwise until it can be removed from the antenna base.
13   Installation is the reverse of removal.

---

14   Headlight bulb - replacement

Refer to illustrations 14.2 and 14.3
Warning: *Halogen bulbs are gas-filled and under pressure and may shatter if the surface is scratched or the bulb is dropped. Wear eye protection and handle the bulbs carefully, grasping only the base whenever possible. Don't touch the surface of the bulb with your fingers because the oil from your skin could*

14.3 Note the position of the bulb, then remove the bulb from the headlight housing

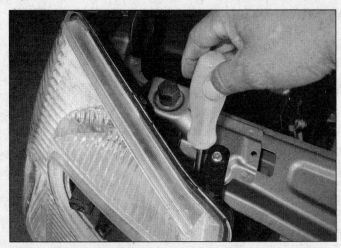

15.1a Use a long Philips screwdriver to access the adjuster . . .

15.1b . . . on the back of the headlight housing

cause it to overheat and fail prematurely. If you do touch the bulb surface, clean it with rubbing alcohol.

Note: *Low and high beam bulbs can be identified by the color of their sockets; low beam bulbs have gray sockets and high beam bulbs have black sockets.*

1   On 2009 and earlier models, remove the headlight housing (see Section 16) and on 2010 and later models, remove the inner fender splash shield (see Chapter 11).

2   Twist the bulb retainer counterclockwise

and withdraw the bulb retainer from the headlight housing **(see illustration)**.

3   Remove the bulb from the headlight housing **(see illustration)**. Handling the new bulb only by the plastic base portion, properly align and insert the new bulb into the headlight housing.

4   Properly align and install the bulb retainer, then twist the retainer clockwise to lock it in place.

5   Installation of the housing is the reverse of the removal procedure.

15   Headlights and fog lights - adjustment

Warning: *The headlights must be aimed correctly. If adjusted incorrectly, they could temporarily blind the driver of an oncoming vehicle and cause an accident or seriously reduce your ability to see the road. The headlights should be checked for proper aim every 12 months and any time a new headlight is installed or front-end bodywork is performed. The following procedure is only an interim step to provide temporary adjustment until the headlights can be adjusted by a properly equipped shop.*

Headlights

Refer to illustrations 15.1a, 15.1b and 15.2

1   These models are equipped with composite headlights with adjustment screws that control up-and-down movement **(see illustrations)**. Left-and-right movement is not adjustable.

2   There are several methods of adjusting the headlights. The simplest method requires a blank wall 25 feet in front of the vehicle and a level floor **(see illustration)**.

3   Position masking tape on the wall in reference to the vehicle centerline and the centerlines of both headlights.

4   Measure the height of the headlight reference marks (in the centers of the headlight lenses) from the ground. Position a horizontal tape line on the wall at the same height as the headlight reference marks. **Note:** *It may be easier to position the tape on the wall with the vehicle parked only a few inches away.*

5   Adjustment should be made with the vehicle sitting level, the gas tank half-full and no unusually heavy load in the vehicle.

6   Turn on the low beams. Turn the adjusting screw **(see illustrations 15.1a and 15.1b)** to position the high intensity zone so it is two inches below the horizontal line.

7   Have the headlights adjusted by a dealer service department at the earliest opportunity.

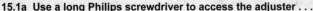

High-Intensity Area

Floor to Center of Headlamp Lens

Center of Vehicle to Center of Headlamp Lens

Vehicle Centerline

25 FT

Front of Headlamp

50029-12-19.3 HAYNES

15.2 Headlight adjustment details

15.10 Fog light adjustment screw location

16.2 Using a trim tool, pry the center trim from the headlight housing, and remove the center portion of the trim by pressing the retainers through the grille to release it - 2009 and earlier

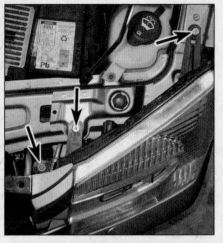

16.3 Remove the headlight housing mounting screws

Fog lights

Refer to illustration 15.10

8   Some models have optional fog lights that can be aimed just like headlights. As with the headlights, there are no left-and-right adjustments.

9   Position tape on a wall 25 feet in front of the vehicle **(see illustration 15.2)**. Tape a horizontal line on the wall that represents the height of the fog light centers, and another tape line four inches below that line.

10   Using the adjusting screws on the fog lights **(see illustration)**, adjust the pattern on the wall so that the top of the fog light beam meets the lower line on the wall.

16   Headlight housing - removal and installation

Refer to illustrations 16.2, 16.3, 16.4 and 16.5

1   Disconnect the cable from the negative battery terminal (see Chapter 5).

2   On 2009 and earlier Equinox models, remove the grille center trim **(see illustration)** and on all 2010 and later models remove the front bumper cover (see Chapter 11).

3   Remove the upper headlight housing mounting bolts **(see illustration)**.

4   Disengage the headlight assembly side pin retainers by firmly pulling forward **(see illustration)**.

5   Maneuver the housing out of the bumper cover **(see illustration)**.

6   Disconnect the electrical connectors from the bulb holders.

7   Installation is the reverse of removal.

17   Bulb replacement

Exterior light bulbs

Front turn signal/parking light bulbs

Refer to illustrations 17.2, 17.3a and 17.3b

1   Remove the headlight housing (see Section 16).

2   Turn the bulb holder counterclockwise and remove it from the headlight housing **(see illustration)**.

3   Pull the bulb straight out of the holder to

16.4 Pull out on the bottom outer corner of the housing to detach the housing lower rear retaining pin

16.5 Push down on the bumper cover and remove the headlight housing

17.2 Rotate the bulb holder counterclockwise and remove it from the housing

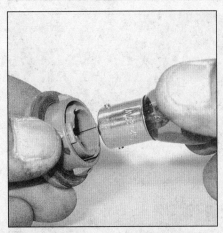

17.3a Push in, rotate counterclockwise and pull out the turn signal bulb. Note that the locating pins are offset; the bulb will fit in the holder only one way

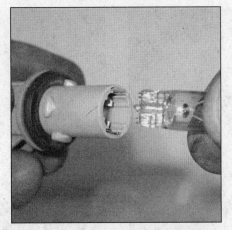

17.3b The side marker bulb pulls straight out of the socket

17.7a Release the connector retaining tabs, disconnect the connector . . .

remove it **(see illustrations)**.

4   Installation is the reverse of removal.

Fog light bulbs

Refer to illustrations 17.7a and 17.7b

Warning: *Halogen bulbs are gas-filled and under pressure and may shatter if the surface is scratched or the bulb is dropped. Wear eye protection and handle the bulbs carefully, grasping only the base whenever possible. Don't touch the surface of the bulb with your fingers because the oil from your skin could cause it to overheat and fail prematurely. If you do touch the bulb surface, clean it with rubbing alcohol.*

5   Raise the front of the vehicle and place it securely on jackstands.

6   Remove the forward fasteners of the fender inner liner/splash shield (see Chapter 11). Pull back the splash shield and reach up through the opening in the bottom of the front bumper cover for access to the bulb.

7   Disconnect the electrical connector from the fog light bulb. Turn the fog light bulb counterclockwise and pull it out of the housing **(see illustrations).**

8   Installation is the reverse of removal.

Center high-mounted stop light

Refer to illustrations 17.9 and 17.10

9   Remove the upper interior trim panel from the liftgate (see Chapter 11), then remove the spoiler mounting screws and tilt the spoiler up **(see illustration)**.

10   Remove the two screws from the bulb assembly and pull it away from the spoiler **(see illustration)**. Disconnect the washer hose.

11   Disconnect the electrical connector from the assembly and remove it.

12   The bulb assembly is a sealed unit. The individual bulbs can't be replaced.

13   Installation is the reverse of removal.

17.7b . . . then twist the fog light bulb holder counterclockwise to remove it

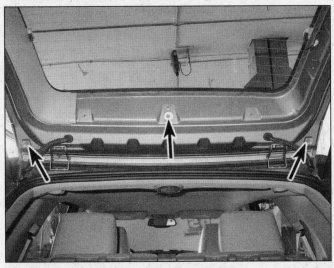

17.9 Rear spoiler mounting screws

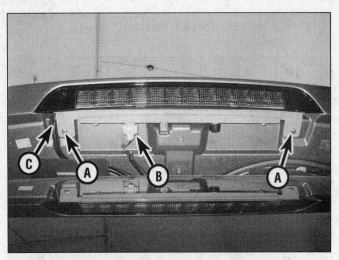

17.10 Center High Mount Stop Light mounting details

A	Mounting screws	C	Washer hose
B	Electrical connector		

17.15 Use a small screwdriver to pry open the trim covers for access to the taillight housing mounting screws

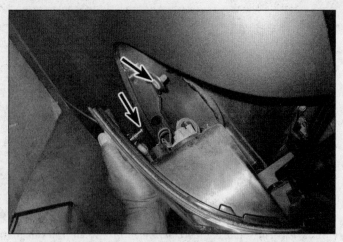

17.16 Disengage the housing retaining pins from the retaining grommets on the body

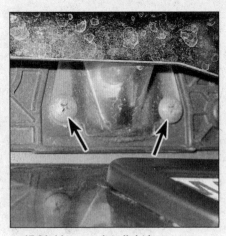

17.21 License plate light lens screws

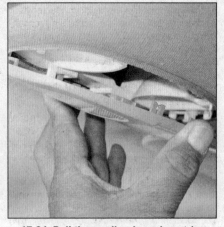

17.24 Pull the reading lamp lens trim assembly down and free the rear retaining tab

17   Turn the bulb holder counterclockwise and remove it from the housing.
18   Remove the bulb from its holder **(see illustrations 17.3a and 17.3b).**
19   Installation is the reverse of removal.

License plate light bulbs

Refer to illustration 17.21

20   Open the liftgate.
21   Remove the license plate light lens screws **(see illustration)** and remove the license plate light lens from its bezel.
22   Pull the bulb from its socket.
23   Installation is the reverse of removal.

Interior light bulbs

Reading light bulbs

Refer to illustrations 17.24 and 17.25

24   Remove the lens from the reading light housing by pulling it straight down **(see illustration).**
25   Pull the bulb straight out of its socket **(see illustration).**
26   Installation is the reverse of removal.

Rear brake/tail, turn signal, brake and back-up light bulbs

Refer to illustrations 17.15 and 17.16

14   Open the liftgate.
15   Remove the taillight housing mounting

screws **(see illustration).**
16   Pull the light assembly outward to disengage the outer retaining pins **(see illustration).**

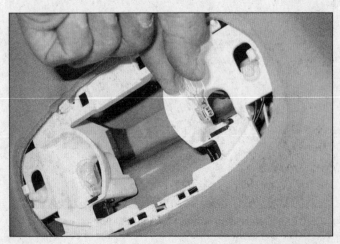

17.25 Pull out the reading lamp bulb from the light assembly

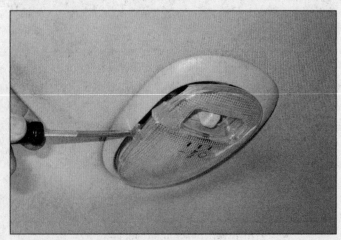

17.27 Using a small screwdriver, pry the lens from the dome light assembly

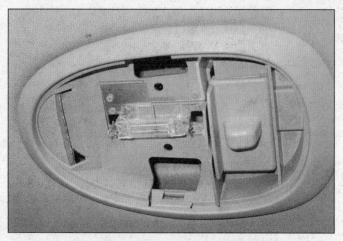

17.28 Remove the bulb from the terminals

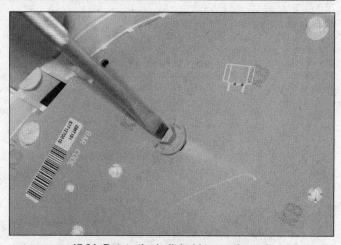

17.31 Rotate the bulb holder to release it

18.1 Remove the nut, mark the wiper arm location and rock the wiper arm to detach it from the shaft - use a small puller if it's stuck

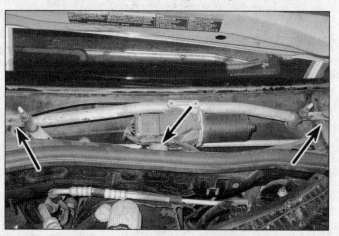

18.4 Remove the wiper linkage mounting bolts

Cargo and dome light bulb

Refer to illustrations 17.27 and 17.28

27 Remove the lens from the light housing **(see illustration)**.
28 Remove the bulb from the terminals **(see illustration)**.
29 Installation is the reverse of removal.

Instrument cluster light bulbs

Refer to illustration 17.31

30 Remove the instrument cluster (see Section 11).
31 Turn the bulb socket using a small screwdriver, then remove it from the instrument cluster **(see illustration)**. **Note:** *It's best to replace all bulbs if one has failed.*
32 Installation is the reverse of removal.

---

18 Wiper motor - replacement

Windshield wiper motor

Refer to illustrations 18.1, 18.4 and 18.6

1 Remove the wiper arm nuts and mark the relationship of the wiper arms to their

shafts **(see illustration)**. Remove both wiper arms.
2 Remove the plastic cowl cover (see Chapter 11).
3 Disconnect the electrical connector from the wiper motor.
4 Remove the three windshield wiper motor and link assembly mounting bolts **(see illustration)** and remove the wiper motor and

link assembly.
5 Using a screwdriver, carefully pry the link rod from the pivot pin on the motor's crank arm.
6 Remove the nut that secures the crank arm to the motor shaft **(see illustration)**.
7 Mark the relationship of the crank arm to the motor shaft and remove the crank arm.

18.6 Remove the nut to separate the wiper motor from the linkage assembly

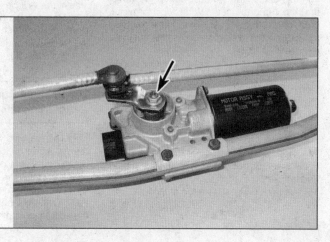

18.10 Remove the nut, mark the wiper arm location and rock the wiper arm to detach it from the shaft - use a small puller if it's stuck. Then remove the round bezel

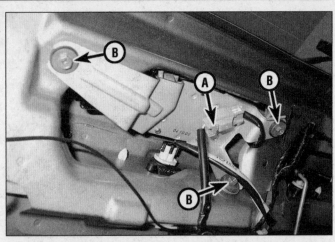

18.14 Disconnect the electrical connector (A) at the rear wiper motor, then remove the mounting bolts (B)

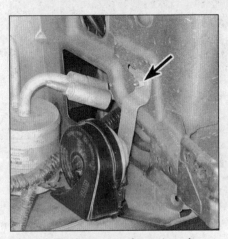

19.2 The horn mounting nut can be accessed after the left headlight is removed

8   Remove the two motor mounting bolts and remove the motor from its mounting bracket.

9   Installation is the reverse of removal.

Rear wiper motor

Refer to illustrations 18.10 and 18.14

10   On 2009 and earlier models, remove the trim cap and unscrew the wiper arm nut **(see illustration)**. On 2010 and later models, rotate the trim cover up to access the nut then remove it.

11   Mark the relationship of the rear wiper arm to the motor shaft **(see illustration 18.1)**, then remove the arm.

12   Remove the trim panel from the rear hatch (see Chapter 11).

13   Disconnect the electrical connector from the motor.

14   Remove the rear wiper motor mounting bolts **(see illustration)** and remove the motor and bracket.

15   Installation is the reverse of removal.

19   Horn - replacement

Refer to illustration 19.2

Note: *These models are equipped with a Body Control Module (BCM). Several systems are linked to a centralized control module that allows simple and accurate troubleshooting, but only with a professional-grade scan tool. The Body Control Module governs the door locks, the power windows, the ignition lock and security system, the interior lights, the Daytime Running Lights system, the horn, the windshield wipers, the heating/air conditioning system and the power mirrors. In the event of a malfunction with this system, have the vehicle diagnosed by a dealership service department or other qualified automotive repair facility.*

1   On 2009 and earlier models, remove the left (driver's side) headlight housing (see Section 16). On 2010 through 2012 models, remove the left (driver's side) inner fender splash shield. On 2013 and later models, remove the left or right inner fender splash shield (see Chapter 11) to replace the appropriate horn.

Note: *On 2012 and earlier models, the horns are mounted on one side and on 2013 and later models, one horn is mounted on each side of the vehicle.*

2   Remove the mounting nut **(see illustration)**, then detach the horn and disconnect the electrical connector.

3   Installation is the reverse of removal.

20   Daytime Running Lights (DRL) - general information

The Daytime Running Lights (DRL) system used on all models illuminates the low beam headlights at reduced intensity whenever the ignition is On. The only exception is with the engine running and the shift lever in Park. Once the parking brake is released or the shift lever is moved, the lights will remain on as long as the ignition switch is on.

21   Rear window defogger - check and repair

1   The rear window defogger consists of a number of horizontal elements baked onto the glass surface.

2   Small breaks in the element can be repaired without removing the rear window.

Check

Refer to illustrations 21.4, 21.5 and 21.7

3   Turn the ignition switch and defogger system switches to the ON position. Using a voltmeter, place the positive probe against the defogger grid positive terminal and the negative probe against the ground terminal. If battery voltage is not indicated, check the fuse, defogger switch and related wiring. If voltage is indicated, but all or part of the

21.4 When measuring the voltage at the rear window defogger grid, wrap a piece of aluminum foil around the positive probe of the voltmeter and press the foil against the wire with your finger

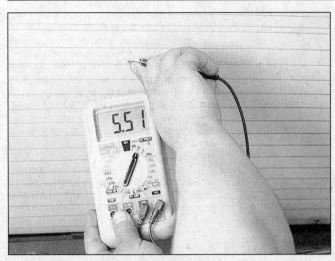

21.5 To determine if a heating element has broken, check the voltage at the center of each element - if the voltage is 6-volts, the element is unbroken

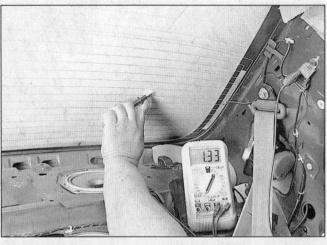

21.7 To find the break, place the voltmeter negative lead against the defogger ground terminal, place the voltmeter positive lead with the foil strip against the heat wire at the positive terminal end and slide it toward the negative terminal end - the point at which the voltmeter deflects from several volts to zero volts is the point at which the wire is broken

defogger doesn't heat, proceed with the following tests.

4   When measuring voltage during the next two tests, wrap a piece of aluminum foil around the tip of the voltmeter positive probe and press the foil against the heating element with your finger **(see illustration)**. Place the negative probe on the defogger grid ground terminal.

5   Check the voltage at the center of each heating element **(see illustration)**. If the voltage is 5 or 6-volts, the element is okay (there is no break). If the voltage is zero, the element is broken between the center of the element and the positive end. If the voltage is 10 to 12-volts, the element is broken between the center of the element and ground. Check each heating element.

6   Connect the negative lead to a good body ground. The reading should stay the same. If it doesn't, the ground connection is

bad.

7   To find the break, place the voltmeter negative probe against the defogger ground terminal. Place the voltmeter positive probe with the foil strip against the heating element at the positive terminal end and slide it toward the negative terminal end. The point at which the voltmeter deflects from several volts to zero is the point at which the heating element is broken **(see illustration)**.

Repair

Refer to illustration 21.13

8   Repair the break in the element using a repair kit specifically recommended for this purpose, available at most auto parts stores. Included in this kit is plastic conductive epoxy.

9   Prior to repairing a break, turn off the system and allow it to cool off for a few minutes.

10   Lightly buff the element area with fine steel wool, then clean it thoroughly with rubbing alcohol.

11   Use masking tape to mask off the area being repaired.

12   Thoroughly mix the epoxy, following the instructions provided with the repair kit.

13   Apply the epoxy material to the slit in the masking tape, overlapping the undamaged area about 3/4-inch on either end **(see illustration)**.

14   Allow the repair to cure for 24 hours before removing the tape and using the system.

22   Cruise control system - description and check

1   The cruise control system maintains vehicle speed with the Powertrain Control Module (PCM), throttle actuator control motor, brake switch, control switches and associ-

ated wiring. There is no mechanical connection, such as a vacuum servo or cable. Some features of the system require special testers and diagnostic procedures that are beyond the scope of the home mechanic. Listed below are some general procedures that may be used to locate common problems.

2   Check the fuses (see Section 3).

3   The Brake Pedal Position (BPP) switch (or brake light switch) deactivates the cruise control system. Have an assistant press the brake pedal while you check the brake light operation.

4   If the brake lights do not operate properly, correct the problem and retest the cruise control.

5   Check the wiring between the PCM and throttle actuator motor for opens or shorts and repair as necessary.

6   The cruise control system uses information from the PCM, including the Vehicle Speed Sensor (VSS), which is located in the transmission or transfer case. Refer to Chapter 6 for more information on the VSS.

7   Test drive the vehicle to determine if the cruise control is now working. If it isn't, take it to a dealer service department or other qualified repair shop for further diagnosis.

23   Power window system - description and check

Note: *These models are equipped with a Body Control Module (BCM). Several systems are linked to a centralized control module that allows simple and accurate troubleshooting, but only with a professional-grade scan tool. The Body Control Module governs the door locks, the power windows, the ignition lock and security system, the interior lights, the Daytime Running Lights system, the horn, the windshield wipers, the heating/air conditioning*

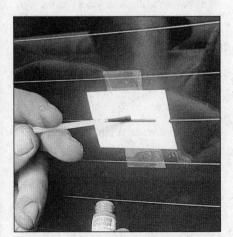

21.13 To use a defogger repair kit, apply masking tape to the inside of the window at the damaged area, then brush on the special conductive coating

24.14 Carefully separate the case halves . . .

24.15 . . . then pry out the battery

system and the power mirrors. In the event of malfunction with this system, have the vehicle diagnosed by a dealership service department or other qualified automotive repair facility.

1   The power window system operates electric motors, mounted in the doors, which lower and raise the windows. The system consists of the control switches, the motors, regulators, glass mechanisms, the Body Control Module (BCM) and associated wiring.

2   The power windows can be lowered and raised from the master control switch by the driver or by remote switches located at the individual windows. Each window has a separate motor that is reversible. The position of the control switch determines the polarity and therefore the direction of operation.

3   The circuit is protected by a fuse and a circuit breaker. Each motor is also equipped with an internal circuit breaker; this prevents one stuck window from disabling the whole system.

4   The power window system will only operate when the ignition switch is ON. In addition, many models have a window lockout switch at the master control switch which, when activated, disables the switches at the rear windows and, sometimes, the switch at the passenger's window also. Always check these items before troubleshooting a window problem.

5   These procedures are general in nature, so if you can't find the problem using them, take the vehicle to a dealer service department or other properly equipped repair facility.

6   If the power windows won't operate, always check the fuse and circuit breaker first.

7   If only the rear windows are inoperative, or if the windows only operate from the master control switch, check the rear window lockout switch for continuity in the unlocked position. Replace it if it doesn't have continuity.

8   Check the wiring between the switches and fuse panel for continuity. Repair the wiring, if necessary.

9   If only one window is inoperative from the master control switch, try the other con-

trol switch at the window. **Note:** *This doesn't apply to the driver's door window.*

10   If the same window works from one switch, but not the other, check the switch for continuity.

11   If the switch tests OK, check for a short or open in the circuit between the affected switch and the window motor.

12   If one window is inoperative from both switches, remove the switch panel from the affected door. Check for voltage at the switch and at the motor (refer to Chapter 11 for door panel removal) while the switch is operated.

13   If voltage is reaching the motor, disconnect the glass from the regulator (see Chapter 11). Move the window up and down by hand while checking for binding and damage. Also check for binding and damage to the regulator. If the regulator is not damaged and the window moves up and down smoothly, replace the motor. If there's binding or damage, lubricate, repair or replace parts, as necessary.

14   If voltage isn't reaching the motor, check the wiring in the circuit for continuity between the switches and the body control module, and between the body control module and the motors. You'll need to consult the wiring diagram at the end of this Chapter. If the circuit is equipped with a relay, check that the relay is grounded properly and receiving voltage.

15   Test the windows after you are done to confirm proper repairs.

---

24   Power door lock and keyless entry system - description and check

Note: *These models are equipped with a Body Control Module (BCM). Several systems are linked to a centralized control module that allows simple and accurate troubleshooting, but only with a professional-grade scan tool. The Body Control Module governs the door locks, the power windows, the ignition lock and security system, the interior lights, the Daytime Running Lights system, the horn, the windshield wipers, the heating/air conditioning*

system and the power mirrors. In the event of malfunction with this system, have the vehicle diagnosed by a dealership service department or other qualified automotive repair facility.

1   The power door lock system operates the door lock actuators mounted in each door. The system consists of the switches, actuators, Body Control Module (BCM) and associated wiring. Diagnosis can usually be limited to simple checks of the wiring connections and actuators for minor faults that can be easily repaired.

2   Power door lock systems are operated by bi-directional solenoids located in the doors. The lock switches have two operating positions: Lock and Unlock. These switches send a signal to the BCM, which in turn sends a signal to the door lock solenoids.

3   If you are unable to locate the trouble using the following general steps, consult your dealer service department.

4   Always check the circuit protection first. Some vehicles use a combination of circuit breakers and fuses. Refer to the wiring diagrams at the end of this Chapter.

5   Check for voltage at the switches. If no voltage is present, check the wiring between the fuse panel and the switches for shorts and opens.

6   If voltage is present, test the switch for continuity. Replace it if there's not continuity in both switch positions. To remove the switch, use a flat-bladed trim tool to pry out the door/window switch assembly (see Chapter 11).

7   If the switch has continuity, check the wiring between the switch and door lock solenoid.

8   If all but one lock solenoids operate, remove the trim panel from the affected door (see Chapter 11) and check for voltage at the solenoid while the lock switch is operated. One of the wires should have voltage in the Lock position; the other should have voltage in the Unlock position.

9   If the inoperative solenoid is receiving voltage, replace the solenoid.

10   If the inoperative solenoid isn't receiving voltage, check for an open or short in the wire between the lock solenoid and the relay. **Note:** *It's common for wires to break in the portion of the harness between the body and door (opening and closing the door fatigues and eventually breaks the wires).*

11   On the models covered by this manual, power door lock system communication goes through the Body Control Module. If the above tests do not pinpoint a problem, take the vehicle to a dealer or qualified shop with the proper scan tool to retrieve trouble codes from the BCM.

Keyless entry system

12   The keyless entry system consists of a remote control transmitter that sends a coded infrared signal to a receiver, which then operates the door lock system.

13   Replace the battery when the transmitter doesn't operate the locks at a distance of ten feet. Normal range should be about 30 feet.

Key remote control battery replacement

Refer to illustrations 24.14 and 24.15

14 Use a coin to carefully separate the case halves **(see illustration)**.

15 Replace the battery **(see illustration)**.

16 Snap the case halves together.

Transmitter programming

17 Programming replacement transmitters requires the use of a specialized scan tool. Take the vehicle and the transmitter(s) to a dealer service department or other qualified repair shop equipped with the necessary tool to have the transmitter(s) programmed to the vehicle.

25 Electric side view mirrors - description

Note: *These models are equipped with a Body Control Module (BCM). Several systems are linked to a centralized control module that allows simple and accurate troubleshooting, but only with a professional-grade scan tool. The Body Control Module governs the door locks, the power windows, the ignition lock and security system, the interior lights, the Daytime Running Lights system, the horn, the windshield wipers, the heating/air conditioning system and the power mirrors. In the event of malfunction with this system, have the vehicle diagnosed by a dealership service department or other qualified automotive repair facility.*

1 The electric rear view mirrors use two motors to move the glass; one for up and down adjustments and one for left-right adjustments.

2 The control switch has a selector portion which sends voltage to the left or right side mirror. With the ignition in the ACC position and the engine OFF, roll down the windows and operate the mirror control switch through all functions (left-right and up-down) for both the left and right side mirrors.

3 Listen carefully for the sound of the electric motors running in the mirrors.

4 If the motors can be heard but the mirror glass doesn't move, there's probably a problem with the drive mechanism inside the mirror. Power mirrors have no user-serviceable parts inside - a defective mirror must be replaced as a unit (see Chapter 11).

5 If the mirrors don't operate and no sound comes from the mirrors, check the fuses (see Section 3).

6 If the fuses are OK, remove the mirror control switch. Have the switch continuity checked by a dealer service department or other qualified shop.

7 Check the ground connections.

8 If the mirror still doesn't work, remove the mirror and check the wires at the mirror for voltage.

9 If there's not voltage in each switch position, check the circuit between the mirror and control switch for opens and shorts.

10 If there's voltage, remove the mirror and test it off the vehicle with jumper wires. Replace the mirror if it fails this test.

26 Power seats - description

1 These models feature a six-way seat that goes forward and backward, up and down and tilts forward and backward. The seats are powered by three reversible motors, mounted in one housing, that are controlled by switches on the side of the seat. Each switch changes the direction of seat travel by reversing polarity to the drive motor.

2 Diagnosis is usually a simple matter, using the following procedures.

3 Look under the seat for any object which may be preventing the seat from moving.

4 If the seat won't work at all, check the circuit breaker. See Section 4 for circuit breaker testing.

5 With the engine off to reduce the noise level, operate the seat controls in all directions and listen for sound coming from the seat motors.

6 If the motors make noise or don't work, check for voltage at the motors while an assistant operates the switch. With the door open, try the seat switch again. If the dome light dims while trying to operate the seat, this indicates something may be jammed in the seat tracks.

7 If the motor is getting voltage but doesn't run, test it off the vehicle with jumper wires. If it still doesn't work, replace it.

8 If the motor isn't getting voltage, remove the switch and check for voltage. If there's no voltage to the switch, check the wiring between the fuse block and the switch. If there's battery voltage at the switch, check the other terminals for voltage while moving the switch around. If the switch is OK, check for a short or open in the wiring between switch and motor.

9 Test the completed repairs.

27 Data Link Communication system - description

1 The vehicles covered by this manual have a complex electrical system, encompassing many power accessories, and a number of separate electronic modules.

2 The Powertrain Control Module (PCM) is mainly responsible for engine and transaxle control, but also communicates with other modules around the vehicle through a Data Link Communication system, which sends serial port data very quickly between the various modules. Many of the computer functions involved in the operation of body systems are routed through the Body Control Module (BCM), which communicates with the PCM.

3 Among the modules in the Data Link system besides the BCM and PCM are the Sensing Diagnostic Module (airbag system), the Electronic Brake Control Module and the instrument panel cluster. The BCM further communicates with various body subsystems.

4 All of the modules in the vehicle have associated trouble codes. When other troubleshooting procedures fail to pinpoint the problem, check the wiring diagrams at the end of this Chapter to see if the BCM or PCM are involved in the circuit. If so, bring your vehicle to a dealer service department or other qualified repair shop with the proper diagnostic tools to extract the trouble codes.

28 Airbag system - general information

1 All models are equipped with a Supplemental Restraint System (SRS), more commonly known as the airbag system. The airbag system is designed to protect the driver and the front seat passenger from serious injury in the event of a head-on or frontal collision. It consists of the impact sensors, a driver's airbag module in the center of the steering wheel, a passenger's airbag module in the glove box area of the instrument panel and a sensing/diagnostic module mounted under the center console. Some models are also equipped with side-curtain airbags and seat belt pre-tensioners.

Airbag modules

Driver's airbag

2 The airbag inflator module contains a housing incorporating the cushion (airbag) and inflator unit, mounted in the center of the steering wheel. The inflator assembly is mounted on the back of the housing over a hole through which gas is expelled, inflating the bag almost instantaneously when an electrical signal is sent from the system. A spiral cable (or clockspring) assembly on the steering column under the steering wheel carries this signal to the module. This clockspring can transmit an electrical signal regardless of steering wheel position.

Passenger's airbag

3 The airbag is mounted inside the right side of the instrument panel, in the area above the glove box. It's similar in design to the driver's airbag, except that it's larger than the steering wheel unit. The trim cover (on the side of the instrument panel that faces toward the passenger) is textured and colored to match the instrument panel and has a molded seam that splits open when the bag inflates.

Side curtain airbags

4 In addition to the side-impact airbags, extra side-impact protection is also provided by side-curtain airbags on some models. These are long airbags that, in the event of a side impact, come out of the headliner at each side of the car and come down between the side windows and the seats. They are designed to protect the heads of both front seat and rear seat passengers.

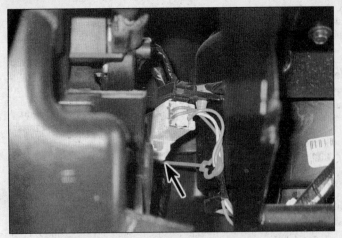

28.20 Pull out the connector lock, then disconnect the passenger's airbag connector behind the glove box area

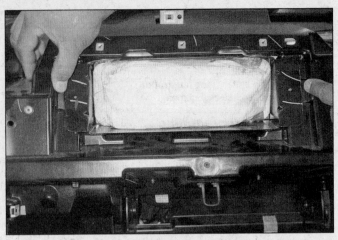

28.22 Separate the airbag module retaining tabs and remove the airbag

Sensing and diagnostic module

5   The sensing and diagnostic module supplies the current to the airbag system in the event of the collision, even if battery power is cut off. It checks this system every time the vehicle is started, causing the "AIR BAG" light to go on then off, if the system is operating properly. If there is a fault in the system, the light will go on and stay on, flash, or the dash will make a beeping sound. If this happens, the vehicle should be taken to your dealer immediately for service. This module is mounted under the center console. There is also a roll-over sensor located directly behind it on later models.

Seat belt pre-tensioners

6   Some models are equipped with pyrotechnic (explosive) units in the front seat belt retracting mechanisms. During an impact that would trigger the airbag system, the airbag control unit also triggers the seat belt retractors. When the pyrotechnic charges go off, they accelerate the retractors to instantly take up any slack in the seat belt system to more fully prepare the driver and front seat passenger for impact.
7   The airbag system should be disabled any time work is done to or around the seats. **Warning:** *Never strike the pillars or floorpan with a hammer or use an impact-driver tool in these areas unless the system is disabled.*

Disarming the system and other precautions

Warning: *Failure to follow these precautions could result in accidental deployment of the airbag and personal injury.*
8   Whenever working in the vicinity of the steering wheel, steering column or any of the other SRS system components, the system must be disarmed.

To disarm the airbag system:

9   Point the wheels straight ahead and turn the key to the Lock position.
10   Disconnect the cable from the negative battery terminal.
11   Wait at least two minutes for the back-up power supply to be depleted.

Whenever handling an airbag module:

12   Always keep the airbag opening (the trim side) pointed away from your body. Never place the airbag module on a bench or other surface with the airbag opening facing the surface. Always place the airbag module in a safe location with the airbag opening facing up.
13   Never measure the resistance of any SRS component. An ohmmeter has a built-in battery supply that could accidentally deploy the airbag.
14   Never use electrical welding equipment on a vehicle equipped with an airbag without first disconnecting the electrical connector for each airbag.
15   Never dispose of a live airbag module. Return it to a dealer service department or other qualified repair shop for safe deployment and disposal.

Component removal and installation

Driver's side airbag module and spiral cable

16   Refer to Chapter 10, *Steering wheel - removal and installation*, for the driver's side airbag module and clockspring removal and installation procedures.

Passenger's airbag module

Refer to illustrations 28.20 and 28.22
17   Disarm the airbag system as described earlier in this Section.
18   Remove the radio (see Section 12). Remove the glove box, the right side lower panel and the instrument panel upper trim panel (see Chapter 11).
19   Remove the air outlet duct from the pas-

senger's side of the instrument panel.
20   Disconnect the airbag module electrical connector **(see illustration)**.
21   Working through the radio and air outlet duct openings, remove the bolts holding the airbag module to the instrument panel reinforcement.
22   Disengage the airbag module from the instrument panel by spreading the retainers apart from each other **(see illustration)** and remove the module from the instrument panel.

Other airbag modules

23   We don't recommend removing any of the other airbag modules. These jobs are best left to a professional. Should you ever have to remove the instrument panel, you'll have to disconnect the electrical connector for the passenger airbag module and remove the fasteners that secure the passenger airbag module to the instrument panel reinforcement bracket.

---

29   Wiring diagrams - general information

Since it isn't possible to include all wiring diagrams for every year and model covered by this manual, the following diagrams are those that are typical and most commonly needed.

Prior to troubleshooting any circuits, check the fuse and circuit breakers (if equipped) to make sure they're in good condition. Make sure the battery is properly charged and check the cable connections (see Chapter 1).

When checking a circuit, make sure that all connectors are clean, with no broken or loose terminals. When disconnecting a connector, do not pull on the wires. Pull only on the connector housings themselves.

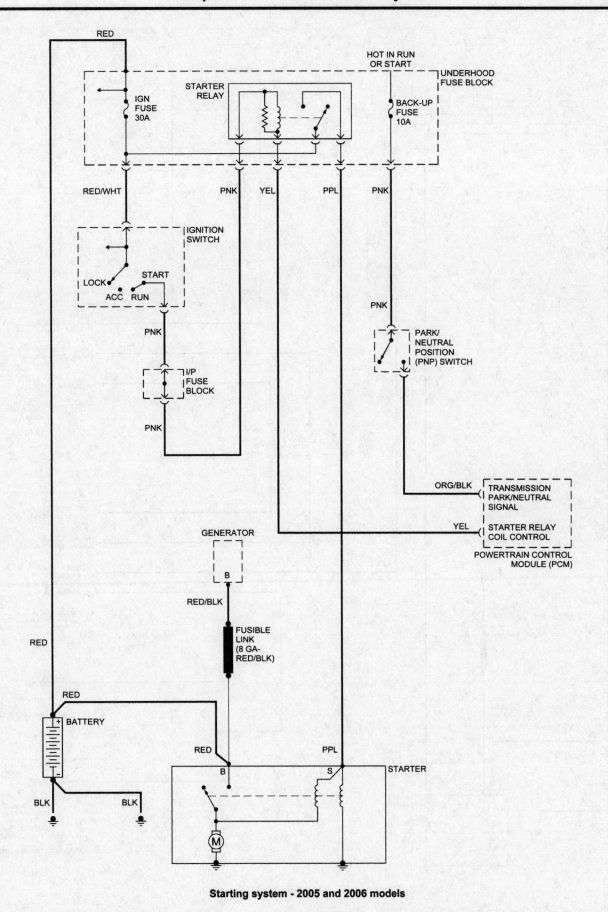

Starting system - 2005 and 2006 models

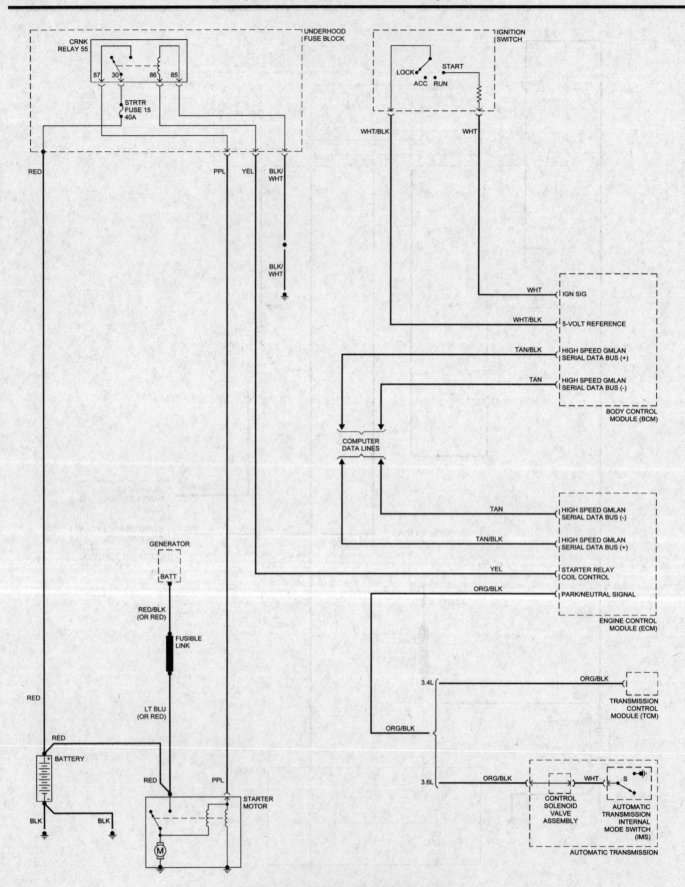

Starting system - 2007 through 2009 models

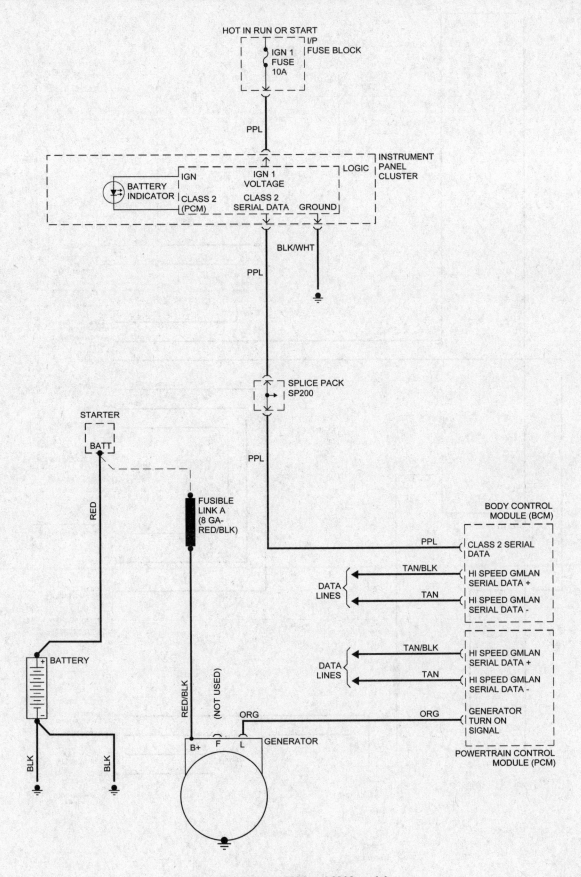

Charging system - 2005 and 2006 models

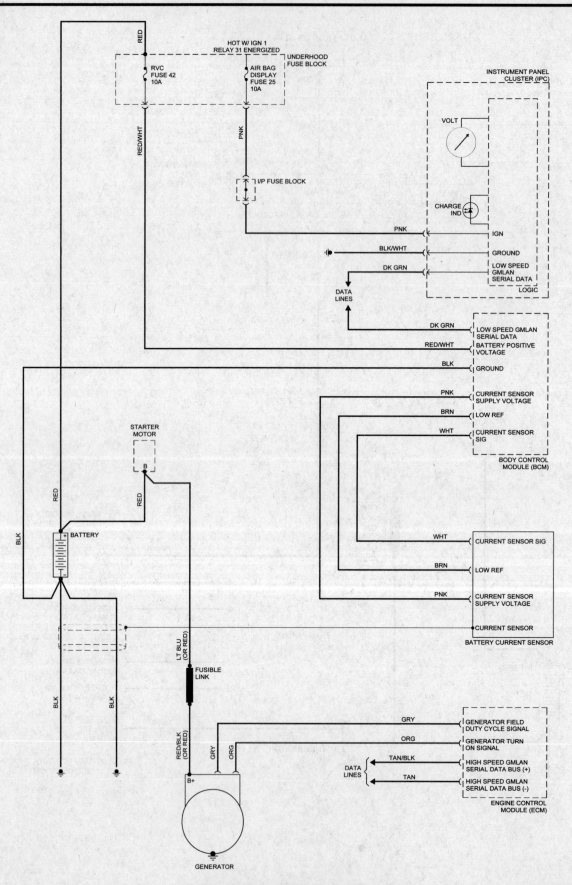

Charging system - 2007 through 2009 models

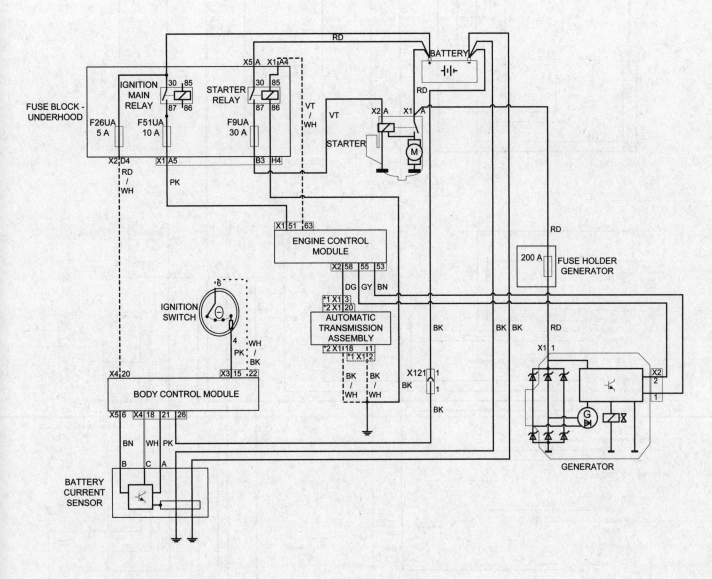

*1 Transmission code: MH7, MHC
*2 Transmission code: MH2, MH4

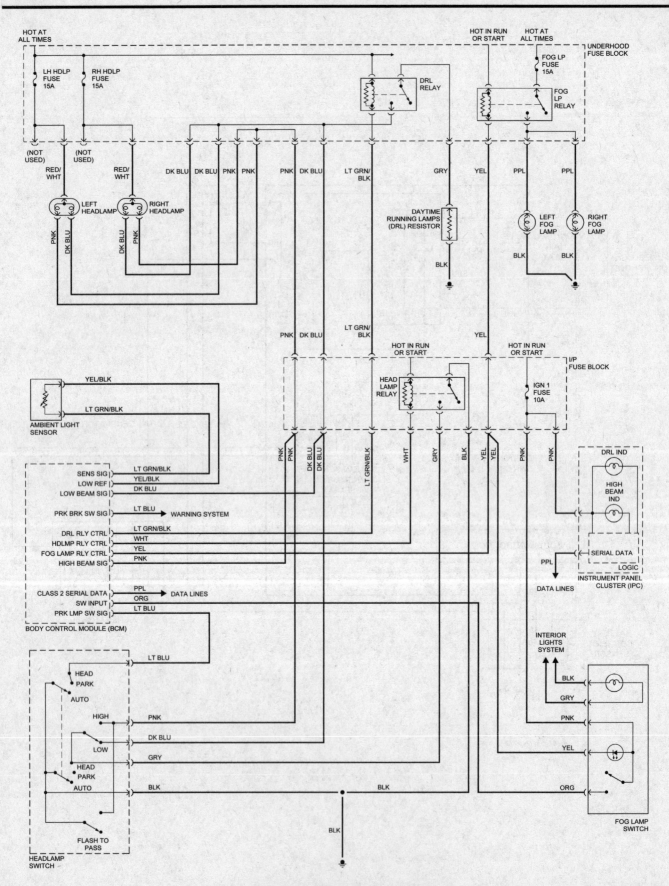

Headlight system - 2005 and 2006 models

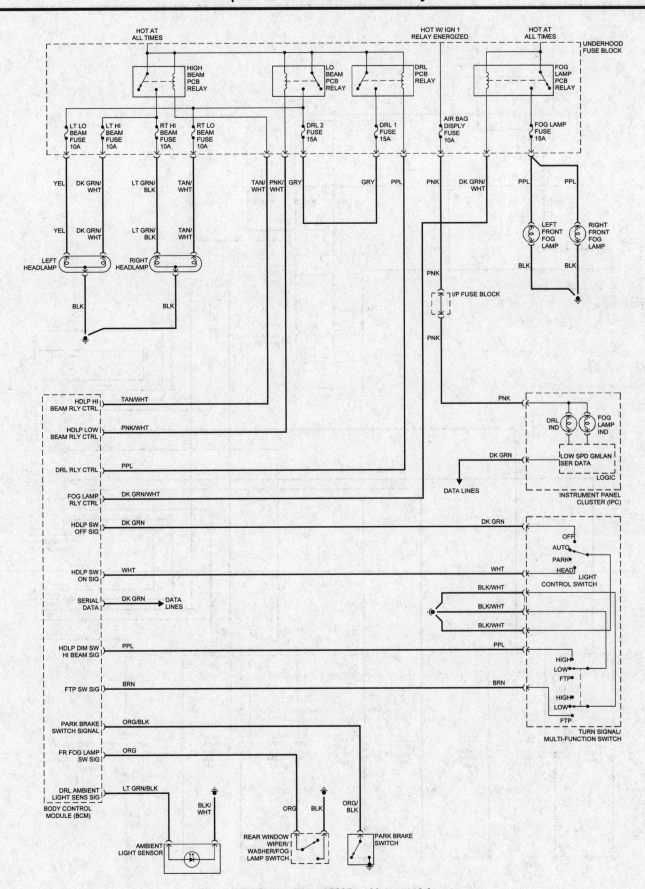

Headlight system - 2007 and later models

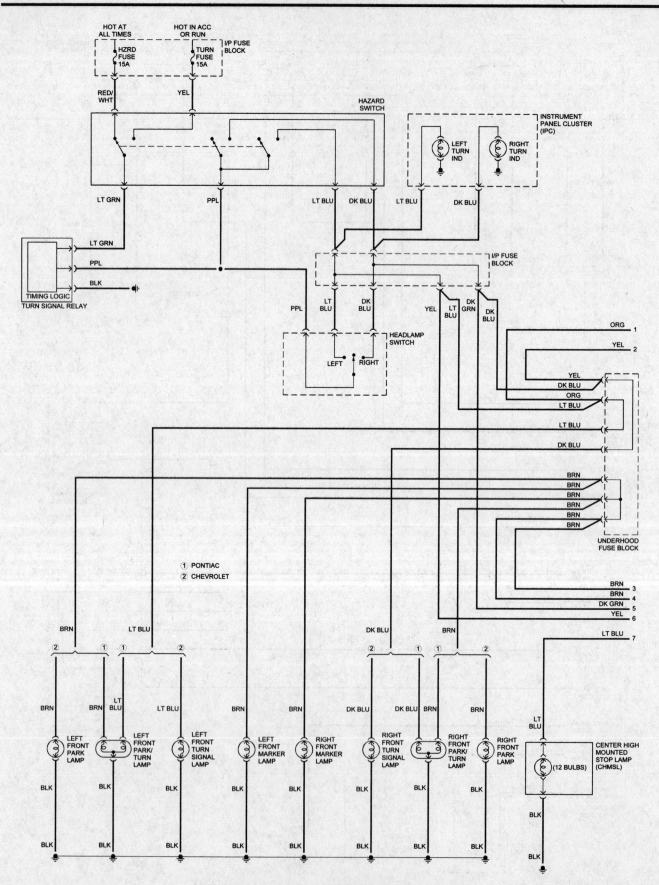

Exterior lighting system - 2005 and 2006 models (1 of 2)

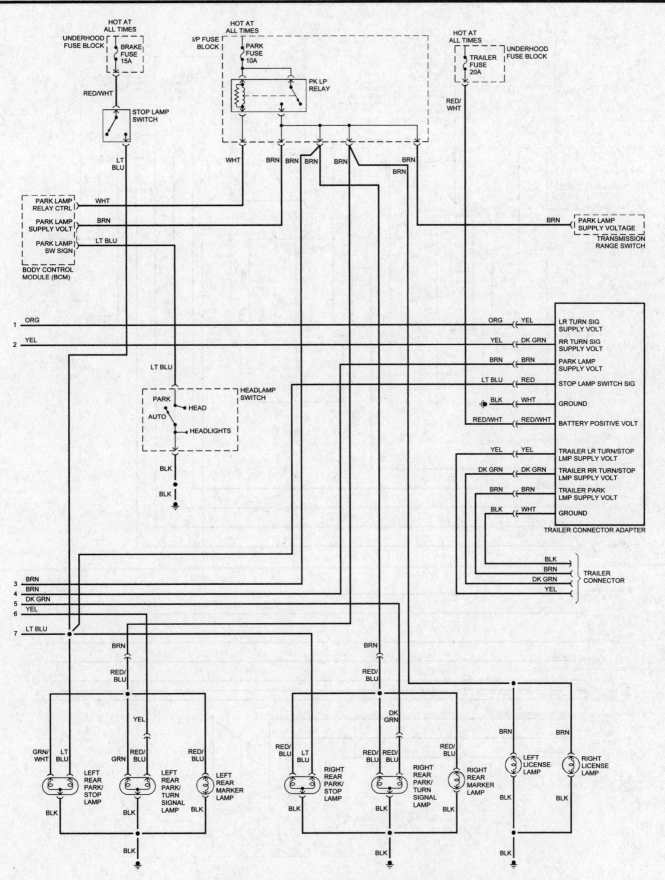

Exterior lighting system - 2005 and 2006 models (2 of 2)

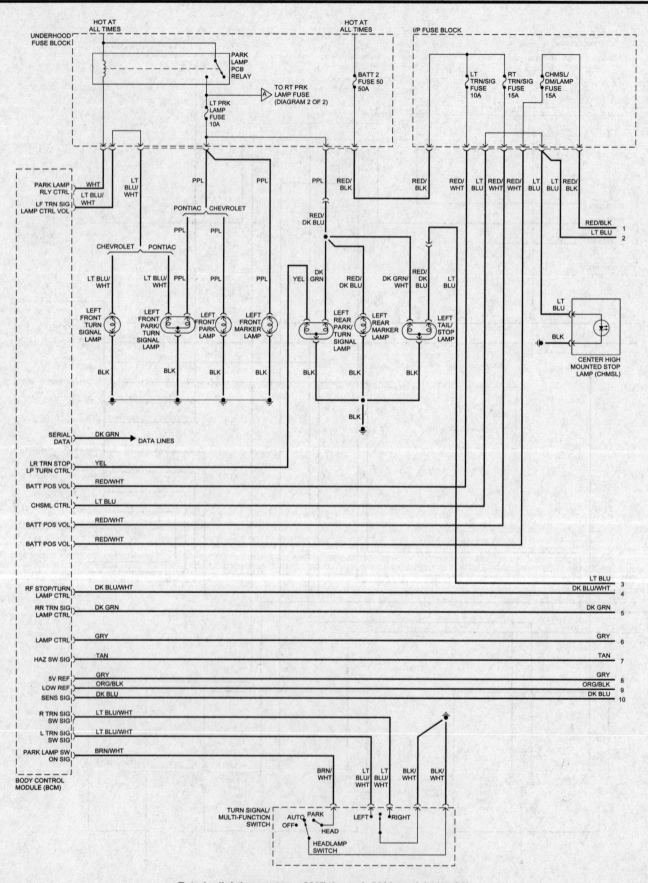

Exterior lighting system - 2007 through 2009 models (1 of 2)

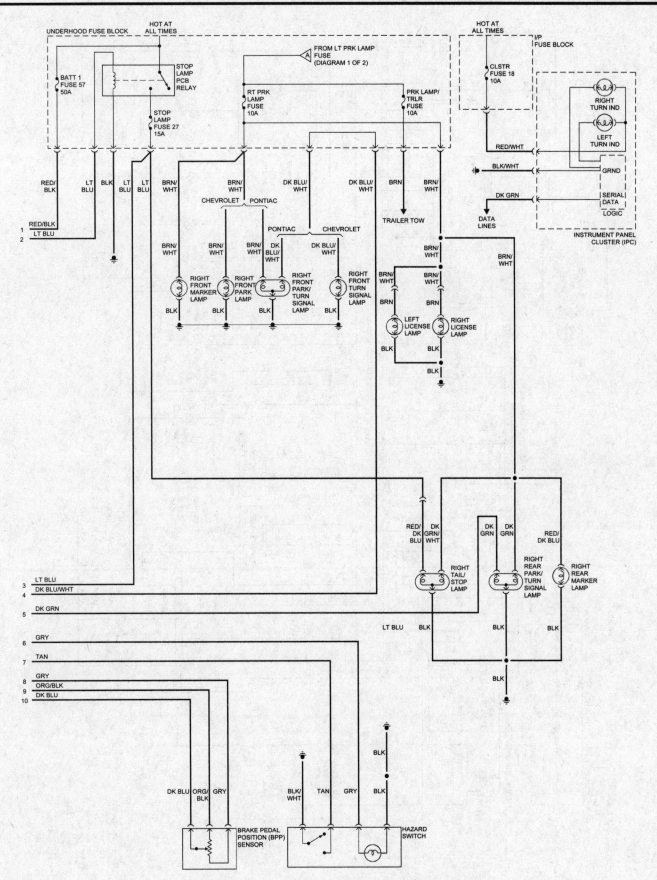

Exterior lighting system - 2007 through 2009 models (2 of 2)

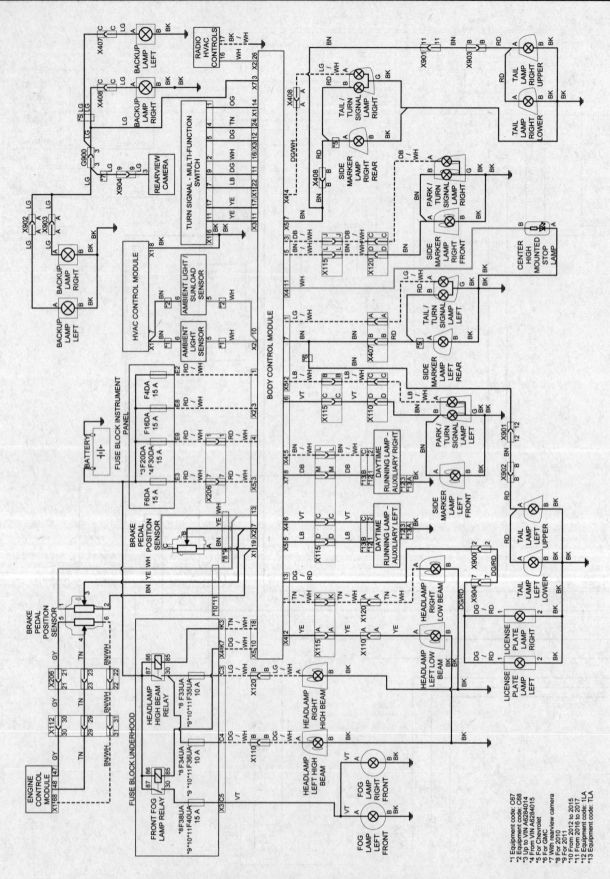

Headlights and exterior lighting system - 2010 and later models

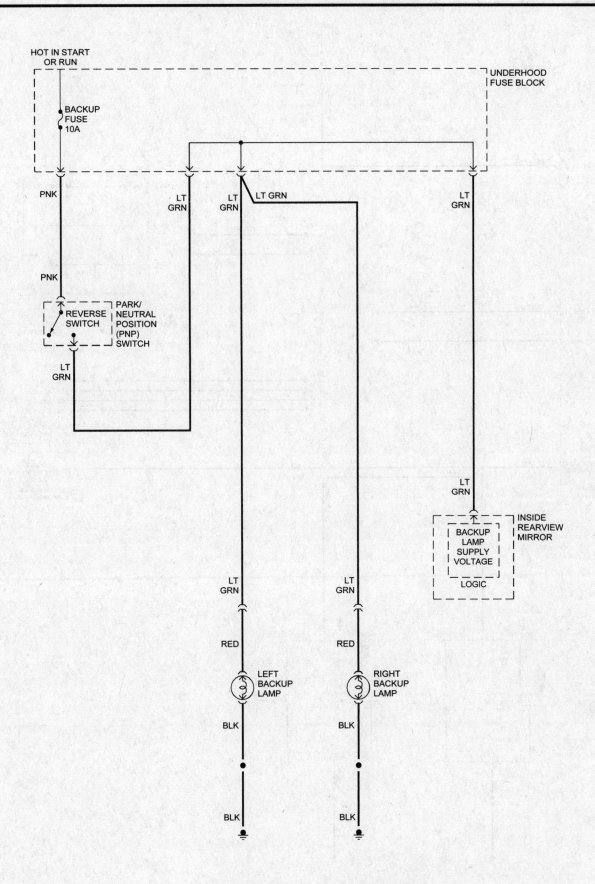

Back-up light system - 2005 and 2006 models

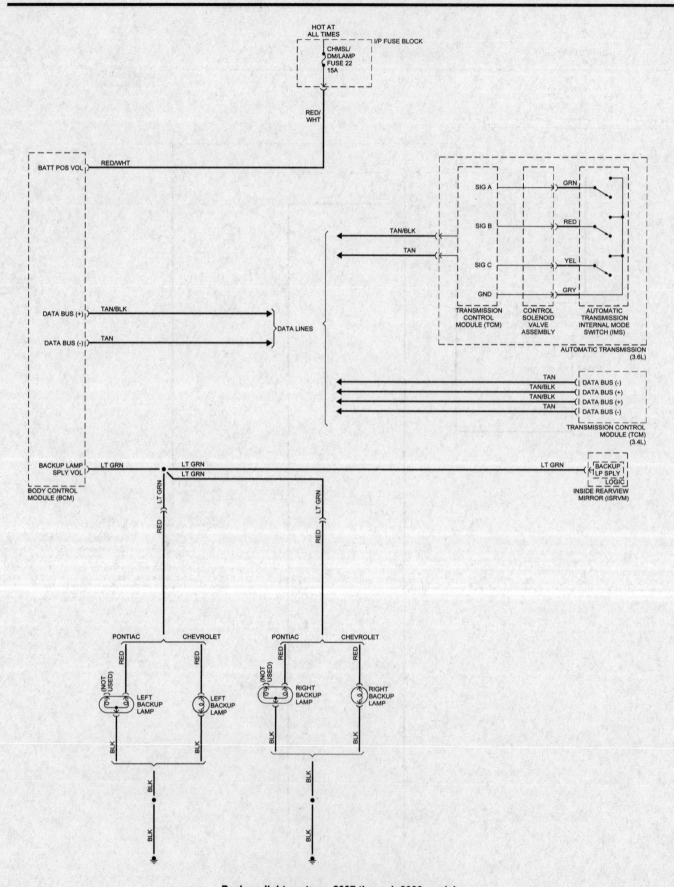

Back-up light system - 2007 through 2009 models

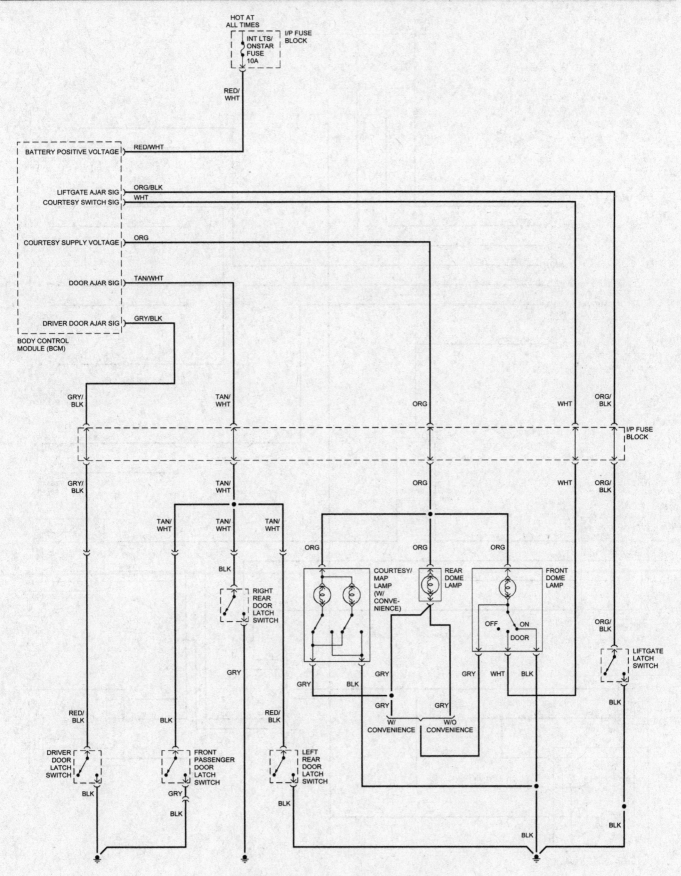

Courtesy lighting system - 2005 and 2006 models

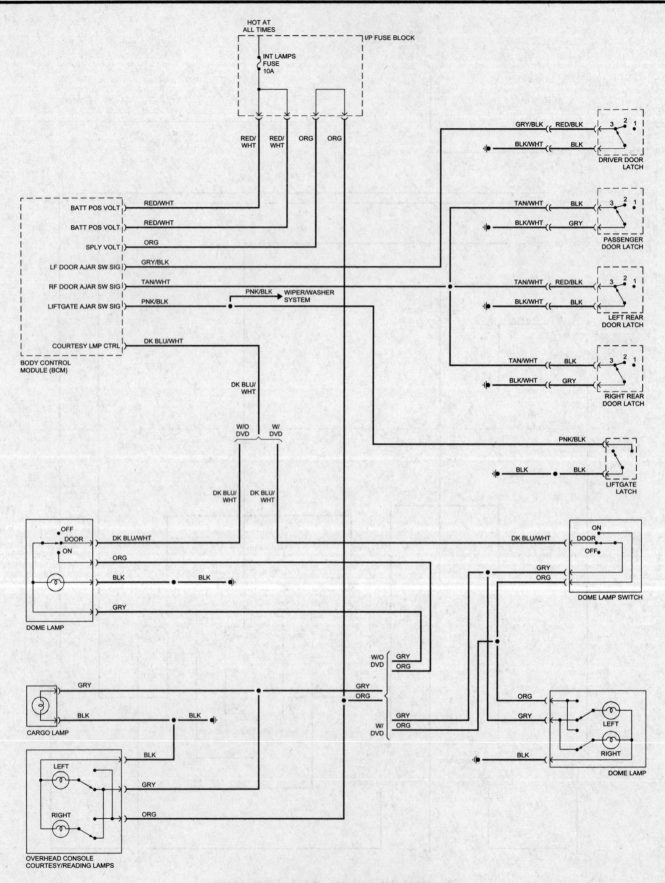

Courtesy lighting system - 2007 through 2009 models

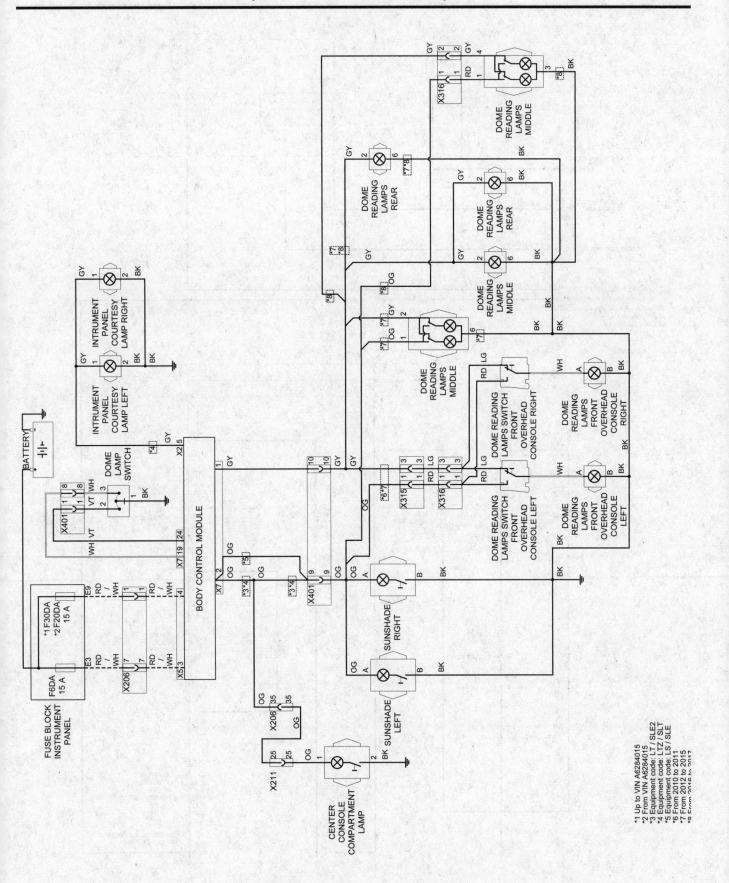

Interior lighting system - 2010 and later models

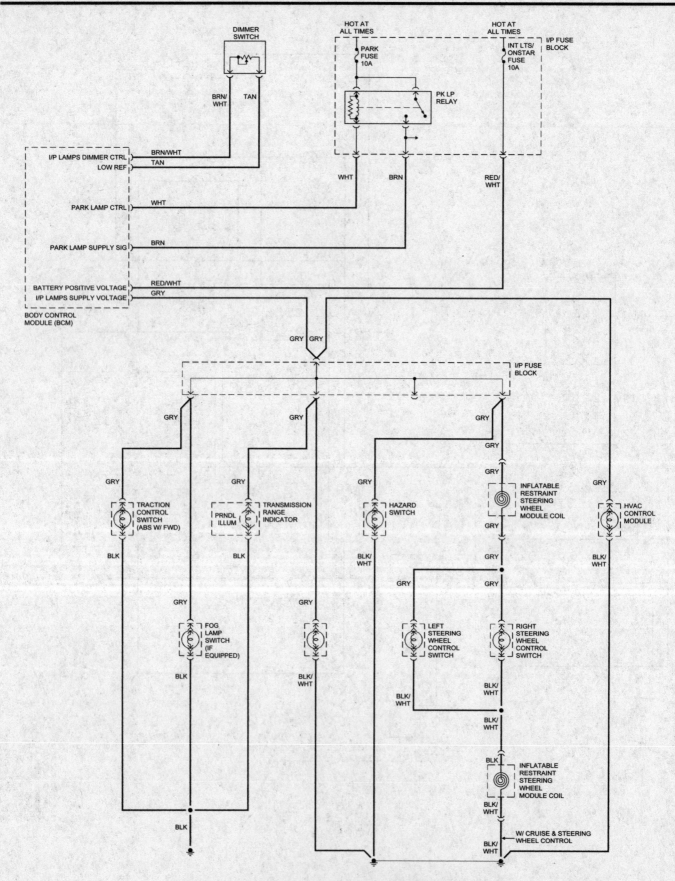

Instrument and switch illumination system - 2005 and 2006 models

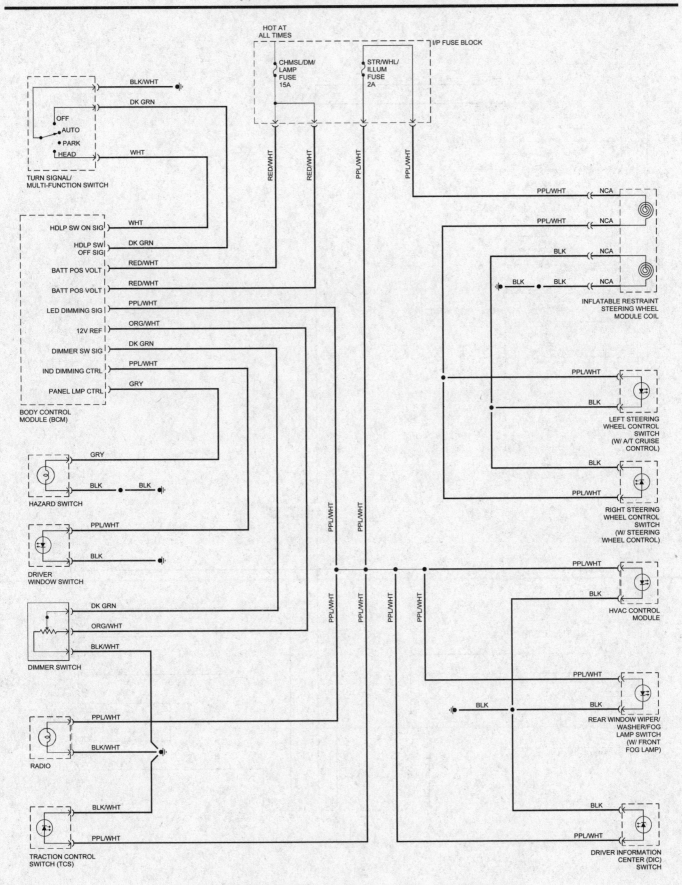

Instrument and switch illumination system - 2007 through 2009 models

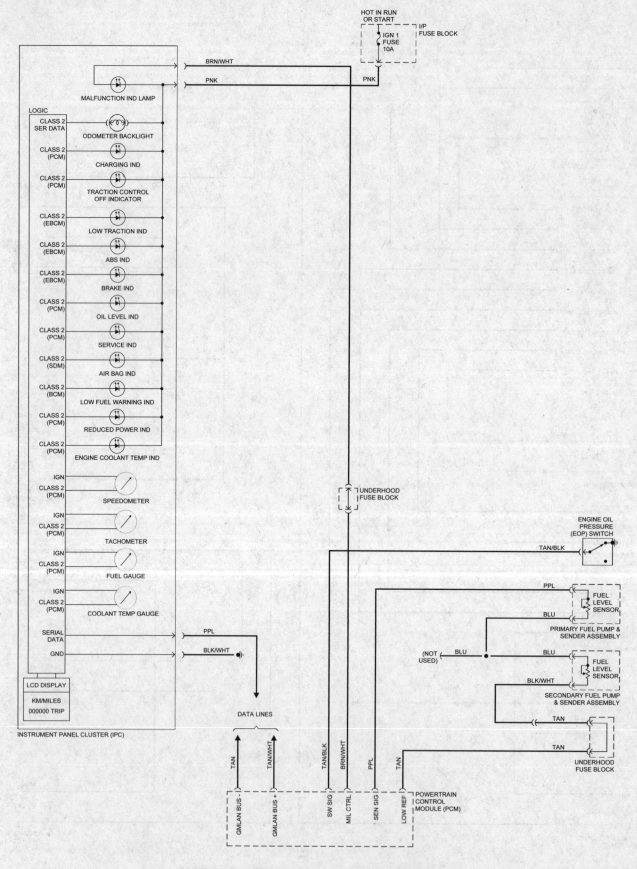

Gauges and warning lights system - 2005 and 2006 models

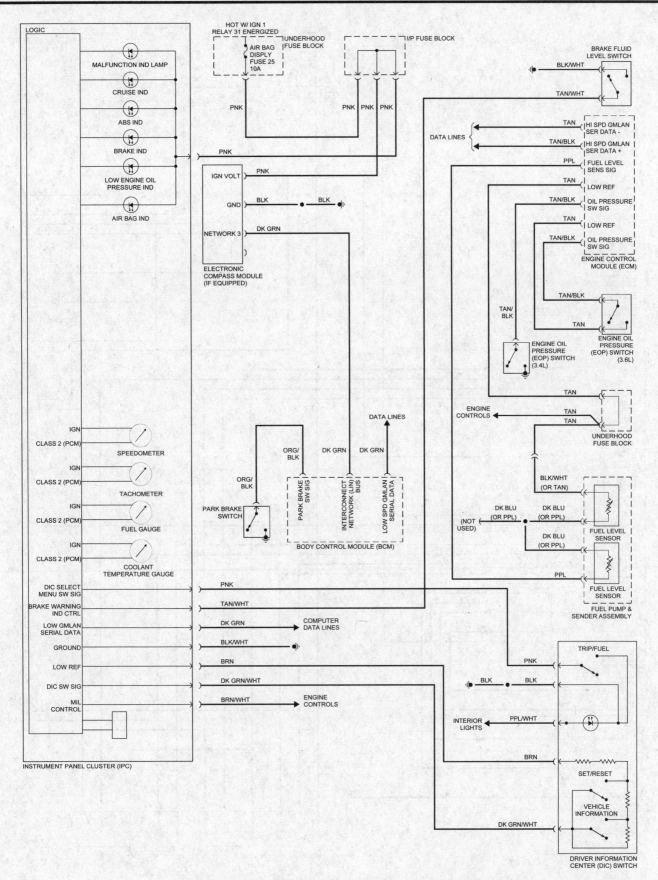

Gauges and warning lights system - 2007 through 2009 models

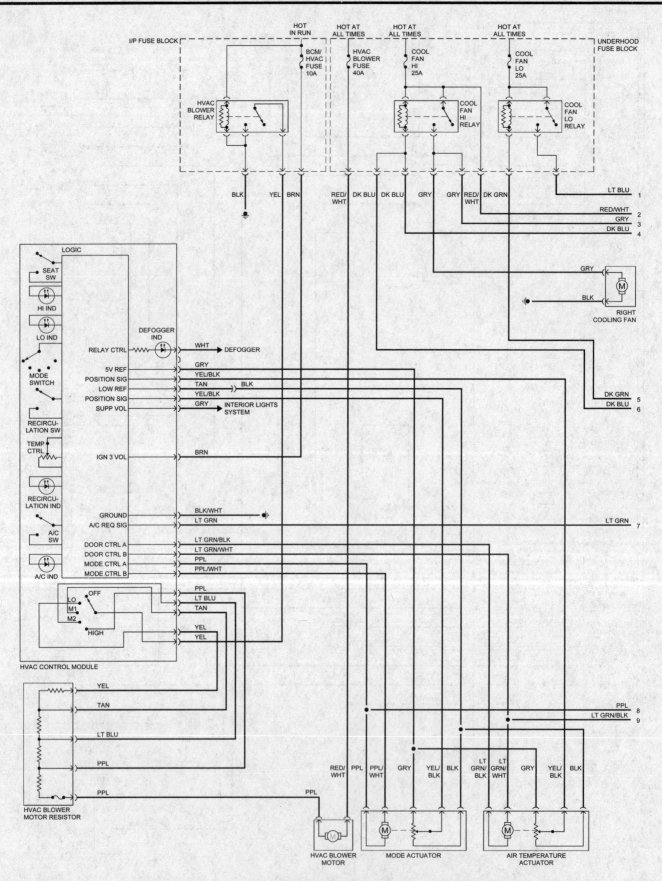

Air conditioning and engine cooling fans system - 2005 and 2006 models (1 of 2)

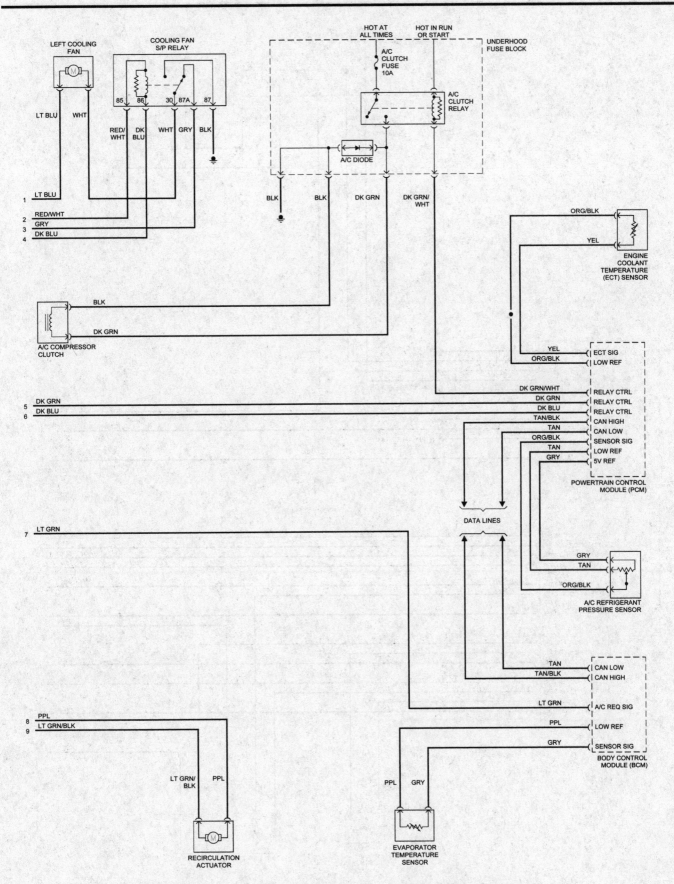

Air conditioning and engine cooling fans system - 2005 and 2006 models (2 of 2)

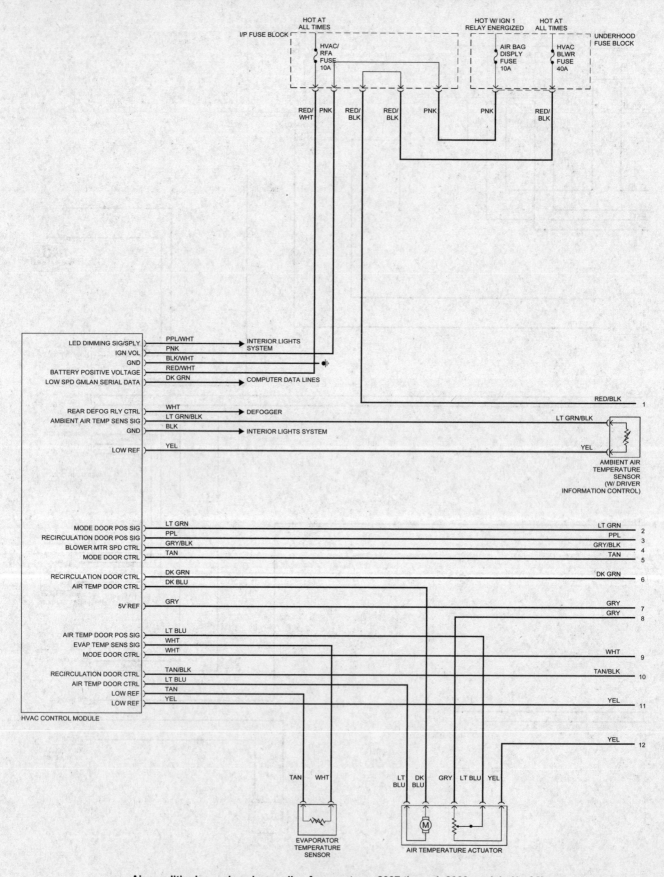

Air conditioning and engine cooling fans system - 2007 through 2009 models (1 of 2)

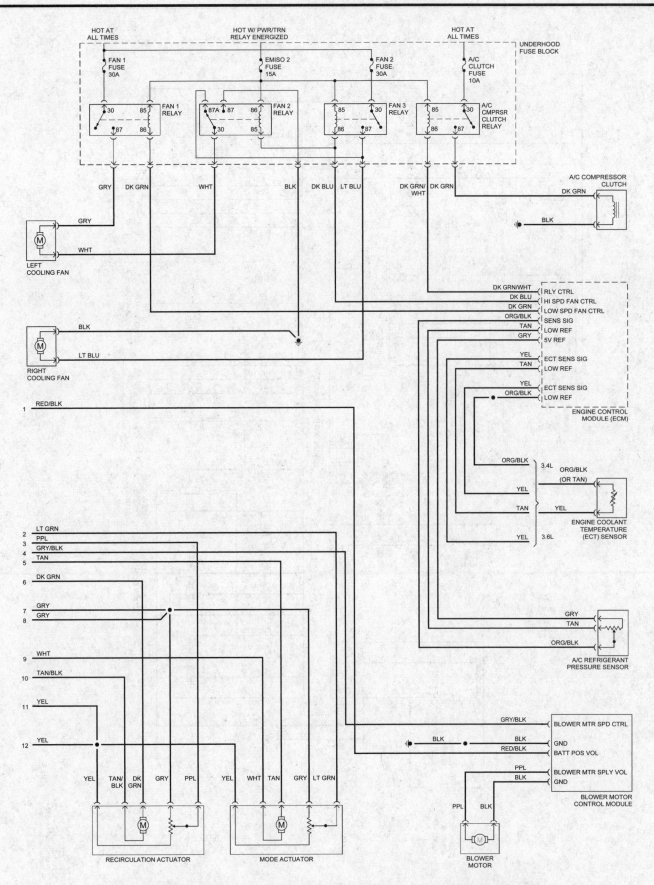

Air conditioning and engine cooling fans system - 2007 through 2009 models (2 of 2)

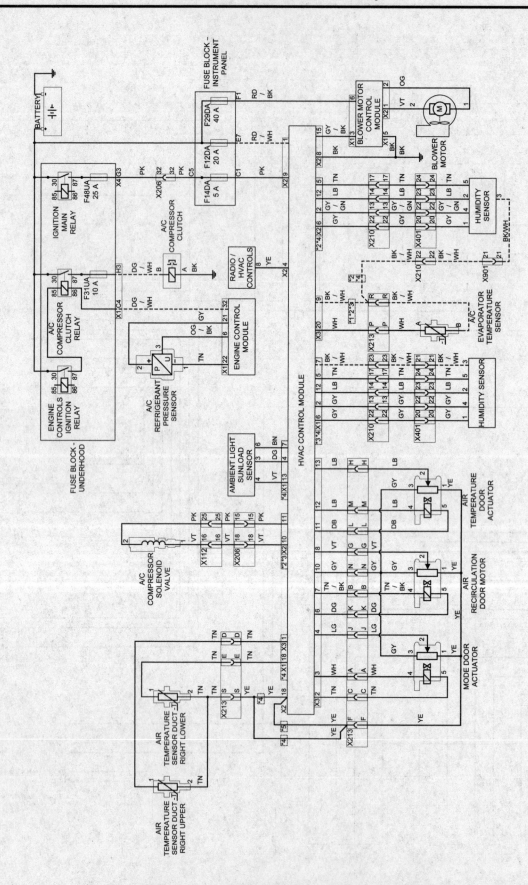

Heating and air conditioning system - 2010 and later models

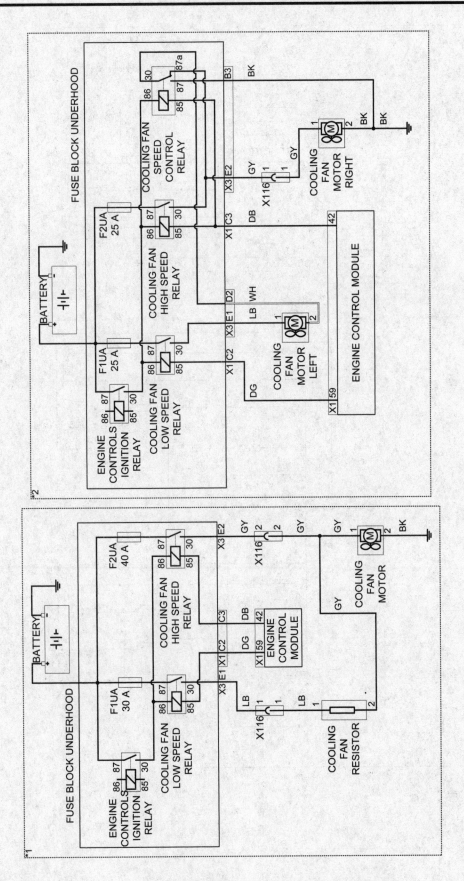

Engine cooling fan system - 2010 and later models

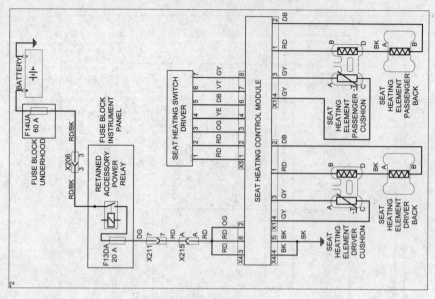

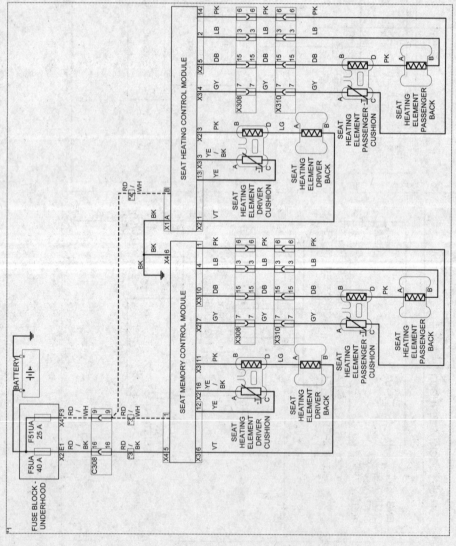

Seat heater system - 2010 and later models

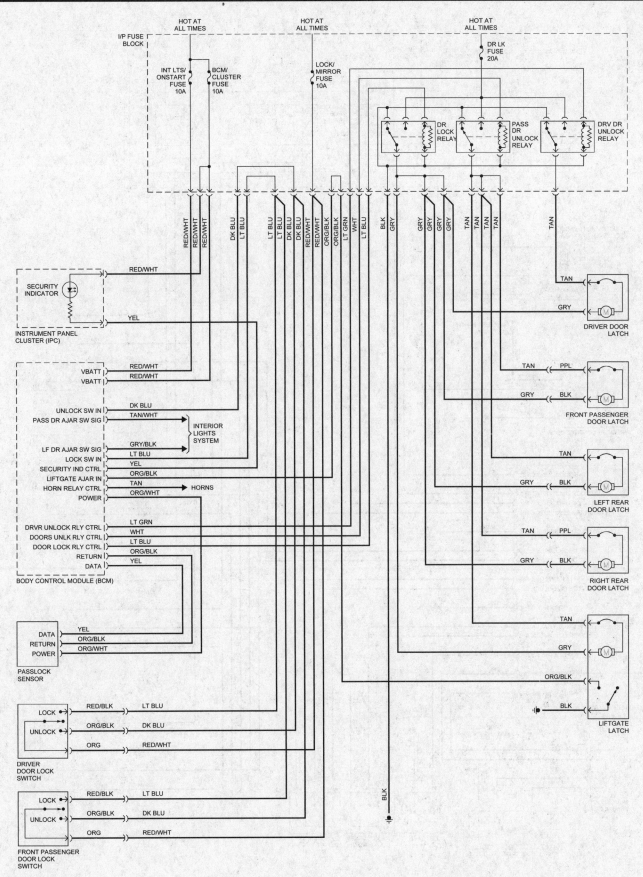

Power door locks system - 2005 and 2006 models

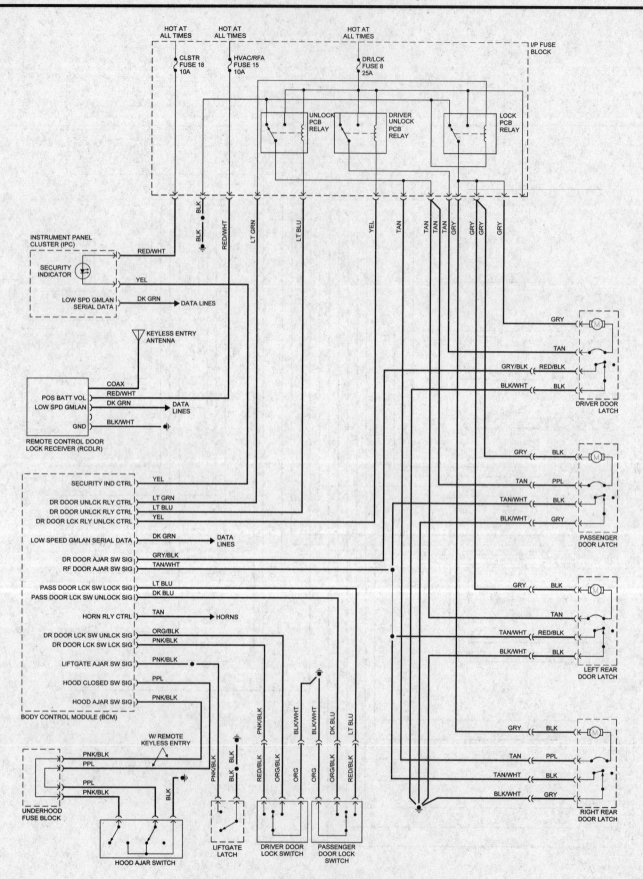

Power door locks system - 2007 through 2009 models

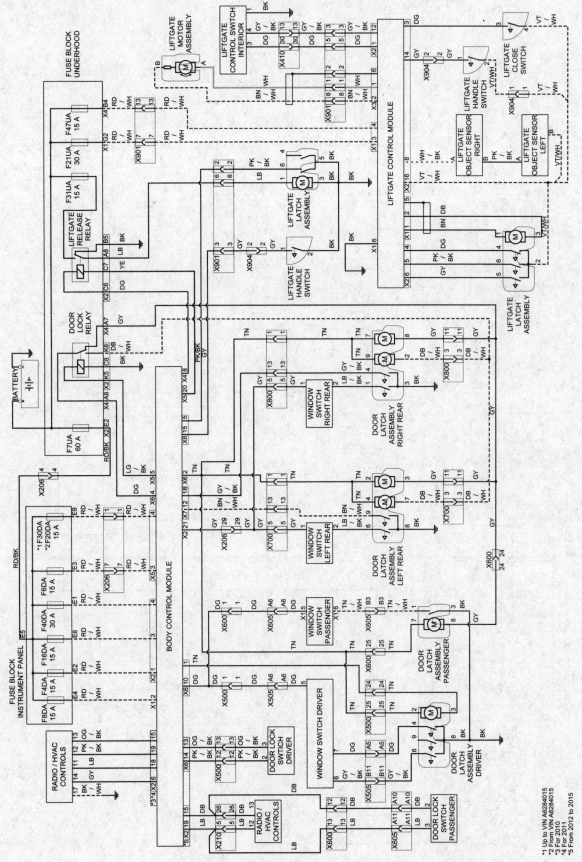

Power door locks system - 2010 through 2015 models

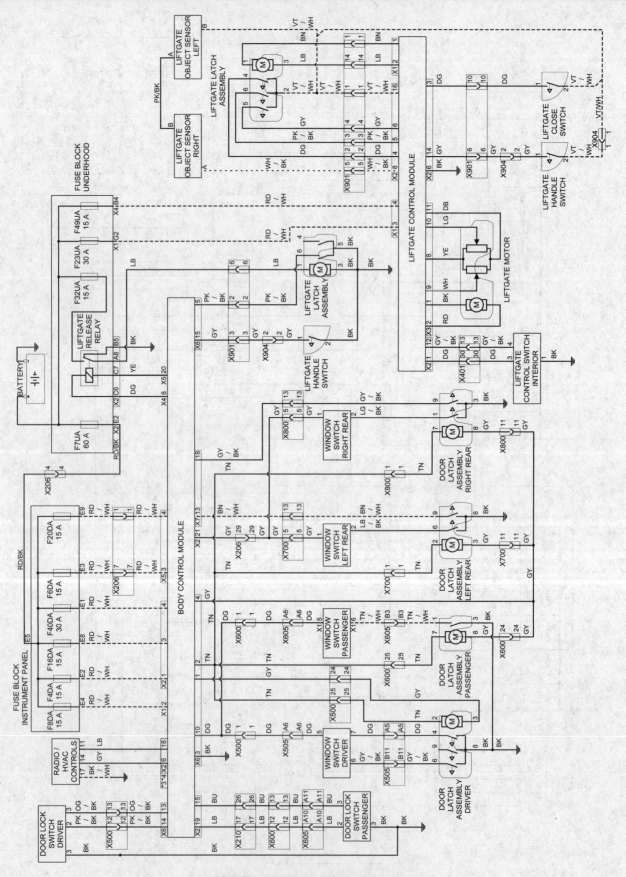

Power door locks system - 2016 and later models

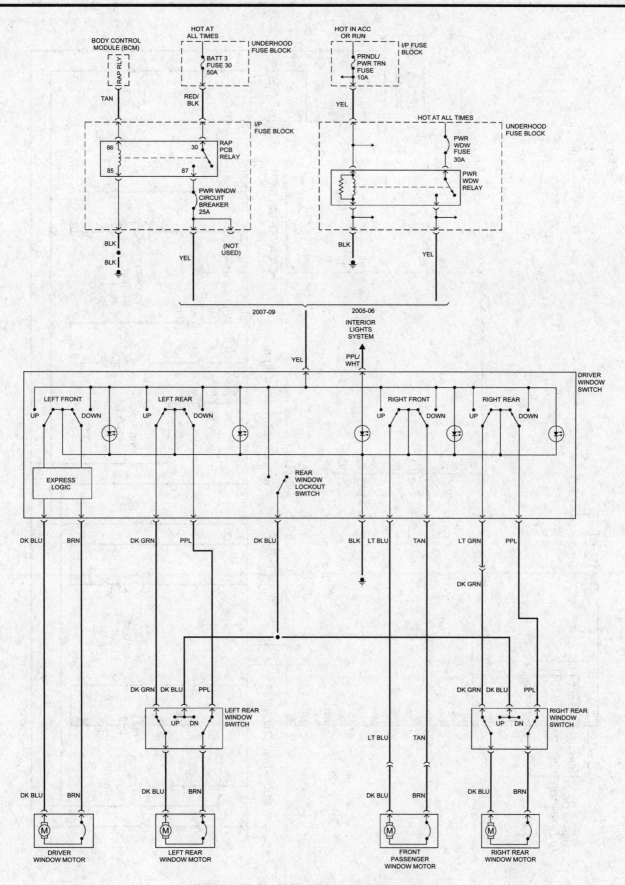

Power windows system - 2009 and earlier models

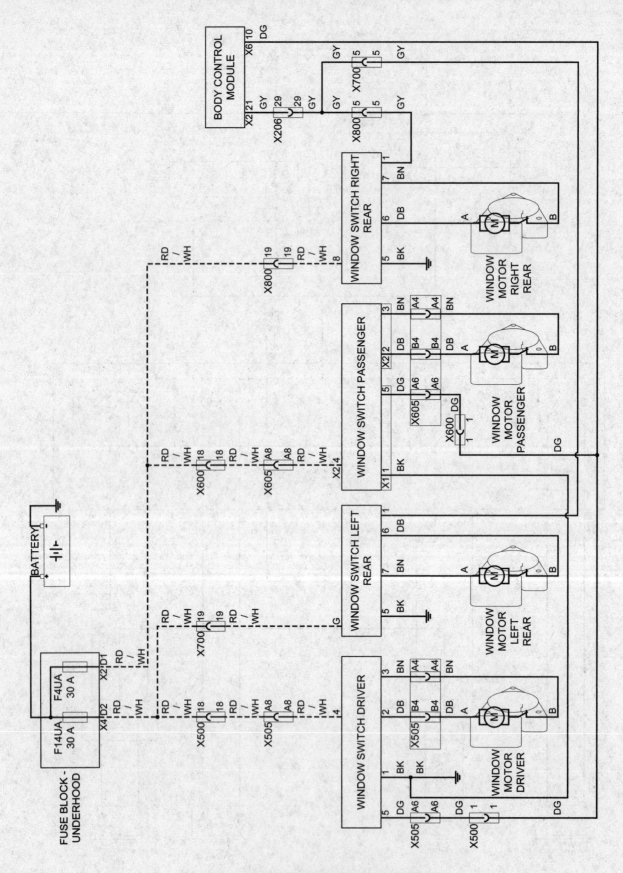

Power windows system - 2010 and later models

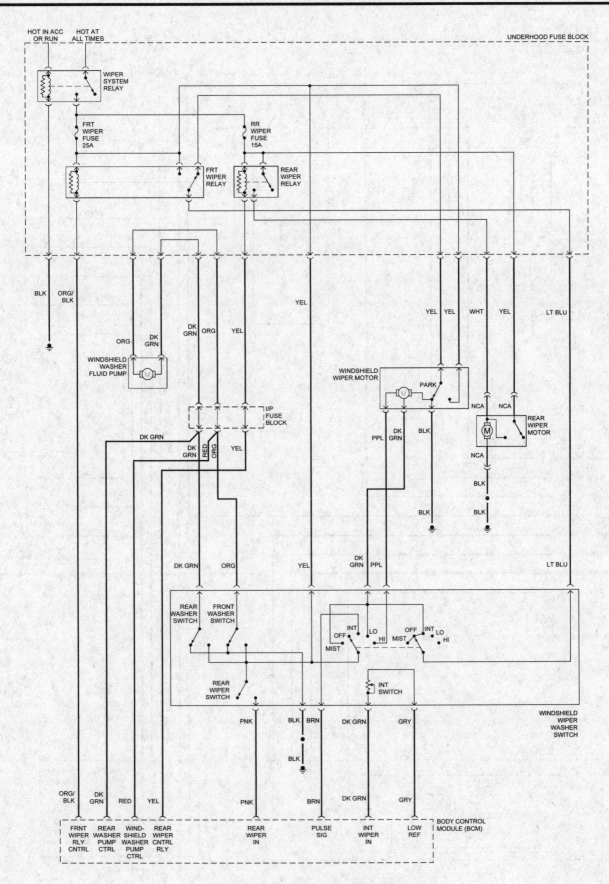

Wiper/washer system - 2005 and 2006 models

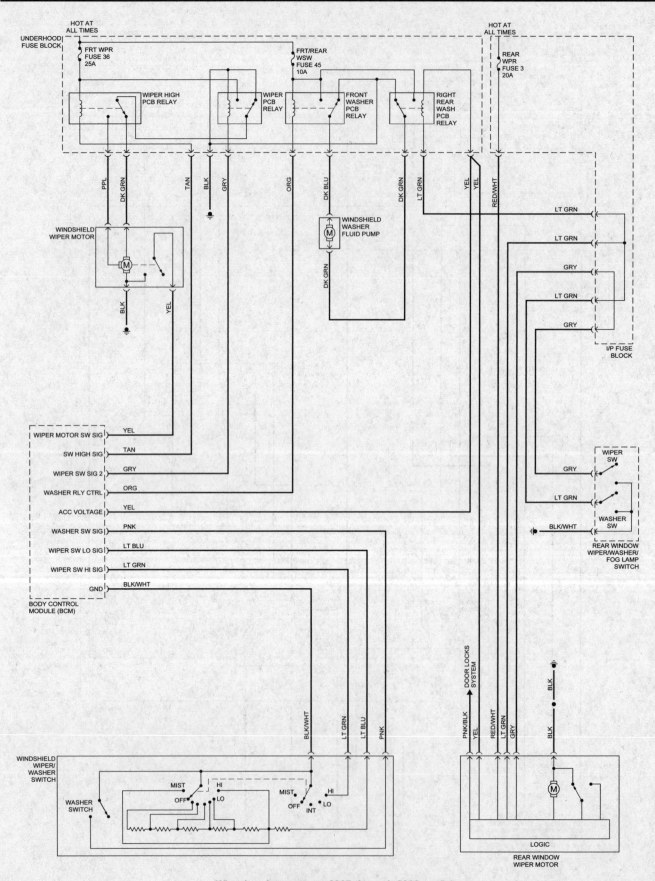

Wiper/washer system - 2007 through 2009 models

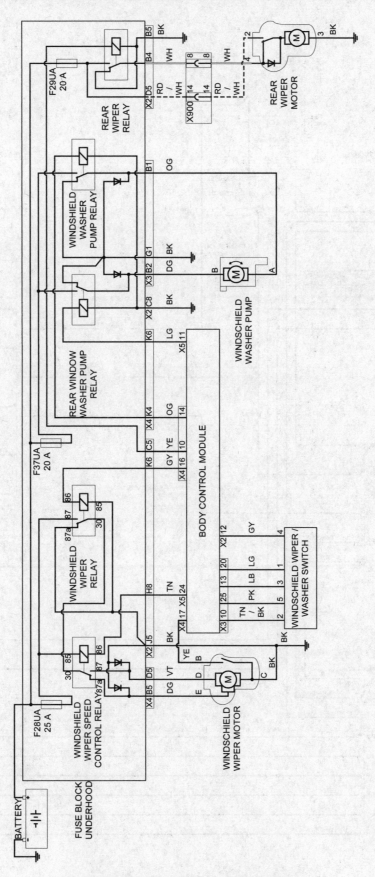

Wiper/washer system - 2010 and later models

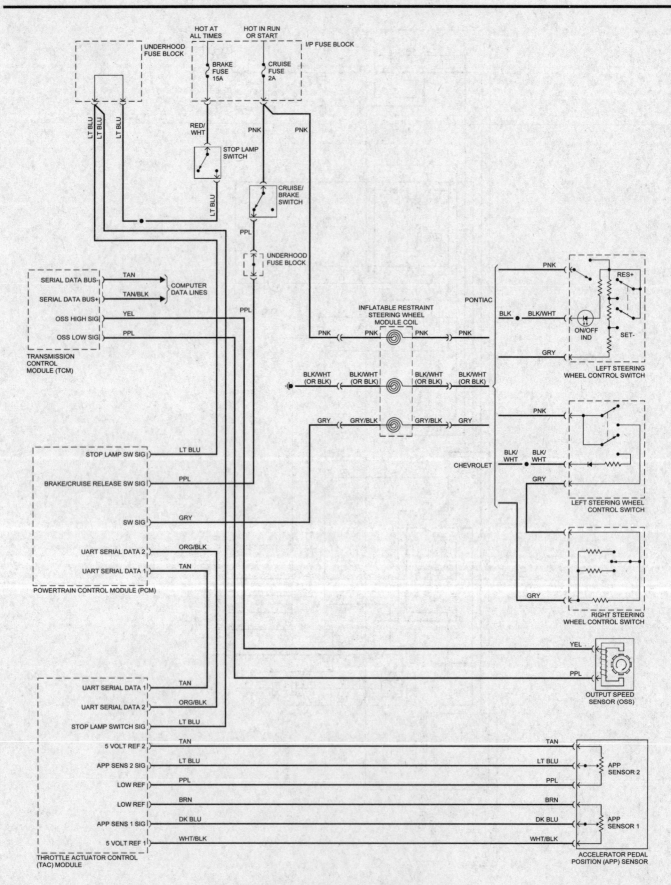

Cruise control system - 2005 and 2006 models

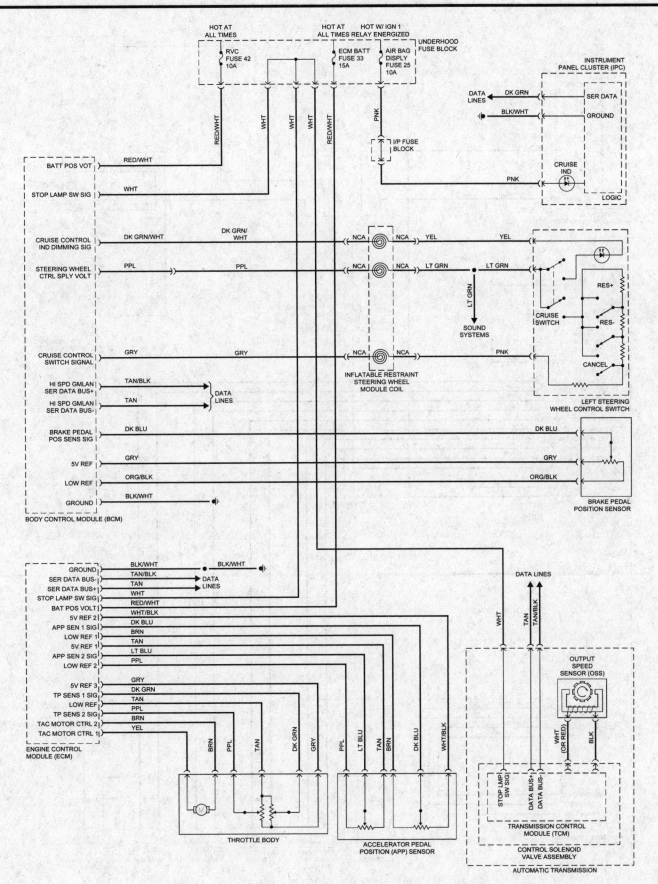

Cruise control system - 2007 through 2009 (later models similar)

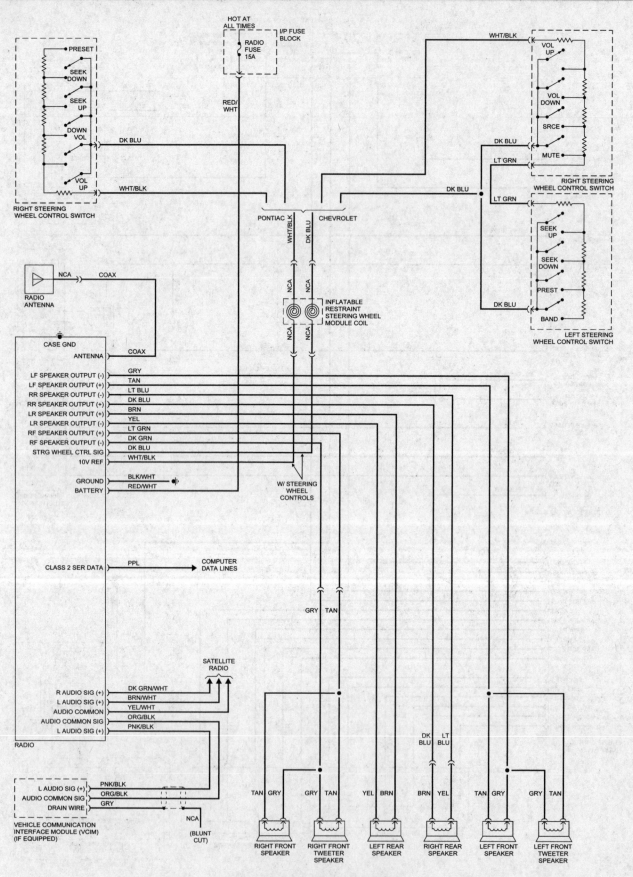

Audio system - 2005 and 2006 models

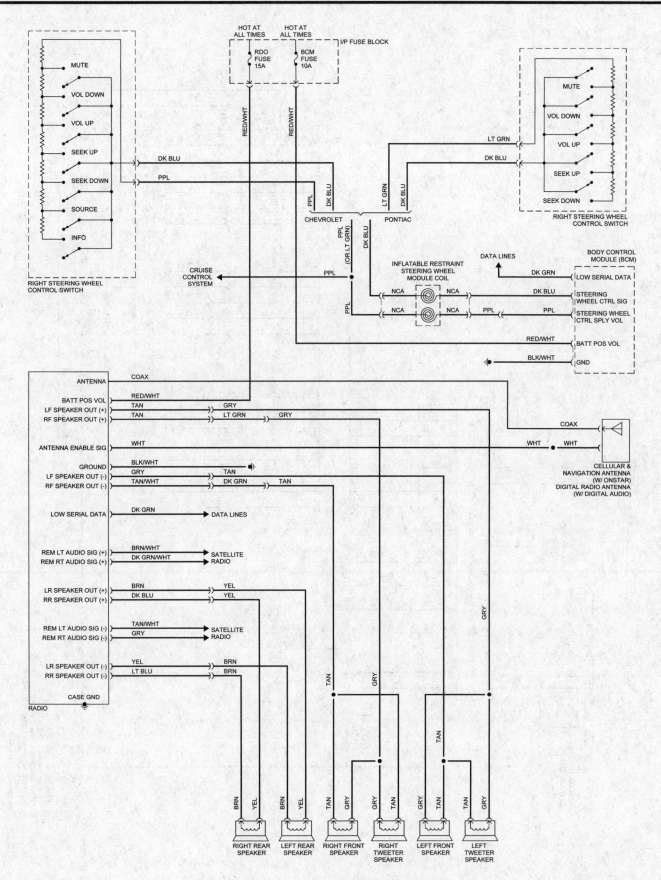

Audio system - 2007 through 2009 models

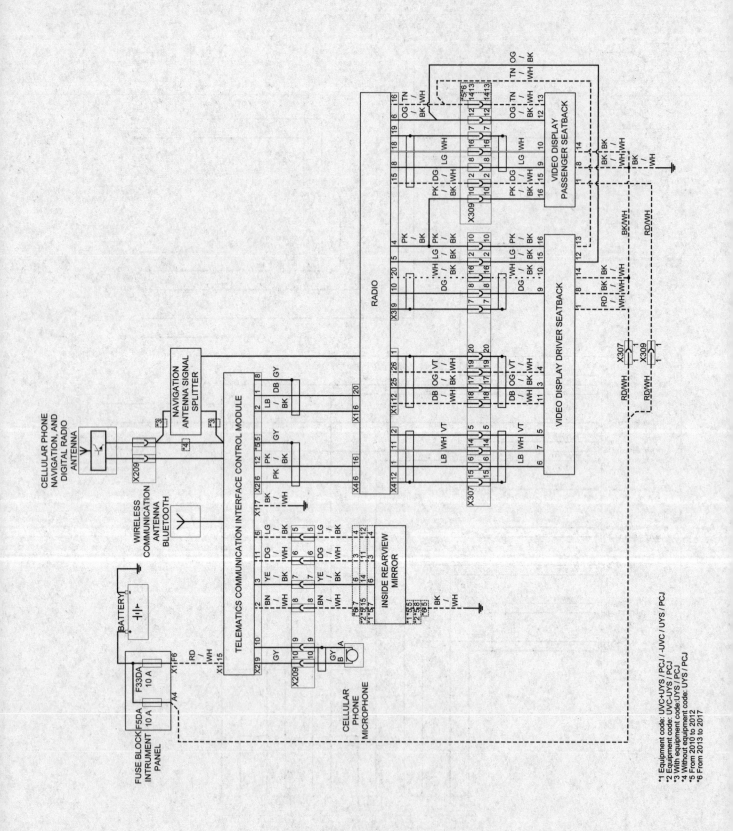

Entertainment systems (OnStar, telematics, video) - 2010 and later models

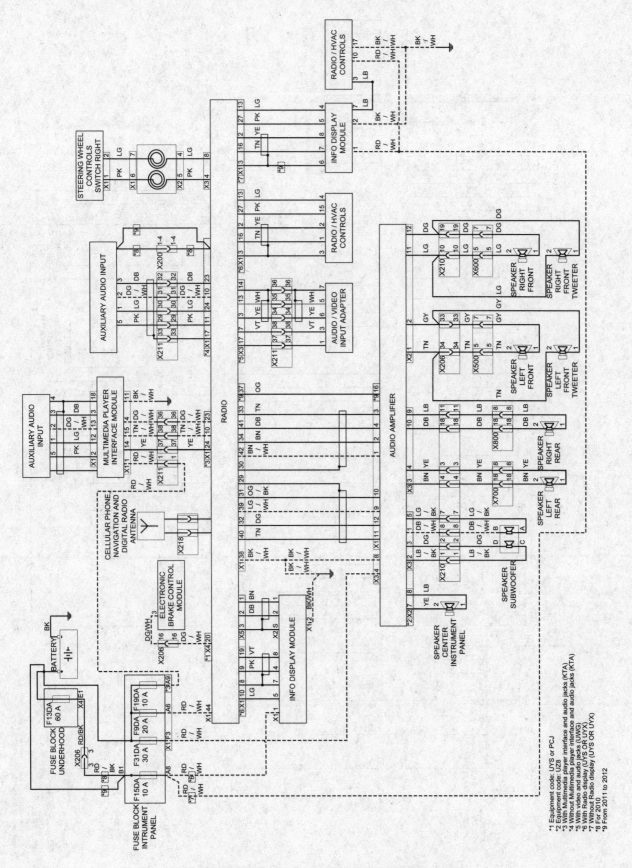

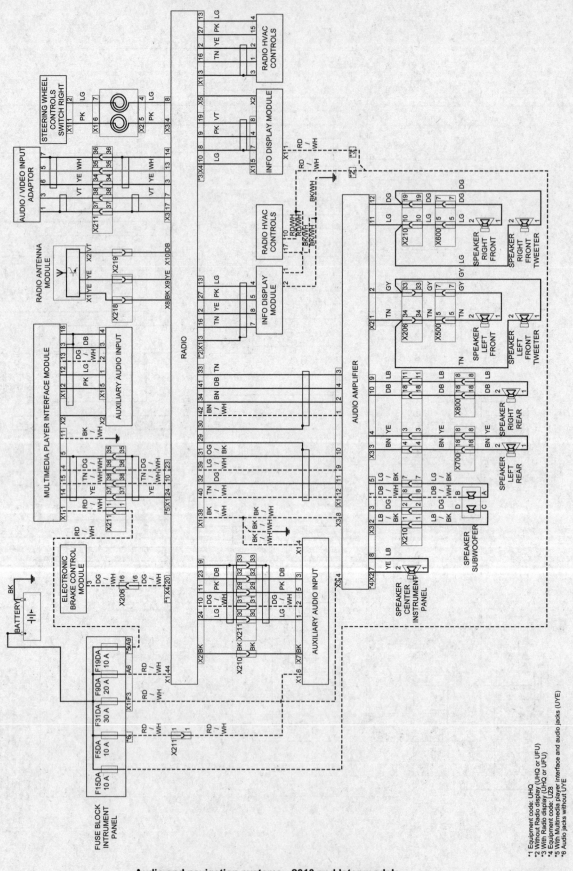

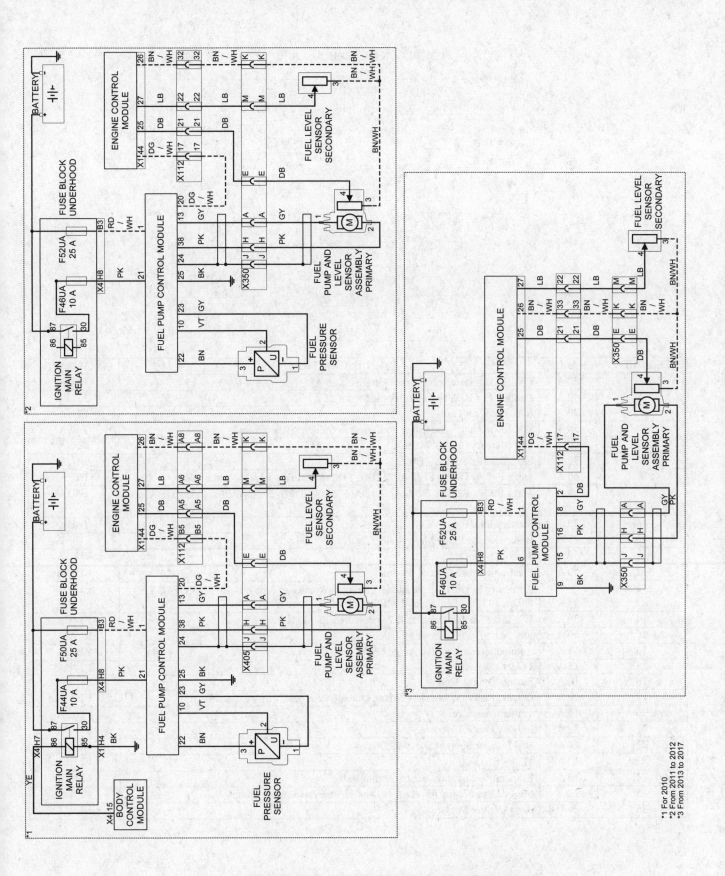

Fuel pump system - 2010 and later models (earlier models similar)

FUSE BLOCK UNDERHOOD 2010

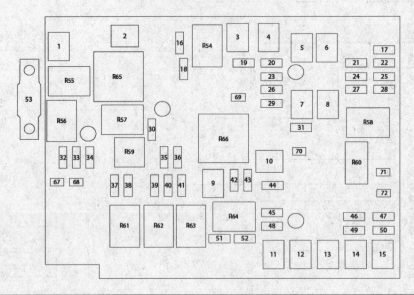

FUSE/RELAY	VALUE	DESCRIPTION	OEM NAME
1	30 A	Cooling Fan Low Speed Relay (also used 25 A)	F1UA
2	40 A	Cooling Fan High Speed Relay (also used 25 A)	F2UA
3	30 A	Rear Defogger Grid	F3UA
4	30 A	Window Switch – Passenger and Window Switch – Right Rear	F4UA
5	40 A	Seat Memory Control Module	F5UA
6	30 A	Seat Adjuster Switch – Driver	F6UA
7	60 A	Fuses F4DA, F8DA, F10DA, F12DA, F16DA, F30DA, F34DA, and F40DA	F7UA
8	60 A	Fuses F6DA, F20DA, F29DA, F31DA, F32DA, F33DA, F35DA, F36DA, F37DA, and F38DA	F8UA
9	30 A	Starter Motor	F9UA
10	25 A	Brake Booster Pump Motor Relay	F10UA
11	30 A	Sunroof Motor	F11UA
12	40 A	Electronic Brake Control Module	F12UA
13		Accessory Relay, Logistic Mode 1 Relay or Accessory Relay, Fuses F5DA, F7DA, F9DA, F11DA, F15DA, and F19DA	F13UA
14	30 A	Window Switch – Driver and Window Switch – Left Rear	F14UA
15	20 A	Electronic Brake Control Module	F15UA
16	15 A	Automatic Transmission Assembly	F16UA
17	15 A	Trailer Connector	F17UA
18	10 A	Engine Control Module	F18UA
19	10 A	Outside Rearview Mirror – Driver, Outside Rearview Mirror – Passenger	F19UA
20	10 A	Trailer Connector	F20UA
21	30 A	Liftgate Control Module	F21UA
22	20 A	Seat Lumbar Support Switch – Driver	F22UA
23	10 A	Trailer Connector	F23UA
24	10 A	Evaporative Emission Vent Solenoid Valve	F24UA
25	7.5 a	Mirror Control Module – Left or Outside Rearview Mirror Switch	F25UA
26	5 A	Body Control Module	F26UA

FUSE/RELAY	VALUE	DESCRIPTION	OEM NAME
27	20 A	Accessory Power Receptacle – Rear	F27UA
28	25 A	Windshield Wiper Speed Control Relay, Windshield Wiper Relay	F28UA
29	20 A	Rear Wiper Relay	F29UA
30	10 A	A/C Compressor Clutch	F30UA
31	15 A	Liftgate Release Relay	F31UA
32	15 A	Horn – High Note, Horn – Low Note	F32UA
33	10 A	Headlamp – Right High Beam	F33UA
34	10 A	Headlamp – Left High Beam	F34UA
35	20 A	Engine Control Module, Ignition Coil 2, Ignition Coil 4, and Ignition Coil 6	F35UA
36	20 A	Engine Control Module, Ignition Coil 1, Ignition Coil 3, and Ignition Coil 5	F36UA
37	20 A	Rear Window Washer Pump Relay, Windshield Washer Pump Relay	F37UA
38	15 A	Fog Lamp – Left Front, Fog Lamp – Right Front	F38UA
39	15 A	Heated Oxygen Sensor Bank 1 Sensor 2, Heated Oxygen Sensor Bank 2 Sensor 2	F39UA
40	20 A	Engine Control Module	F40UA
41	15 A	Evaporative Emission Canister Purge Solenoid Valve, Heated Oxygen Sensor 1, Heated Oxygen Sensor 2 , Heated Oxygen Sensor Bank 1 Sensor 1 , Heated Oxygen Sensor Bank 2 Sensor 1 , Mass Air Flow (MAF) Sensor/Intake Air Temperature (IAT) Sensor	F41UA
42	15 A	Automatic Transmission	F42UA
43	7.5 A	Inflatable Restraint Passenger Air Bag On/Off Indicator, Inside Rearview Mirror	F43UA
44	10 A	Fuel Pump Flow Control Module	F44UA
45	25 A	Spare fuse	-
46	15 A	Rear Differential Clutch Control Module	F46UA
47	15 A	Liftgate Module	F47UA
48	25 A	HVAC, MIL, and SDM MDL IGN Fuses	F48UA
49	25 A	Heated Seat Control Module or Memory Seat Module	F49UA
50	25 A	Fuel Pump Flow Control Module	F50UA
51	10 A	Engine Control Module	F51UA
52	5 A	Rearview Camera	F52UA
53	80 A	Power Steering Control Module	F53UA
R54	-	Rear Defogger Relay	KR5
R55	-	Cooling Fan Low Speed Relay	KR20C
R56	-	Headlamp High Beam Relay	KR48
R57	-	Cooling Fan Speed Control Relay	KR20E
R58	-	Windshield Wiper Relay	KR12B
R59	-	A/C Compressor Clutch Relay	KR29
R60	-	Windshield Wiper Speed Control Relay	KR12C
R61	-	Front Fog Lamp Relay	KR46
R62	-	Engine Controls Ignition Relay	KR75
R63	-	Starter Relay	KR27
R64	-	Ignition Main Relay	KR73
R65	-	Cooling Fan High Speed Relay	KR20D
R66	-	Brake booster pump motor relay	KR14
Items listed below are diagnostic test points			
67	-	Windshield Washer Pump	228A
68	-	Windshield Washer Pump	392
69	-	Rear Wiper Motor	393
70	-	Liftgate Latch Assembly	6128
71	-	Door Latch Assembly - Left Rear, Door Latch Assembly - Passenger, and Door Latch Assembly - Right Rear	295

FUSE/RELAY	VALUE	DESCRIPTION	OEM NAME
72	-	Door Lock Relay	3271
		Relays listed below are non-serviceable	
-	-	Horn Relay	KR3
-	-	Rear Window Washer Pump Relay	KR6
-	-	Rear Wiper Relay	KR7
-	-	Windshield Washer Pump Relay	KR11
-	-	Park Lamp Relay	KR53
-	-	Trailer Stop / Turn Signal Lamp Relay - Left	KR63L
-	-	Trailer Stop / Turn Signal Lamp Relay - Right	KR63R
-	-	Liftgate Release Relay	KR63A
-	-	Door Lock Relay	KR97

FUSE BLOCK UNDERHOOD FROM 2011 TO 2017

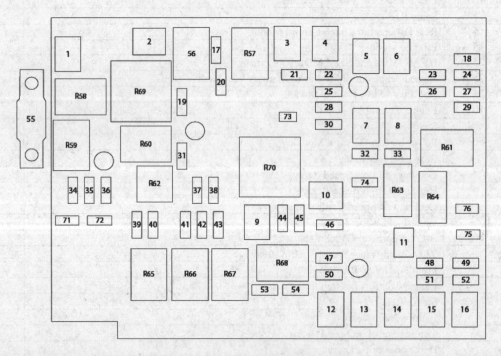

FUSE/RELAY	VALUE	DESCRIPTION	OEM NAME
1	30 A	Cooling Fan Low Speed Relay (also used 25 A)	F1UA
2	40 A	Cooling Fan High Speed Relay (also used 25 A)	F2UA
3	25 A	Brake Booster Pump Motor Relay	F3UA
4	30 A	Window Switch – Passenger, Window Switch – Right Rear	F4UA
5	40 A	Seat Memory Control Module	F5UA
6	30 A	Seat Adjuster Switch – Driver	F6UA

FUSE/RELAY	VALUE	DESCRIPTION	OEM NAME
7	60 A	Fuses F4DA, F8DA, F10DA, F12DA, F16DA, F30DA, F34DA, and F40DA	F7UA
8	30 A	Rear Defogger Grid	F8UA
9	30 A	Starter Motor	F9UA
10	-	Not used	-
11	60 A	Fuses F6DA, F20DA, F29DA, F31DA, F32DA, F33DA, F35DA, F36DA, F37DA, and F38DA	F11UA
12	30 A	Sunroof Motor	F12UA
13	40 A	Electronic Brake Control Module	F13UA
14	60 A	Accessory Relay, Battery Saver Relay 1 or Accessory Relay, Fuses F5DA, F7DA, F9DA, F11DA, F15DA, and F19DA	F14UA
15	30 A	Window Switch – Driver, Window Switch – Left Rear	F15UA
16	20 A	Electronic Brake Control Module	F16UA
17	15 A	Automatic Transmission Assembly	F17UA
18	15 A	Trailer Connector	F18UA
19	-	Not used	-
20	10 A	Engine Control Module	F20UA
21	10 A	Evaporative Emission Vent Solenoid Valve	F21UA
22	10 A	Trailer Connector	F22UA
23	30 A	Liftgate Control Module	F23UA
24	20 A	Seat Lumbar Support Switch – Driver	F24UA
25	10 A	Trailer Connector	F25UA
26	20 A	Accessory Power Receptacle – Rear Compartment	F26UA
27	7.5 A	Mirror Control Module – Left or Outside Rearview Mirror Switch	F27UA
28	5 A	Body Control Module	F28UA
29	25 A	Windshield Wiper Speed Control Relay, Windshield Wiper Relay	F29UA
30	20 A	Rear Wiper Relay	F30UA
31	10 A	A/C Compressor Clutch	F31UA
32	15 A	Liftgate Unlatch Relay	F32UA
33	10 A	Outside Rearview Mirror – Driver, Outside Rearview Mirror – Passenger	F33UA
34	15 A	Horn – High Note, Horn – Low Note	F34UA
35	10 A	Headlamp – Right High Beam	F35UA
36	10 A	Headlamp – Left High Beam	F36UA
37	20 A	Engine Control Module, Ignition Coil 2, T8D Ignition Coil 4, T8F Ignition Coil 6	F37UA
38	20 A	Engine Control Module, Ignition Coil 1, Ignition Coil 2, Ignition Coil 3, Ignition Coil 4, Ignition Coil 5	F38UA
39	20 A	Rear Window Washer Pump Relay, Windshield Washer Pump Relay	F39UA
40	15 A	Fog Lamp – Left Front, Fog Lamp – Right Front	F40UA
41	15 A	Heated Oxygen Sensor Bank 1 Sensor 2 , Heated Oxygen Sensor Bank 2 Sensor 2 (LFW)	F41UA
42	20 A	Engine Control Module	F42UA
43	15 A	Evaporative Emission Canister Purge Solenoid Valve, Heated Oxygen Sensor 1, Heated Oxygen Sensor 2, Heated Oxygen Sensor Bank 1 Sensor 1, Heated Oxygen Sensor Bank 2 Sensor 1, Mass Air Flow/Intake Air Temperature Sensor, Fuel Composition Sensor	F43UA
44	15 A	Automatic Transmission Assembly	F44UA
45	7.5 A	Passenger Air Bag Disable Indicator, Inside Rearview Mirror	F45UA
46	10 A	Fuel Pump Control Module	F46UA
47	-	Not used	-
48	15 A	Rear Differential Clutch Control Module	F48UA

Fuses and relays - 2010 and later models (4 of 7)

FUSE/RELAY	VALUE	DESCRIPTION	OEM NAME
49	15 A	Liftgate Control Module	F49UA
50	25 A	HVAC, MIL, and SDM MDL IGN Fuses	F50UA
51	25 A	Seat Heating Control Module , Seat Memory Control Module	F51UA
52	25 A	Fuel Pump Control Module	F52UA
53	10 A	Engine Control Module	F53UA
54	5 A	Rearview Camera	F54UA
55	80 A	Power Steering Control Module	F55UA
56	-	Not used	-
R57	-	Brake Booster Pump Motor Relay	KR14
R58	-	Cooling Fan Low Speed Relay	KR20
R59	-	Headlamp High Beam Relay	KR48
R60	-	Cooling Fan Speed Control Relay	KR20E
R61	-	Windshield Wiper Relay	KR12B
R62	-	A/C Compressor Clutch Relay	KR29
R63	-	Rear Defogger Relay	KR5
R64	-	Windshield Wiper Speed Control Relay	KR12C
R65	-	Front Fog Lamp Relay	KR46
R66	-	Engine Controls Ignition Relay	KR75
R67	-	Starter Relay	KR27
R68	-	Ignition Main Relay	KR73
R69	-	Cooling Fan High Speed Relay	KR20D
R70	-	Not used	-
Items listed below are diagnostic test points			
T71	-	Windshield Washer Pump	228A
T72	-	Windshield Washer Pump	392
T73	-	Rear Wiper Motor	393
T74	-	Liftgate Latch Assembly	6128
T75	-	Door Latch Assembly – Left Rear, Door Latch Assembly – Passenger, Door Latch Assembly – Right Rear	295
T76	-	Door Lock Relay	3271
Relays listed below are non-serviceable			
-	-	Horn Relay	KR3
-	-	Rear Window Washer Pump Relay	KR63R
-	-	Rear Wiper Relay	KR7
-	-	Windshield Washer Pump Relay	KR11
-	-	Park Lamp Relay	KR53
-	-	Trailer Stop / Turn Signal Lamp Relay - Left	KR63L
-	-	Trailer Stop / Turn Signal Lamp Relay - Right	KR63R
-	-	Liftgate Release Relay	KR95A
-	-	Door Lock Relay	KR97

FUSE BLOCK INSTRUMENT PANEL

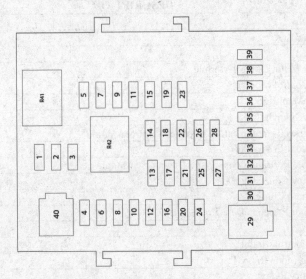

FUSE/RELAY	VALUE	DESCRIPTION	OEM NAME
1	2 A	Steering Wheel Control Switch - Left, Steering Wheel Control Switch - Right	F1DA
2	-	Not used	-
3	-	Not used	-
4	15 A	Body Control Module	F4DA
5	10 A	Video Display - Driver Seatback , Video Display - Passenger Seatback	F5DA
6	15 A	Body Control Module	F6DA
7	5 A	Active Noise Cancellation Module	F7DA
8	15 A	Body Control Module	F8DA
9	20 A	Radio	F9DA
10	20 A	Spare fuse	F10DA
11	10 A	Object Alarm Control Module or Rear parking Assist Control Module	F11DA
12	20 A	HVAC Control Module	F12DA
13	20 A	Accessory Power Receptacle - Instrument Panel	F13DA
14	5 A	HVAC Control Module	F14DA
15	10 A	Info Display Module, Radio/HVAC Control	F15DA
16	15 A	Body Control Module	F16DA
17	20 A	Accessory Power Receptacle - Center Console 1, Accessory Power Receptacle - Center Console 2	F17DA
18	5 A	Instrument Cluster	F18DA
19	10 A	Multimedia Player Interface Module	F19DA
20	15 A	Body Control Module	F20DA
21	20 A	Spare fuse	F21DA
22	10 A	Inflatable Restraint Sensing and Diagnostic Module	F22DA
23	-	Not used	F23DA
24	-	Not used	F24DA
25	10 A	Transmission Shift Lever Position Indicator	F25DA

FUSE/RELAY	VALUE	DESCRIPTION	OEM NAME
26	-	Not used	F26DA
27	-	Not used	F27DA
28	-	Not used	F28DA
29	40 A	Blower Motor Control Module	F29DA
30	15 A	Body Control Module	F30DA
31	30 A	Audio Amplifier	F31DA
32	5 A	Ignition Switch	F32DA
33	10 A	Telematics Communication Interface Control Module	F33DA
34	15 A	Body Control Module	F34DA
35	10 A	Inflatable Restraint Sensing and Diagnostic Module	F35DA
36	7.5 A	Data Link Connector	F36DA
37	10 A	Instrument Cluster	F37DA
38	5 A	Passenger Presence Detection Sensor	F38DA
39	-	Not used	F39DA
40	30 A	Body Control Module	F40DA
R41	-	Logistic Mode Relay 1	KR104A
R42	-	Accessory Relay	KR80

Index

F

Notes

Notes

Haynes Automotive Manuals

NOTE: If you do not see a listing for your vehicle, consult your local Haynes dealer for the latest product information.

ACURA

- 12020 Integra '86 thru '89 & Legend '86 thru '90
- 12021 Integra '90 thru '93 & Legend '91 thru '95
  Integra '94 thru '00 - see HONDA Civic (42025)
  MDX '01 thru '07 - see HONDA Pilot (42037)
- 12050 Acura TL all models '99 thru '08

AMC

- Jeep CJ - see JEEP (50020)
- 14020 Mid-size models '70 thru '83
- 14025 (Renault) Alliance & Encore '83 thru '87

AUDI

- 15020 4000 all models '80 thru '87
- 15025 5000 all models '77 thru '83
- 15026 5000 all models '84 thru '88
  Audi A4 '96 thru '01 - see VW Passat (96023)
- 15030 Audi A4 '02 thru '08

AUSTIN-HEALEY

- Sprite - see MG Midget (66015)

BMW

- 18020 3/5 Series '82 thru '92
- 18021 3-Series incl. Z3 models '92 thru '98
- 18022 3-Series incl. Z4 models '99 thru '05
- 18023 3-Series '06 thru '10
- 18025 320i all 4 cyl models '75 thru '83
- 18050 1500 thru 2002 except Turbo '59 thru '77

BUICK

- 19010 Buick Century '97 thru '05
  Century (front-wheel drive) - see GM (38005)
- 19020 Buick, Oldsmobile & Pontiac Full-size
  (Front-wheel drive) '85 thru '05
  Buick Electra, LeSabre and Park Avenue;
  Oldsmobile Delta 88 Royale, Ninety Eight
  and Regency; Pontiac Bonneville
- 19025 Buick, Oldsmobile & Pontiac Full-size
  (Rear wheel drive) '70 thru '90
  Buick Estate, Electra, LeSabre, Limited,
  Oldsmobile Custom Cruiser, Delta 88,
  Ninety-eight, Pontiac Bonneville,
  Catalina, Grandville, Parisienne
- 19030 Mid-size Regal & Century all rear-drive
  models with V6, V8 and Turbo '74 thru '87
  Regal - see GENERAL MOTORS (38010)
  Riviera - see GENERAL MOTORS (38030)
  Roadmaster - see CHEVROLET (24046)
  Skyhawk - see GENERAL MOTORS (38015)
  Skylark - see GM (38020, 38025)
  Somerset - see GENERAL MOTORS (38025)

CADILLAC

- 21015 CTS & CTS-V '03 thru '12
- 21030 Cadillac Rear Wheel Drive '70 thru '93
  Cimarron - see GENERAL MOTORS (38015)
  DeVille - see GM (38031 & 38032)
  Eldorado - see GM (38030 & 38031)
  Fleetwood - see GM (38031)
  Seville - see GM (38030, 38031 & 38032)

CHEVROLET

- 10305 Chevrolet Engine Overhaul Manual
- 24010 Astro & GMC Safari Mini-vans '85 thru '05
- 24015 Camaro V8 all models '70 thru '81
- 24016 Camaro all models '82 thru '92
- 24017 Camaro & Firebird '93 thru '02
  Cavalier - see GENERAL MOTORS (38016)
  Celebrity - see GENERAL MOTORS (38005)
- 24020 Chevelle, Malibu & El Camino '69 thru '87
- 24024 Chevette & Pontiac T1000 '76 thru '87
  Citation - see GENERAL MOTORS (38020)
- 24027 Colorado & GMC Canyon '04 thru '10
- 24032 Corsica/Beretta all models '87 thru '96
- 24040 Corvette all V8 models '68 thru '82
- 24041 Corvette all models '84 thru '96
- 24045 Full-size Sedans Caprice, Impala, Biscayne,
  Bel Air & Wagons '69 thru '90
- 24046 Impala SS & Caprice and Buick Roadmaster
  '91 thru '96
  Impala '00 thru '05 - see LUMINA (24048)
- 24047 Impala & Monte Carlo all models '06 thru '11
  Lumina '90 thru '94 - see GM (38010)
- 24048 Lumina & Monte Carlo '95 thru '05
  Lumina APV - see GM (38035)
- 24050 Luv Pick-up all 2WD & 4WD '72 thru '82
  Malibu '97 thru '03 - see GM (38026)
- 24055 Monte Carlo all models '70 thru '88
  Monte Carlo '95 thru '01 - see LUMINA (24048)
- 24059 Nova all V8 models '69 thru '79
- 24060 Nova and Geo Prizm '85 thru '92
- 24064 Pick-ups '67 thru '87 - Chevrolet & GMC
- 24065 Pick-ups '88 thru '98 - Chevrolet & GMC

- 24066 Pick-ups '99 thru '06 - Chevrolet & GMC
- 24067 Chevrolet Silverado & GMC Sierra '07 thru '12
- 24070 S-10 & S-15 Pick-ups '82 thru '93,
  Blazer & Jimmy '83 thru '94,
- 24071 S-10 & Sonoma Pick-ups '94 thru '04, including Blazer, Jimmy & Hombre
- 24072 Chevrolet TrailBlazer, GMC Envoy &
  Oldsmobile Bravada '02 thru '09
- 24075 Sprint '85 thru '88 & Geo Metro '89 thru '01
- 24080 Vans - Chevrolet & GMC '68 thru '96
- 24081 Chevrolet Express & GMC Savana
  Full-size Vans '96 thru '10

CHRYSLER

- 10310 Chrysler Engine Overhaul Manual
- 25015 Chrysler Cirrus, Dodge Stratus,
  Plymouth Breeze '95 thru '00
- 25020 Full-size Front-Wheel Drive '88 thru '93
  K-Cars - see DODGE Aries (30008)
  Laser - see DODGE Daytona (30030)
- 25025 Chrysler LHS, Concorde, New Yorker,
  Dodge Intrepid, Eagle Vision, '93 thru '97
- 25026 Chrysler LHS, Concorde, 300M,
  Dodge Intrepid, '98 thru '04
- 25027 Chrysler 300, Dodge Charger &
  Magnum '05 thru '09
- 25030 Chrysler & Plymouth Mid-size
  front wheel drive '82 thru '95
  Rear-wheel Drive - see Dodge (30050)
- 25035 PT Cruiser all models '01 thru '10
- 25040 Chrysler Sebring '95 thru '06, Dodge Stratus
  '01 thru '06, Dodge Avenger '95 thru '00

DATSUN

- 28005 200SX all models '80 thru '83
- 28007 B-210 all models '73 thru '78
- 28009 210 all models '79 thru '82
- 28012 240Z, 260Z & 280Z Coupe '70 thru '78
- 28014 280ZX Coupe & 2+2 '79 thru '83
  300ZX - see NISSAN (72010)
- 28018 510 & PL521 Pick-up '68 thru '73
- 28020 510 all models '78 thru '81
- 28022 620 Series Pick-up all models '73 thru '79
  720 Series Pick-up - see NISSAN (72030)
- 28025 810/Maxima all gasoline models '77 thru '84

DODGE

- 400 & 600 - see CHRYSLER (25030)
- 30008 Aries & Plymouth Reliant '81 thru '89
- 30010 Caravan & Plymouth Voyager '84 thru '95
- 30011 Caravan & Plymouth Voyager '96 thru '02
- 30012 Challenger/Plymouth Saporro '78 thru '83
- 30013 Caravan, Chrysler Voyager, Town &
  Country '03 thru '07
- 30016 Colt & Plymouth Champ '78 thru '87
- 30020 Dakota Pick-ups all models '87 thru '96
- 30021 Durango '98 & '99, Dakota '97 thru '99
- 30022 Durango '00 thru '03 Dakota '00 thru '04
- 30023 Durango '04 thru '09, Dakota '05 thru '11
- 30025 Dart, Demon, Plymouth Barracuda,
  Duster & Valiant 6 cyl models '67 thru '76
- 30030 Daytona & Chrysler Laser '84 thru '89
  Intrepid - see CHRYSLER (25025, 25026)
- 30034 Neon all models '95 thru '99
- 30035 Omni & Plymouth Horizon '78 thru '90
- 30036 Dodge and Plymouth Neon '00 thru '05
- 30040 Pick-ups all full-size models '74 thru '93
- 30041 Pick-ups all full-size models '94 thru '01
- 30042 Pick-ups full-size models '02 thru '08
- 30045 Ram 50/D50 Pick-ups & Raider and
  Plymouth Arrow Pick-ups '79 thru '93
- 30050 Dodge/Plymouth/Chrysler RWD '71 thru '89
- 30055 Shadow & Plymouth Sundance '87 thru '94
- 30060 Spirit & Plymouth Acclaim '89 thru '95
- 30065 Vans - Dodge & Plymouth '71 thru '03

EAGLE

- Talon - see MITSUBISHI (68030, 68031)
- Vision - see CHRYSLER (25025)

FIAT

- 34010 124 Sport Coupe & Spider '68 thru '78
- 34025 X1/9 all models '74 thru '80

FORD

- 10320 Ford Engine Overhaul Manual
- 10355 Ford Automatic Transmission Overhaul
- 11500 Mustang '64-1/2 thru '70 Restoration Guide
- 36004 Aerostar Mini-vans all models '86 thru '97
- 36006 Contour & Mercury Mystique '95 thru '00
- 36008 Courier Pick-up all models '72 thru '82
- 36012 Crown Victoria & Mercury Grand
  Marquis '88 thru '10
- 36016 Escort/Mercury Lynx all models '81 thru '90
- 36020 Escort/Mercury Tracer '91 thru '02

- 36022 Escape & Mazda Tribute '01 thru '11
- 36024 Explorer & Mazda Navajo '91 thru '01
- 36025 Explorer/Mercury Mountaineer '02 thru '10
- 36028 Fairmont & Mercury Zephyr '78 thru '83
- 36030 Festiva & Aspire '88 thru '97
- 36032 Fiesta all models '77 thru '80
- 36034 Focus all models '00 thru '11
- 36036 Ford & Mercury Full-size '75 thru '87
- 36044 Ford & Mercury Mid-size '75 thru '86
- 36045 Fusion & Mercury Milan '06 thru '10
- 36048 Mustang V8 all models '64-1/2 thru '73
- 36049 Mustang II 4 cyl, V6 & V8 models '74 thru '78
- 36050 Mustang & Mercury Capri '79 thru '93
- 36051 Mustang all models '94 thru '04
- 36052 Mustang '05 thru '10
- 36054 Pick-ups & Bronco '73 thru '79
- 36058 Pick-ups & Bronco '80 thru '96
- 36059 F-150 & Expedition '97 thru '09, F-250 '97
  thru '99 & Lincoln Navigator '98 thru '09
- 36060 Super Duty Pick-ups, Excursion '99 thru '10
- 36061 F-150 full-size '04 thru '10
- 36062 Pinto & Mercury Bobcat '75 thru '80
- 36066 Probe all models '89 thru '92
  Probe '93 thru '97 - see MAZDA 626 (61042)
- 36070 Ranger/Bronco II gasoline models '83 thru '92
- 36071 Ranger '93 thru '10 & Mazda Pick-ups '94 thru '09
- 36074 Taurus & Mercury Sable '86 thru '95
- 36075 Taurus & Mercury Sable '96 thru '05
- 36078 Tempo & Mercury Topaz '84 thru '94
- 36082 Thunderbird/Mercury Cougar '83 thru '88
- 36086 Thunderbird/Mercury Cougar '89 thru '97
- 36090 Vans all V8 Econoline models '69 thru '91
- 36094 Vans full size '92 thru '10
- 36097 Windstar Mini-van '95 thru '07

GENERAL MOTORS

- 10360 GM Automatic Transmission Overhaul
- 38005 Buick Century, Chevrolet Celebrity,
  Oldsmobile Cutlass Ciera & Pontiac 6000
  all models '82 thru '96
- 38010 Buick Regal, Chevrolet Lumina,
  Oldsmobile Cutlass Supreme &
  Pontiac Grand Prix (FWD) '88 thru '07
- 38015 Buick Skyhawk, Cadillac Cimarron,
  Chevrolet Cavalier, Oldsmobile Firenza &
  Pontiac J-2000 & Sunbird '82 thru '94
- 38016 Chevrolet Cavalier &
  Pontiac Sunfire '95 thru '05
- 38017 Chevrolet Cobalt & Pontiac G5 '05 thru '11
- 38020 Buick Skylark, Chevrolet Citation,
  Olds Omega, Pontiac Phoenix '80 thru '85
- 38025 Buick Skylark & Somerset,
  Oldsmobile Achieva & Calais and
  Pontiac Grand Am all models '85 thru '98
- 38026 Chevrolet Malibu, Olds Alero & Cutlass,
  Pontiac Grand Am '97 thru '03
- 38027 Chevrolet Malibu '04 thru '10
- 38030 Cadillac Eldorado, Seville, Oldsmobile
  Toronado, Buick Riviera '71 thru '85
- 38031 Cadillac Eldorado & Seville, DeVille, Fleetwood
  & Olds Toronado, Buick Riviera '86 thru '93
- 38032 Cadillac DeVille '94 thru '05 & Seville '92 thru '04
  Cadillac DTS '06 thru '10
- 38035 Chevrolet Lumina APV, Olds Silhouette
  & Pontiac Trans Sport all models '90 thru '96
- 38036 Chevrolet Venture, Olds Silhouette,
  Pontiac Trans Sport & Montana '97 thru '05
  General Motors Full-size
  Rear-wheel Drive - see BUICK (19025)
- 38040 Chevrolet Equinox '05 thru '09 Pontiac
  Torrent '06 thru '09
- 38070 Chevrolet HHR '06 thru '11

GEO

- Metro - see CHEVROLET Sprint (24075)
- Prizm - '85 thru '92 see CHEVY (24060),
  '93 thru '02 see TOYOTA Corolla (92036)
- 40030 Storm all models '90 thru '93
  Tracker - see SUZUKI Samurai (90010)

GMC

- Vans & Pick-ups - see CHEVROLET

HONDA

- 42010 Accord CVCC all models '76 thru '83
- 42011 Accord all models '84 thru '89
- 42012 Accord all models '90 thru '93
- 42013 Accord all models '94 thru '97
- 42014 Accord all models '98 thru '02
- 42015 Accord '03 thru '07
- 42020 Civic 1200 all models '73 thru '79
- 42021 Civic 1300 & 1500 CVCC '80 thru '83
- 42022 Civic 1500 CVCC all models '75 thru '79

(Continued on other side)

Haynes North America, Inc., 859 Lawrence Drive, Newbury Park, CA 91320-1514 • (805) 498-6703 • http://www.haynes.com

Haynes Automotive Manuals (continued)

NOTE: If you do not see a listing for your vehicle, consult your local Haynes dealer for the latest product information.

42023 Civic all models '84 thru '91
42024 Civic & del Sol '92 thru '95
42025 Civic '96 thru '00, CR-V '97 thru '01, Acura Integra '94 thru '00
42026 Civic '01 thru '10, CR-V '02 thru '09
42035 Odyssey all models '99 thru '10
Passport - *see ISUZU Rodeo (47017)*
42037 Honda Pilot '03 thru '07, Acura MDX '01 thru '07
42040 Prelude CVCC all models '79 thru '89

HYUNDAI
43010 Elantra all models '96 thru '10
43015 Excel & Accent all models '86 thru '09
43050 Santa Fe all models '01 thru '06
43055 Sonata all models '99 thru '08

INFINITI
G35 '03 thru '08 - *see NISSAN 350Z (72011)*

ISUZU
Hombre - *see CHEVROLET S-10 (24071)*
47017 Rodeo, Amigo & Honda Passport '89 thru '02
47020 Trooper & Pick-up '81 thru '93

JAGUAR
49010 XJ6 all 6 cyl models '68 thru '86
49011 XJ6 all models '88 thru '94
49015 XJ12 & XJS all 12 cyl models '72 thru '85

JEEP
50010 Cherokee, Comanche & Wagoneer Limited all models '84 thru '01
50020 CJ all models '49 thru '86
50025 Grand Cherokee all models '93 thru '04
50026 Grand Cherokee '05 thru '09
50029 Grand Wagoneer & Pick-up '72 thru '91
Grand Wagoneer '84 thru '91, Cherokee & Wagoneer '72 thru '83, Pick-up '72 thru '88
50030 Wrangler all models '87 thru '11
50035 Liberty '02 thru '07

KIA
54050 Optima '01 thru '10
54070 Sephia '94 thru '01, Spectra '00 thru '09, Sportage '05 thru '10

LEXUS
ES 300/330 - *see TOYOTA Camry (92007) (92008)*
RX 330 - *see TOYOTA Highlander (92095)*

LINCOLN
Navigator - *see FORD Pick-up (36059)*
59010 Rear-Wheel Drive all models '70 thru '10

MAZDA
61010 GLC Hatchback (rear-wheel drive) '77 thru '83
61011 GLC (front-wheel drive) '81 thru '85
61012 Mazda3 '04 thru '11
61015 323 & Protegé '90 thru '03
61016 MX-5 Miata '90 thru '09
61020 MPV all models '89 thru '98
Navajo - *see Ford Explorer (36024)*
61030 Pick-ups '72 thru '93
Pick-ups '94 thru '00 - *see Ford Ranger (36071)*
61035 RX-7 all models '79 thru '85
61036 RX-7 all models '86 thru '91
61040 626 (rear-wheel drive) all models '79 thru '82
61041 626/MX-6 (front-wheel drive) '83 thru '92
61042 626, MX-6/Ford Probe '93 thru '02
61043 Mazda6 '03 thru '11

MERCEDES-BENZ
63012 123 Series Diesel '76 thru '85
63015 190 Series four-cyl gas models, '84 thru '88
63020 230/250/280 6 cyl sohc models '68 thru '72
63025 280 123 Series gasoline models '77 thru '81
63030 350 & 450 all models '71 thru '80
63040 C-Class: C230/C240/C280/C320/C350 '01 thru '07

MERCURY
64200 Villager & Nissan Quest '93 thru '01
All other titles, see FORD Listing.

MG
66010 MGB Roadster & GT Coupe '62 thru '80
66015 MG Midget, Austin Healey Sprite '58 thru '80

MINI
67010 Mini '02 thru '11

MITSUBISHI
68020 Cordia, Tredia, Galant, Precis & Mirage '83 thru '93
68030 Eclipse, Eagle Talon & Ply. Laser '90 thru '94
68031 Eclipse '95 thru '05, Eagle Talon '95 thru '98
68035 Galant '94 thru '10
68040 Pick-up '83 thru '96 & Montero '83 thru '93

NISSAN
72010 300ZX all models including Turbo '84 thru '89
72011 350Z & Infiniti G35 all models '03 thru '08
72015 Altima all models '93 thru '06
72016 Altima '07 thru '10
72020 Maxima all models '85 thru '92
72021 Maxima all models '93 thru '04
72025 Murano '03 thru '10
72030 Pick-up '80 thru '97 Pathfinder '87 thru '95
72031 Frontier Pick-up, Xterra, Pathfinder '96 thru '04
72032 Frontier & Xterra '05 thru '11
72040 Pulsar all models '83 thru '86
Quest - *see MERCURY Villager (64200)*
72050 Sentra all models '82 thru '94
72051 Sentra & 200SX all models '95 thru '06
72060 Stanza all models '82 thru '90
72070 Titan pick-ups '04 thru '10 Armada '05 thru '10

OLDSMOBILE
73015 Cutlass V6 & V8 gas models '74 thru '88
For other OLDSMOBILE titles, see BUICK, CHEVROLET or GENERAL MOTORS listing.

PLYMOUTH
For PLYMOUTH titles, see DODGE listing.

PONTIAC
79008 Fiero all models '84 thru '88
79018 Firebird V8 models except Turbo '70 thru '81
79019 Firebird all models '82 thru '92
79025 G6 all models '05 thru '09
79040 Mid-size Rear-wheel Drive '70 thru '87
Vibe '03 thru '11 - *see TOYOTA Matrix (92060)*
For other PONTIAC titles, see BUICK, CHEVROLET or GENERAL MOTORS listing.

PORSCHE
80020 911 except Turbo & Carrera 4 '65 thru '89
80025 914 all 4 cyl models '69 thru '76
80030 924 all models including Turbo '76 thru '82
80035 944 all models including Turbo '83 thru '89

RENAULT
Alliance & Encore - *see AMC (14020)*

SAAB
84010 900 all models including Turbo '79 thru '88

SATURN
87010 Saturn all S-series models '91 thru '02
87011 Saturn Ion '03 thru '07
87020 Saturn all L-series models '00 thru '04
87040 Saturn VUE '02 thru '07

SUBARU
89002 1100, 1300, 1400 & 1600 '71 thru '79
89003 1600 & 1800 2WD & 4WD '80 thru '94
89100 Legacy all models '90 thru '99
89101 Legacy & Forester '00 thru '06

SUZUKI
90010 Samurai/Sidekick & Geo Tracker '86 thru '01

TOYOTA
92005 Camry all models '83 thru '91
92006 Camry all models '92 thru '96
92007 Camry, Avalon, Solara, Lexus ES 300 '97 thru '01
92008 Toyota Camry, Avalon and Solara and Lexus ES 300/330 all models '02 thru '06
92009 Camry '07 thru '11
92015 Celica Rear Wheel Drive '71 thru '85
92020 Celica Front Wheel Drive '86 thru '99
92025 Celica Supra all models '79 thru '92
92030 Corolla all models '75 thru '79
92032 Corolla all rear wheel drive models '80 thru '87
92035 Corolla all front wheel drive models '84 thru '92
92036 Corolla & Geo Prizm '93 thru '02
92037 Corolla models '03 thru '11
92040 Corolla Tercel all models '80 thru '82
92045 Corona all models '74 thru '82
92050 Cressida all models '78 thru '82
92055 Land Cruiser FJ40, 43, 45, 55 '68 thru '82
92056 Land Cruiser FJ60, 62, 80, FZJ80 '80 thru '96
92060 Matrix & Pontiac Vibe '03 thru '11
92065 MR2 all models '85 thru '87
92070 Pick-up all models '69 thru '78
92075 Pick-up all models '79 thru '95
92076 Tacoma, 4Runner, & T100 '93 thru '04
92077 Tacoma all models '05 thru '09
92078 Tundra '00 thru '06 & Sequoia '01 thru '07
92079 4Runner all models '03 thru '09
92080 Previa all models '91 thru '95
92081 Prius all models '01 thru '08
92082 RAV4 all models '96 thru '10
92085 Tercel all models '87 thru '94
92090 Sienna all models '98 thru '09
92095 Highlander & Lexus RX-330 '99 thru '07

TRIUMPH
94007 Spitfire all models '62 thru '81
94010 TR7 all models '75 thru '81

VW
96008 Beetle & Karmann Ghia '54 thru '79
96009 New Beetle '98 thru '11
96016 Rabbit, Jetta, Scirocco & Pick-up gas models '75 thru '92 & Convertible '80 thru '92
96017 Golf, GTI & Jetta '93 thru '98, Cabrio '95 thru '02
96018 Golf, GTI, Jetta '99 thru '05
96019 Jetta, Rabbit, GTI & Golf '05 thru '11
96020 Rabbit, Jetta & Pick-up diesel '77 thru '84
96023 Passat '98 thru '05, Audi A4 '96 thru '01
96030 Transporter 1600 all models '68 thru '79
96035 Transporter 1700, 1800 & 2000 '72 thru '79
96040 Type 3 1500 & 1600 all models '63 thru '73
96045 Vanagon all air-cooled models '80 thru '83

VOLVO
97010 120, 130 Series & 1800 Sports '61 thru '73
97015 140 Series all models '66 thru '74
97020 240 Series all models '76 thru '93
97040 740 & 760 Series all models '82 thru '88
97050 850 Series all models '93 thru '97

TECHBOOK MANUALS
10205 Automotive Computer Codes
10206 OBD-II & Electronic Engine Management
10210 Automotive Emissions Control Manual
10215 Fuel Injection Manual '78 thru '85
10220 Fuel Injection Manual '86 thru '99
10225 Holley Carburetor Manual
10230 Rochester Carburetor Manual
10240 Weber/Zenith/Stromberg/SU Carburetors
10305 Chevrolet Engine Overhaul Manual
10310 Chrysler Engine Overhaul Manual
10320 Ford Engine Overhaul Manual
10330 GM and Ford Diesel Engine Repair Manual
10333 Engine Performance Manual
10340 Small Engine Repair Manual, 5 HP & Less
10341 Small Engine Repair Manual, 5.5 - 20 HP
10345 Suspension, Steering & Driveline Manual
10355 Ford Automatic Transmission Overhaul
10360 GM Automatic Transmission Overhaul
10405 Automotive Body Repair & Painting
10410 Automotive Brake Manual
10411 Automotive Anti-lock Brake (ABS) Systems
10415 Automotive Detailing Manual
10420 Automotive Electrical Manual
10425 Automotive Heating & Air Conditioning
10430 Automotive Reference Manual & Dictionary
10435 Automotive Tools Manual
10440 Used Car Buying Guide
10445 Welding Manual
10450 ATV Basics
10452 Scooters 50cc to 250cc

SPANISH MANUALS
98903 Reparación de Carrocería & Pintura
98904 Manual de Carburador Modelos Holley & Rochester
98905 Códigos Automotrices de la Computadora
98906 OBD-II & Sistemas de Control Electrónico del Motor
98910 Frenos Automotriz
98913 Electricidad Automotriz
98915 Inyección de Combustible '86 al '99
99040 Chevrolet & GMC Camionetas '67 al '87
99041 Chevrolet & GMC Camionetas '88 al '98
99042 Chevrolet & GMC Camionetas Cerradas '68 al '95
99043 Chevrolet/GMC Camionetas '94 al '04
99048 Chevrolet/GMC Camionetas '99 al '06
99055 Dodge Caravan & Plymouth Voyager '84 al '95
99075 Ford Camionetas y Bronco '80 al '94
99076 Ford F-150 '97 al '09
99077 Ford Camionetas Cerradas '69 al '91
99088 Ford Modelos de Tamaño Mediano '75 al '86
99089 Ford Camionetas Ranger '93 al '10
99091 Ford Taurus & Mercury Sable '86 al '95
99095 GM Modelos de Tamaño Grande '70 al '90
99100 GM Modelos de Tamaño Mediano '70 al '88
99106 Jeep Cherokee, Wagoneer & Comanche '84 al '00
99110 Nissan Camioneta '80 al '96, Pathfinder '87 al '95
99118 Nissan Sentra '82 al '94
99125 Toyota Camionetas y 4Runner '79 al '95

Over 100 Haynes motorcycle manuals also available

7-12